D0311859

521 586 39 X

To Explain the World

STEVEN WEINBERG

To Explain the World

The Discovery of Modern Science

ALLEN LANE
an imprint of
PENGUIN BOOKS

ALLEN LANE

UK | USA | Canada | Ireland | Australia
India | New Zealand | South Africa

Penguin Books is part of the Penguin Random House group of companies whose addresses can be found at global.penguinrandomhouse.com.

First published in the United States of America by HarperCollins 2015
First published in Great Britain by Allen Lane 2015
001

Printed in Great Britain by Clays Ltd, St Ives plc

A CIP catalogue record for this book is available from the British Library

ISBN: 978–0–241–19662–5

www.greenpenguin.co.uk

To Louise, Elizabeth, and Gabrielle

These three hours that we have spent,

Walking here, two shadows went

Along with us, which we ourselves produced;

But, now the sun is just above our head,

We do those shadows tread;

And to brave clearness all things are reduced.

John Donne, "A Lecture upon the Shadow"

Contents

Preface

I am a physicist, not a historian, but over the years I have become increasingly fascinated by the history of science. It is an extraordinary story, one of the most interesting in human history. It is also a story in which scientists like myself have a personal stake. Today's research can be aided and illuminated by a knowledge of its past, and for some scientists knowledge of the history of science helps to motivate present work. We hope that our research may turn out to be a part, however small, of the grand historical tradition of natural science.

Where my own past writing has touched on history, it has been mostly the modern history of physics and astronomy, roughly from the late nineteenth century to the present. Although in this era we have learned many new things, the goals and standards of physical science have not materially changed. If physicists of 1900 were somehow taught today's Standard Model of cosmology or of elementary particle physics, they would have found much to amaze them, but the idea of seeking mathematically formulated and experimentally validated impersonal principles that explain a wide variety of phenomena would have seemed quite familiar.

A while ago I decided that I needed to dig deeper, to learn more about an earlier era in the history of science, when the goals and standards of science had not yet taken their present shape. As is natural for an academic, when I want to learn about something, I volunteer to teach a course on the subject. Over the past decade at the University of Texas, I have from time to time taught undergraduate courses on the history of physics and astronomy to students who had no special background in science, mathematics, or history. This book grew out of the lecture notes for those courses.

But as the book has developed, perhaps I have been able to offer something that goes a little beyond a simple narrative: it is the perspective of a modern working scientist on the science of the past. I have taken this opportunity to explain my views about the nature of physical science, and about its continued tangled relations with religion, technology, philosophy, mathematics, and aesthetics.

Before history there was science, of a sort. At any moment nature presents us with a variety of puzzling phenomena: fire, thunderstorms, plagues, planetary motion, light, tides, and so on. Observation of the world led to useful generalizations: fires are hot; thunder presages rain; tides are highest when the Moon is full or new, and so on. These became part of the common sense of mankind. But here and there, some people wanted more than just a collection of facts. They wanted to explain the world.

It was not easy. It is not only that our predecessors did not know what we know about the world—more important, they did not have anything like our ideas of what there was to know about the world, and how to learn it. Again and again in preparing the lectures for my course I have been impressed with how different the work of science in past centuries was from the science of my own times. As the much quoted lines of a novel of L. P. Hartley put it, “The past is a foreign country; they do things differently there.” I hope that in this book I have been able to give the reader not only an idea of what happened in the history of the exact sciences, but also a sense of how hard it has all been.

So this book is not solely about how we came to learn various things about the world. That is naturally a concern of any history of science. My focus in this book is a little different—it is how we came to learn how to learn about the world.

I am not unaware that the word "explain" in the title of this book raises problems for philosophers of science. They have pointed out the difficulty in drawing a precise distinction between explanation and description. (I will have a little to say about this in Chapter 8.) But this is a work on the history rather than the philosophy of science. By explanation I mean something admittedly imprecise, the same as is meant in ordinary life when we try to explain why a horse has won a race or why an airplane has crashed.

The word "discovery" in the subtitle is also problematic. I had thought of using *The Invention of Modern Science* as a subtitle. After all, science could hardly exist without human beings to practice it. I chose "Discovery" instead of "Invention" to suggest that science is the way it is not so much because of various adventitious historic acts of invention, but because of the way nature is. With all its imperfections, modern science is a technique that is sufficiently well tuned to nature so that it works—it is a practice that allows us to learn reliable things about the world. In this sense, it is a technique that was waiting for people to discover it.

Thus one can talk about the discovery of science in the way that a historian can talk about the discovery of agriculture. With all its variety and imperfections, agriculture is the way it is because its practices are sufficiently well tuned to the realities of biology so that it works—it allows us to grow food.

I also wanted with this subtitle to distance myself from the few remaining social constructivists: those sociologists, philosophers, and historians who try to explain not only the process but even the results of science as products of a particular cultural milieu.

Among the branches of science, this book will emphasize physics and astronomy. It was in physics, especially as applied to astronomy, that science first took a modern form. Of course there are limits to the extent to which sciences like biology, whose principles depend so much on historical accidents, can or should be modeled on physics. Nevertheless, there is a sense in which the development of scientific biology as well as chemistry in the nineteenth and twentieth centuries followed the model of the revolution in physics of the seventeenth century.

Science is now international, perhaps the most international aspect of our civilization, but the discovery of modern science happened in what may loosely be called the West. Modern science learned its methods from research done in Europe during the scientific revolution, which in turn evolved from work done in Europe and in Arab countries during the Middle Ages, and ultimately from the precocious science of the Greeks. The West borrowed much scientific knowledge from elsewhere—geometry from Egypt, astronomical data from Babylon, the techniques of arithmetic from Babylon and India, the magnetic compass from China, and so on—but as far as I know, it did not import the *methods* of modern science. So this book will emphasize the West (including medieval Islam) in just the way that was deplored by Oswald Spengler and Arnold Toynbee: I will have little to say about science outside the West, and nothing at all to say about the interesting but entirely isolated progress made in pre-Columbian America.

In telling this story, I will be coming close to the dangerous ground that is most carefully avoided by contemporary historians, of judging the past by the standards of the present. This is an irreverent history; I am not unwilling to criticize the methods and theories of the past from a modern viewpoint. I have even taken some pleasure in uncovering a few errors made by scientific heroes that I have not seen mentioned by historians.

A historian who devotes years to study the works of some great man of the past may come to exaggerate what his hero has accomplished. I have seen this in particular in works on Plato, Aristotle, Avicenna, Grosseteste, and Descartes. But it is not my purpose here to accuse some past natural philosophers of stupidity. Rather, by showing how far these very intelligent individuals were from our present conception of science, I want to show how difficult was the discovery of modern science, how far from obvious are its practices and standards. This also serves as a warning, that science may not yet be in its final form. At several points in this book I suggest that, as great as is the progress that has been made in the methods of science, we may today be repeating some of the errors of the past.

Some historians of science make a shibboleth of not referring to present scientific knowledge in studying the science of the past. I will instead make a point of using present knowledge to clarify past science. For instance, though it might be an interesting intellectual exercise to try to understand how the Hellenistic astronomers Apollonius and Hipparchus developed the theory that the planets go around the Earth on looping epicyclic orbits by using only the data that had been available to them, this is impossible, for much of the data they used is lost. But we do know that in ancient times the Earth and planets went around the Sun on nearly circular orbits, just as they do today, and by using this knowledge we will be able to understand how the data available to ancient astronomers could have suggested to them their theory of epicycles. In any case, how can anyone today, reading about ancient astronomy, forget our present knowledge of what actually goes around what in the solar system?

For readers who want to understand in greater detail how the work of past scientists fits in with what actually exists in nature, there are "technical notes" at the back of the book. It is not necessary to read these notes to follow the book's main text, but some readers may learn a few odd bits of physics and astronomy from them, as I did in preparing them.

Science is not now what it was at its start. Its results are impersonal. Inspiration and aesthetic judgment are important in the development of scientific theories, but the verification of these theories relies finally on impartial experimental tests of their predictions. Though mathematics is used in the formulation of physical theories and in working out their consequences, science is not a branch of mathematics, and scientific theories cannot be deduced by purely mathematical reasoning. Science and technology benefit each other, but at its most fundamental level science is not undertaken for any practical reason. Though science has nothing to say one way or the other about the existence of God or an afterlife, its goal is to find explanations of natural phenomena that are purely naturalistic. Science is cumulative; each new theory incorporates successful earlier theories as approximations, and even explains why these approximations work, when they do work.

None of this was obvious to the scientists of the ancient world or the Middle Ages, and all of it was learned only with great difficulty in the scientific revolution of the sixteenth and seventeenth centuries. Nothing like modern science was a goal from the beginning. How then did we get to the scientific revolution, and beyond it to where we are now? That is what we must try to learn as we explore the discovery of modern science.

PART I

GREEK PHYSICS

During or before the flowering of Greek science, significant contributions to technology, mathematics, and astronomy were being made by the Babylonians, Chinese, Egyptians, Indians, and other peoples. Nevertheless, it was from Greece that Europe drew its model and its inspiration, and it was in Europe that modern science began, so the Greeks played a special role in the discovery of science.

One can argue endlessly about why it was the Greeks who accomplished so much. It may be significant that Greek science began when Greeks lived in small independent city-states, many of them democracies. But as we shall see, the Greeks made their most impressive scientific achievements after these small states had been absorbed into great powers: the Hellenistic kingdoms, and then the Roman Empire. The Greeks in Hellenistic and Roman times made contributions to science and mathematics that were not significantly surpassed until the scientific revolution of the sixteenth and seventeenth centuries in Europe.

This part of my account of Greek science deals with physics, leaving Greek astronomy to be discussed in Part II. I have divided Part I into five chapters, dealing in more or less chronological order with five modes of thought with which science has had to come to terms: poetry, mathematics, philosophy, technology, and religion. The theme of the relationship of science to these five intellectual neighbors will recur throughout this book.

1

Matter and Poetry

First, to set the scene. By the sixth century BC the western coast of what is now Turkey had for some time been settled by Greeks, chiefly speaking the Ionian dialect. The richest and most powerful of the Ionian cities was Miletus, founded at a natural harbor near where the river Meander flows into the Aegean Sea. In Miletus, over a century before the time of Socrates, Greeks began to speculate about the fundamental substance of which the world is made.

I first learned about the Milesians as an undergraduate at Cornell, taking courses on the history and philosophy of science. In lectures I heard the Milesians called "physicists." At the same time, I was also attending classes on physics, including the modern atomic theory of matter. There seemed to me to be very little in common between Milesian and modern physics. It was not so much that the Milesians were wrong about the nature of matter, but rather that I could not understand how they could have reached their conclusions. The historical record concerning Greek thought before the time of Plato is fragmentary, but I was pretty sure that during the Archaic and Classical eras (roughly from 600 to 450 BC and from 450 to 300 BC, respectively) neither the Milesians nor any of the other Greek students of nature were reasoning in anything like the way scientists reason today.

The first Milesian of whom anything is known was Thales,

who lived about two centuries before the time of Plato. He was supposed to have predicted a solar eclipse, one that we know did occur in 585 BC and was visible from Miletus. Even with the benefit of Babylonian eclipse records it's unlikely that Thales could have made this prediction, because any solar eclipse is visible from only a limited geographic region, but the fact that Thales was credited with this prediction shows that he probably flourished in the early 500s BC. We don't know if Thales put any of his ideas into writing. In any case, nothing written by Thales has survived, even as a quotation by later authors. He is a legendary figure, one of those (like his contemporary Solon, who was supposed to have founded the Athenian constitution) who were conventionally listed in Plato's time as the "seven sages" of Greece. For instance, Thales was reputed to have proved or brought from Egypt a famous theorem of geometry (see Technical Note 1). What matters to us here is that Thales was said to hold the view that all matter is composed of a single fundamental substance. According to Aristotle's *Metaphysics*, "Of the first philosophers, most thought the principles which were of the nature of matter were the only principles of all things. . . . Thales, the founder of this school of philosophy, says the principle is water."[1] Much later, Diogenes Laertius (fl. AD 230), a biographer of the Greek philosophers, wrote, "His doctrine was that water is the universal primary substance, and that the world is animate and full of divinities."[2]

By "universal primary substance" did Thales mean that all matter is composed of water? If so, we have no way of telling how he came to this conclusion, but if someone is convinced that all matter is composed of a single common substance, then water is not a bad candidate. Water not only occurs as a liquid but can be easily converted into a solid by freezing or into a vapor by boiling. Water evidently also is essential to life. But we don't know if Thales thought that rocks, for example, are really formed from ordinary water, or only that there is something profound that rock and all other solids have in common with frozen water.

Thales had a pupil or associate, Anaximander, who came to a

different conclusion. He too thought that there is a single fundamental substance, but he did not associate it with any common material. Rather, he identified it as a mysterious substance he called the unlimited, or infinite. On this, we have a description of his views by Simplicius, a Neoplatonist who lived about a thousand years later. Simplicius includes what seems to be a direct quotation from Anaximander, indicated here in italics:

> Of those who say that [the principle] is one and in motion and unlimited, Anaximander, son of Praxiades, a Milesian who became successor and pupil to Thales, said that the unlimited is both principle and element of the things that exist. He says that it is neither water nor any other of the so-called elements, but some other unlimited nature, from which the heavens and the worlds in them come about; and the things from which is the coming into being for the things that exist are also those into which their destruction comes about, in accordance with what must be. *For they give justice and reparation to one another for their offence in accordance with the ordinance of time*—speaking of them thus in rather poetical terms. And it is clear that, having observed the change of the four elements into one another, he did not think fit to make any one of these an underlying stuff, but something else apart from these.[3]

A little later another Milesian, Anaximenes, returned to the idea that everything is made of some one common substance, but for Anaximenes it was not water but air. He wrote one book, of which just one whole sentence has survived: "The soul, being our air, controls us, and breath and air encompass the whole world."[4]

With Anaximenes the contributions of the Milesians came to an end. Miletus and the other Ionian cities of Asia Minor became subject to the growing Persian Empire in about 550 BC. Miletus started a revolt in 499 BC and was devastated by the Persians. It revived later as an important Greek city, but it never again became a center of Greek science.

Concern with the nature of matter continued outside Miletus

among the Ionian Greeks. There is a hint that earth was nominated as the fundamental substance by Xenophanes, who was born around 570 BC at Colophon in Ionia and migrated to southern Italy. In one of his poems, there is the line "For all things come from earth, and in earth all things end."[5] But perhaps this was just his version of the familiar funerary sentiment, "Ashes to ashes, dust to dust." We will meet Xenophanes again in another connection, when we come to religion in Chapter 5.

At Ephesus, not far from Miletus, around 500 BC Heraclitus taught that the fundamental substance is fire. He wrote a book, of which only fragments survive. One of these fragments tells us, "This ordered *kosmos*,* which is the same for all, was not created by any one of the gods or of mankind, but it was ever and is and shall be ever-living Fire, kindled in measure and quenched in measure."[6] Heraclitus elsewhere emphasized the endless changes in nature, so for him it was more natural to take flickering fire, an agent of change, as the fundamental element than the more stable earth, air, or water.

The classic view that all matter is composed not of one but of four elements—water, air, earth, and fire—is probably due to Empedocles. He lived in Acragas, in Sicily (the modern Agrigento), in the mid-400s BC, and he is the first and nearly the only Greek in this early part of the story to have been of Dorian rather than of Ionian stock. He wrote two hexameter poems, of which many fragments have survived. In *On Nature*, we find "how from the mixture of Water, Earth, Aether, and Sun [fire] there came into being the forms and colours of mortal things"[7] and also "fire and water and earth and the endless height of air, and cursed Strife apart from them, balanced in every way, and Love among them, equal in height and breadth."[8]

* As pointed out by Gregory Vlastos, in *Plato's Universe* (University of Washington Press, Seattle, 1975), an adverbial form of the word *kosmos* was used by Homer to mean "socially decent" and "morally proper." This use survives in English in the word "cosmetic." Its use by Heraclitus reflects the Hellenic view that the world is pretty much what it should be. The word appears in English also in the cognates "cosmos" and "cosmology."

It is possible that Empedocles and Anaximander used terms like "love" and "strife" or "justice" and "injustice" only as metaphors for order and disorder, in something like the way Einstein occasionally used "God" as a metaphor for the unknown fundamental laws of nature. But we should not force a modern interpretation onto the pre-Socratics' words. As I see it, the intrusion of human emotions like Empedocles' love and strife, or of values like Anaximander's justice and reparation, into speculations about the nature of matter is more likely to be a sign of the great distance of the thought of the pre-Socratics from the spirit of modern physics.

These pre-Socratics, from Thales to Empedocles, seem to have thought of the elements as smooth undifferentiated substances. A different view that is closer to modern understanding was introduced a little later at Abdera, a town on the seacoast of Thrace founded by refugees from the revolt of the Ionian cities against Persia started in 499 BC. The first known Abderite philosopher is Leucippus, from whom just one sentence survives, suggesting a deterministic worldview: "No thing happens in vain, but everything for a reason and by necessity."[9] Much more is known of Leucippus' successor Democritus. He was born at Miletus, and had traveled in Babylon, Egypt, and Athens before settling in Abdera in the late 400s BC. Democritus wrote books on ethics, natural science, mathematics, and music, of which many fragments survive. One of these fragments expresses the view that all matter consists of tiny indivisible particles called atoms (from the Greek for "uncuttable"), moving in empty space: "Sweet exists by convention, bitter by convention; atoms and Void [alone] exist in reality."[10]

Like modern scientists, these early Greeks were willing to look beneath the surface appearance of the world, pursuing knowledge about a deeper level of reality. The matter of the world does not appear at first glance as if it is all made of water, or air, or earth, or fire, or all four together, or even of atoms.

Acceptance of the esoteric was taken to an extreme by Parmenides of Elea (the modern Velia) in southern Italy, who was greatly admired by Plato. In the early 400s BC Parmenides taught, contra Heraclitus, that the apparent change and variety in nature are an illusion. His ideas were defended by his pupil Zeno of Elea (not to be confused with other Zenos, such as Zeno the Stoic). In his book *Attacks*, Zeno offered a number of paradoxes to show the impossibility of motion. For instance, to traverse the whole course of a racetrack, it is necessary first to cover half the distance, and then half the remaining distance, and so on indefinitely, so that it is impossible ever to traverse the whole track. By the same reasoning, as far as we can tell from surviving fragments, it appeared to Zeno to be impossible ever to travel *any* given distance, so that all motion is impossible.

Of course, Zeno's reasoning was wrong. As pointed out later by Aristotle,[11] there is no reason why we cannot accomplish an infinite number of steps in a finite time, as long as the time needed for each successive step decreases sufficiently rapidly. It is true that an infinite series like $\frac{1}{2} + \frac{1}{3} + \frac{1}{4} + \ldots$ has an infinite sum, but the infinite series $\frac{1}{2} + \frac{1}{4} + \frac{1}{8} + \ldots$ has a finite sum, in this case equal to 1.

What is most striking is not so much that Parmenides and Zeno were wrong as that they did not bother to explain why, if motion is impossible, things appear to move. Indeed, none of the early Greeks from Thales to Plato, in either Miletus or Abdera or Elea or Athens, ever took it on themselves to explain in detail how their theories about ultimate reality accounted for the appearances of things.

This was not just intellectual laziness. There was a strain of intellectual snobbery among the early Greeks that led them to regard an understanding of appearances as not worth having. This is just one example of an attitude that has blighted much of the history of science. At various times it has been thought that circular orbits are more perfect than elliptical orbits, that gold is more noble than lead, and that man is a higher being than his fellow simians.

Are we now making similar mistakes, passing up opportunities for scientific progress because we ignore phenomena that seem unworthy of our attention? One can't be sure, but I doubt it. Of course, we cannot explore everything, but we choose problems that we think, rightly or wrongly, offer the best prospect for scientific understanding. Biologists who are interested in chromosomes or nerve cells study animals like fruit flies and squid, not noble eagles and lions. Elementary particle physicists are sometimes accused of a snobbish and expensive preoccupation with phenomena at the highest attainable energies, but it is only at high energies that we can create and study hypothetical particles of high mass, like the dark matter particles that astronomers tell us make up five-sixths of the matter of the universe. In any case, we give plenty of attention to phenomena at low energies, like the intriguing mass of neutrinos, about a millionth the mass of the electron.

In commenting on the prejudices of the pre-Socratics, I don't mean to say that a priori reasoning has no place in science. Today, for instance, we expect to find that our deepest physical laws satisfy principles of symmetry, which state that physical laws do not change when we change our point of view in certain definite ways. Just like Parmenides' principle of changelessness, some of these symmetry principles are not immediately apparent in physical phenomena—they are said to be spontaneously broken. That is, the equations of our theories have certain simplicities, for instance treating certain species of particles in the same way, but these simplicities are not shared by the solutions of the equations, which govern actual phenomena. Nevertheless, unlike the commitment of Parmenides to changelessness, the a priori presumption in favor of principles of symmetry arose from many years of experience in searching for physical principles that describe the real world, and broken as well as unbroken symmetries are validated by experiments that confirm their consequences. They do not involve value judgments of the sort we apply to human affairs.

With Socrates, in the late fifth century BC, and Plato, some forty years later, the center of the stage for Greek intellectual life moved to Athens, one of the few cities of Ionian Greeks on the Greek mainland. Almost all of what we know about Socrates comes from his appearance in the dialogues of Plato, and as a comic character in Aristophanes' play *The Clouds*. Socrates does not seem to have put any of his ideas into writing, but as far as we can tell he was not very interested in natural science. In Plato's dialogue *Phaedo* Socrates recalls how he was disappointed in reading a book by Anaxagoras (about whom more in Chapter 7) because Anaxagoras described the Earth, Sun, Moon, and stars in purely physical terms, without regard to what is best.[12]

Plato, unlike his hero Socrates, was an Athenian aristocrat. He was the first Greek philosopher from whom many writings have survived pretty much intact. Plato, like Socrates, was more concerned with human affairs than with the nature of matter. He hoped for a political career that would allow him to put his utopian and antidemocratic ideas into practice. In 367 BC Plato accepted an invitation from Dionysius II to come to Syracuse and help reform its government, but, fortunately for Syracuse, nothing came of the reform project.

In one of his dialogues, the *Timaeus*, Plato brought together the idea of four elements with the Abderite notion of atoms. Plato supposed that the four elements of Empedocles consisted of particles shaped like four of the five solid bodies known in mathematics as regular polyhedrons: bodies with faces that are all identical polygons, with all edges identical, coming together at identical vertices. (See Technical Note 2.) For instance, one of the regular polyhedrons is the cube, whose faces are all identical squares, three squares meeting at each vertex. Plato took atoms of earth to have the shape of cubes. The other regular polyhedrons are the tetrahedron (a pyramid with four triangular faces), the eight-sided octahedron, the twenty-sided icosahedron, and the twelve-sided dodecahedron. Plato supposed that the atoms of fire, air, and water have the shapes respectively of the tetrahedron, octahedron, and icosahedron. This left the dodecahedron unaccounted for. Plato regarded it as representing the *kosmos*. Later Aristotle introduced a fifth element, the ether or quintessence, which he supposed filled the space above the orbit of the Moon.

It has been common in writing about these early speculations regarding the nature of matter to emphasize how they prefigure features of modern science. Democritus is particularly admired; one of the leading universities in modern Greece is named Democritus University. Indeed, the effort to identify the fundamental constituents of matter continued for millennia, though with changes from time to time in the menu of elements. By early modern times alchemists had identified three supposed elements: mercury, salt, and sulfur. The modern idea of chemical elements dates from the chemical revolution instigated by Priestley, Lavoisier, Dalton, and others at the end of the eighteenth century, and now incorporates 92 naturally occurring elements, from hydrogen to uranium (including mercury and sulfur but not salt) plus a growing list of artificially created elements heavier than uranium. Under normal conditions, a pure chemical element consists of atoms all of the same type, and the elements are distinguished from one another by the type of atom of which they are composed. Today we look beyond the chemical elements to the elementary particles of which atoms are composed, but one way or another we continue the search, begun at Miletus, for the fundamental constituents of nature.

Nevertheless, I think one should not overemphasize the modern aspects of Archaic or Classical Greek science. There is an important feature of modern science that is almost completely missing in all the thinkers I have mentioned, from Thales to Plato: none of them attempted to verify or even (aside perhaps from Zeno) seriously to justify their speculations. In reading their writings, one continually wants to ask, "How do you know?" This is just as true of Democritus as of the others. Nowhere in the fragments of his books that survive do we see any effort to show that matter really is composed of atoms.

Plato's ideas about the five elements give a good example of his insouciant attitude toward justification. In *Timaeus*, he starts not with regular polyhedrons but with triangles, which he proposes to join together to form the faces of the polyhedrons. What sort of triangles? Plato proposes that these should be the isosceles right triangle, with angles 45°, 45°, and 90°; and the right triangle with angles 30°, 60°, and 90°. The square faces of the cubic atoms of earth can be formed from two isosceles right triangles, and the triangular faces of the tetrahedral, octahedral, and icosahedral atoms of fire, air, and water (respectively) can each be formed from two of the other right triangles. (The dodecahedron, which mysteriously represents the cosmos, cannot be constructed in this way.) To explain this choice, Plato in *Timaeus* says, "If anyone can tell us of a better choice of triangle for the construction of the four bodies, his criticism will be welcome; but for our part we propose to pass over all the rest. . . . It would be too long a story to give the reason, but if anyone can produce a proof that it is not so we will welcome his achievement."[13] I can imagine the reaction today if I supported a new conjecture about matter in a physics article by saying that it would take too long to explain my reasoning, and challenging my colleagues to prove the conjecture is not true.

Aristotle called the earlier Greek philosophers *physiologi,* and this is sometimes translated as "physicists,"[14] but that is misleading. The word *physiologi* simply means students of nature (*physis*), and the early Greeks had very little in common with today's physicists. Their theories had no bite. Empedocles could speculate about the elements, and Democritus about atoms, but their speculations led to no new information about nature—and certainly to nothing that would allow their theories to be tested.

It seems to me that to understand these early Greeks, it is better to think of them not as physicists or scientists or even philosophers, but as poets.

I should be clear about what I mean by this. There is a narrow sense of poetry, as language that uses verbal devices like meter, rhyme, or alliteration. Even in this narrow sense, Xenophanes, Parmenides, and Empedocles all wrote in poetry. After the Dorian invasions and the breakup of the Bronze Age Mycenaean civilization in the twelfth century BC, the Greeks had become largely illiterate. Without writing, poetry is almost the only way that people can communicate to later generations, because poetry can be remembered in a way that prose cannot. Literacy revived among the Greeks sometime around 700 BC, but the new alphabet borrowed from the Phoenicians was first used by Homer and Hesiod to write poetry, some of it the long-remembered poetry of the Greek dark ages. Prose came later.

Even the early Greek philosophers who wrote in prose, like Anaximander, Heraclitus, and Democritus, adopted a poetic style. Cicero said of Democritus that he was more poetic than many poets. Plato when young had wanted to be a poet, and though he wrote prose and was hostile to poetry in the *Republic*, his literary style has always been widely admired.

I have in mind here poetry in a broader sense: language chosen for aesthetic effect, rather than in an attempt to say clearly what one actually believes to be true. When Dylan Thomas writes, "The force that through the green fuse drives the flower drives my green age," we do not regard this as a serious statement about the unification of the forces of botany and zoology, and we do not seek verification; we (or at least I) take it rather as an expression of sadness about age and death.

At times it seems clear that Plato did not intend to be taken literally. One example mentioned above is his extraordinarily weak argument for the choice he made of two triangles as the basis of all matter. As an even clearer example, in the *Timaeus* Plato introduced the story of Atlantis, which supposedly flourished thousands of years before his own time. Plato could not possibly have seriously thought that he really knew anything about what had happened thousands of years earlier.

I don't at all mean to say that the early Greeks decided to write poetically in order to avoid the need to validate their theories. They felt no such need. Today we test our speculations about nature by using proposed theories to draw more or less precise conclusions that can be tested by observation. This did not occur to the early Greeks, or to many of their successors, for a very simple reason: *they had never seen it done.*

There are signs here and there that even when they did want to be taken seriously, the early Greeks had doubts about their own theories, that they felt reliable knowledge was unattainable. I used one example in my 1972 treatise on general relativity. At the head of a chapter about cosmological speculation, I quoted some lines of Xenophanes: "And as for certain truth, no man has seen it, nor will there ever be a man who knows about the gods and about the things I mention. For if he succeeds to the full in saying what is completely true, he himself is nevertheless unaware of it, and opinion is fixed by fate upon all things."[15] In the same vein, in *On the Forms*, Democritus remarked, "We in reality know nothing firmly" and "That in reality we do not know how each thing is or is not has been shown in many ways."[16]

There remains a poetic element in modern physics. We do not write in poetry; much of the writing of physicists barely reaches the level of prose. But we seek beauty in our theories, and use aesthetic judgments as a guide in our research. Some of us think that this works because we have been trained by centuries of success and failure in physics research to anticipate certain aspects of the laws of nature, and through this experience we have come to feel that these features of nature's laws are beautiful.[17] But we do not take the beauty of a theory as convincing evidence of its truth.

For example, string theory, which describes the different species of elementary particles as various modes of vibration of tiny strings, is very beautiful. It appears to be just barely consistent mathematically, so that its structure is not arbitrary, but largely fixed by the requirement of mathematical consistency. Thus it has the beauty of a rigid art form—a sonnet or a sonata. Unfortunately, string theory has not yet led to any predictions that can be tested experimentally, and as a result theorists (at least most of us) are keeping an open mind as to whether the theory actually applies to the real world. It is this insistence on verification that we most miss in all the poetic students of nature, from Thales to Plato.

2

Music and Mathematics

Even if Thales and his successors had understood that from their theories of matter they needed to derive consequences that could be compared with observation, they would have found the task prohibitively difficult, in part because of the limitations of Greek mathematics. The Babylonians had achieved great competence in arithmetic, using a number system based on 60 rather than 10. They had also developed some simple techniques of algebra, such as rules (though these were not expressed in symbols) for solving various quadratic equations. But for the early Greeks, mathematics was largely geometric. As we have seen, mathematicians by Plato's time had already discovered theorems about triangles and polyhedrons. Much of the geometry found in Euclid's *Elements* was already well known before the time of Euclid, around 300 BC. But even by then the Greeks had only a limited understanding of arithmetic, let alone algebra, trigonometry, or calculus.

The phenomenon that was studied earliest using methods of arithmetic may have been music. This was the work of the followers of Pythagoras. A native of the Ionian island of Samos, Pythagoras emigrated to southern Italy around 530 BC. There, in the Greek city of Croton, he founded a cult that lasted until the 300s BC.

The word "cult" seems appropriate. The early Pythagoreans

left no writings of their own, but according to the stories told by other writers[1] the Pythagoreans believed in the transmigration of souls. They are supposed to have worn white robes and forbidden the eating of beans, because that vegetable resembled the human fetus. They organized a kind of theocracy, and under their rule the people of Croton destroyed the neighboring city of Sybaris in 510 BC.

What is relevant to the history of science is that the Pythagoreans also developed a passion for mathematics. According to Aristotle's *Metaphysics,*[2] "the Pythagoreans, as they are called, devoted themselves to mathematics: they were the first to advance this study, and having been brought up in it, they thought its principles were the principles of all things."

Their emphasis on mathematics may have stemmed from an observation about music. They noted that in playing a stringed instrument, if two strings of equal thickness, composition, and tension are plucked at the same time, the sound is pleasant if the lengths of the strings are in a ratio of small whole numbers. In the simplest case, one string is just half the length of the other. In modern terms, we say that the sounds of these two strings are an octave apart, and we label the sounds they produce with the same letter of the alphabet. If one string is two-thirds the length of the other, the two notes produced are said to form a "fifth," a particularly pleasing chord. If one string is three-fourths the length of the other, they produce a pleasant chord called a "fourth." By contrast, if the lengths of the two strings are not in a ratio of small whole numbers (for instance if the length of one string is, say, 100,000/314,159 times the length of the other), or not in a ratio of whole numbers at all, then the sound is jarring and unpleasant. We now know that there are two reasons for this, having to do with the periodicity of the sound produced by the two strings played together, and the matching of the overtones produced by each string (see Technical Note 3). None of this was understood by the Pythagoreans, or indeed by anyone else until the work of the French priest Marin Mersenne in the seventeenth century. Instead, the Pythagoreans according to Aristotle judged "the whole heaven to be a musical scale."[3] This idea had a long afterlife. For instance, Cicero, in *On the Republic*, tells a story in which the ghost of the great Roman general Scipio Africanus introduces his grandson to the music of the spheres.

It was in pure mathematics rather than in physics that the Pythagoreans made the greatest progress. Everyone has heard of *the* Pythagorean theorem, that the area of a square whose edge is the hypotenuse of a right triangle equals the sum of the areas of the two squares whose edges are the other two sides of the triangle. No one knows which if any of the Pythagoreans proved this theorem, or how. It is possible to give a simple proof based on a theory of proportions, a theory due to the Pythagorean Archytas of Tarentum, a contemporary of Plato. (See Technical Note 4. The proof given as Proposition 46 of Book I of Euclid's *Elements* is more complicated.) Archytas also solved a famous outstanding problem: given a cube, use purely geometric methods to construct another cube of precisely twice the volume.

The Pythagorean theorem led directly to another great discovery: geometric constructions can involve lengths that cannot be expressed as ratios of whole numbers. If the two sides of a right triangle adjacent to the right angle each have a length (in some units of measurement) equal to 1, then the total area of the two squares with these edges is $1^2 + 1^2 = 2$, so according to the Pythagorean theorem the length of the hypotenuse must be a number whose square is 2. But it is easy to show that a number whose square is 2 cannot be expressed as a ratio of whole numbers. (See Technical Note 5.) The proof is given in Book X of Euclid's *Elements*, and mentioned earlier by Aristotle in his *Prior Analytics*[4] as an example of a *reductio ad impossibile*, but without giving the original source. There is a legend that this discovery is due to the Pythagorean Hippasus, possibly of Metapontum in southern Italy, and that he was exiled or murdered by the Pythagoreans for revealing it.

We might today describe this as the discovery that numbers like the square root of 2 are irrational—they cannot be expressed as ratios of whole numbers. According to Plato,[5] it was shown by Theodorus of Cyrene that the square roots of 3, 5, 6, . . . , 15, 17, etc. (that is, though Plato does not say so, the square roots of all the whole numbers other than the numbers 1, 4, 9, 16, etc., that are the squares of whole numbers) are irrational in the same sense. But the early Greeks would not have expressed it this way. Rather, as the translation of Plato has it, the sides of squares whose areas are 2, 3, 5, etc., square feet are "incommensurate" with a single foot. The early Greeks had no conception of any but rational numbers, so for them quantities like the square root of 2 could be given only a geometric significance, and this constraint further impeded the development of arithmetic.

The tradition of concern with pure mathematics was continued in Plato's Academy. Supposedly there was a sign over its entrance, saying that no one should enter who was ignorant of geometry. Plato himself was no mathematician, but he was enthusiastic about mathematics, perhaps in part because, during the journey to Sicily to tutor Dionysius the Younger of Syracuse, he had met the Pythagorean Archytas.

One of the mathematicians at the Academy who had a great influence on Plato was Theaetetus of Athens, who was the title character of one of Plato's dialogues and the subject of another. Theaetetus is credited with the discovery of the five regular solids that, as we have seen, provided a basis for Plato's theory of the elements. The proof* offered in Euclid's *Elements* that these are the only possible convex regular solids may be due to Theaetetus, and Theaetetus also contributed to the theory of what are today called irrational numbers.

The greatest Hellenic mathematician of the fourth century BC was probably Eudoxus of Cnidus, a pupil of Archytas and a con-

* In fact (as discussed in Technical Note 2), whatever may have been proved by Theaetetus, *Elements* does not prove what it claims to prove, that there are only five possible convex regular solids. *Elements* does prove that for regular polyhedrons, there are just five combinations of the number of sides of each face of a polyhedron and of the number of faces that meet at each vertex, but it does not prove that for each combination of these numbers there is just one possible convex regular polyhedron.

temporary of Plato. Though resident much of his life in the city of Cnidus on the coast of Asia Minor, Eudoxus was a student at Plato's Academy, and returned later to teach there. No writings of Eudoxus survive, but he is credited with solving a great number of difficult mathematical problems, such as showing that the volume of a cone is one-third the volume of the cylinder with the same base and height. (I have no idea how Eudoxus could have done this without calculus.) But his greatest contribution to mathematics was the introduction of a rigorous style, in which theorems are deduced from clearly stated axioms. It is this style that we find later in the writings of Euclid. Indeed, many of the details in Euclid's *Elements* have been attributed to Eudoxus.

Though a great intellectual achievement in itself, the development of mathematics by Eudoxus and the Pythagoreans was a mixed blessing for natural science. For one thing, the deductive style of mathematical writing, enshrined in Euclid's *Elements*, was endlessly imitated by workers in natural science, where it is not so appropriate. As we will see, Aristotle's writing on natural science involves little mathematics, but at times it sounds like a parody of mathematical reasoning, as in his discussion of motion in *Physics*: "A, then, will move through B in a time C, and through D, which is thinner, in time E (if the length of B is equal to D), in proportion to the density of the hindering body. For let B be water and D be air."[6] Perhaps the greatest work of Greek physics is *On Floating Bodies* by Archimedes, to be discussed in Chapter 4. This book is written like a mathematics text, with unquestioned postulates followed by deduced propositions. Archimedes was smart enough to choose the right postulates, but scientific research is more honestly reported as a tangle of deduction, induction, and guesswork.

More important than the question of style, though related to it, is a false goal inspired by mathematics: to reach certain truth by the unaided intellect. In his discussion of the education of philosopher kings in the *Republic*, Plato has Socrates argue that astronomy should be done in the same way as geometry. According to Socrates, looking at the sky may be helpful as a spur to the intellect, in the same way that looking at a geometric diagram may be helpful in mathematics, but in both cases real knowledge comes solely through thought. Socrates explains in the *Republic* that "we should use the heavenly bodies merely as illustrations to help us study the other realm, as we would if we were faced with exceptional geometric figures."[7]

Mathematics is the means by which we deduce the consequences of physical principles. More than that, it is the indispensable language in which the principles of physical science are expressed. It often inspires new ideas about the natural sciences, and in turn the needs of science often drive developments in mathematics. The work of a theoretical physicist, Edward Witten, has provided so much insight into mathematics that in 1990 he was awarded one of the highest awards in mathematics, the Fields Medal. But mathematics is not a natural science. Mathematics in itself, without observation, cannot tell us anything about the world. And mathematical theorems can be neither verified nor refuted by observation of the world.

This was not clear in the ancient world, nor indeed even in early modern times. We have seen that Plato and the Pythagoreans considered mathematical objects such as numbers or triangles to be the fundamental constituents of nature, and we shall see that some philosophers regarded mathematical astronomy as a branch of mathematics, not of natural science.

The distinction between mathematics and science is pretty well settled. It remains mysterious to us why mathematics that is invented for reasons having nothing to do with nature often turns out to be useful in physical theories. In a famous article,[8] the physicist Eugene Wigner has written of "the unreasonable effectiveness of mathematics." But we generally have no trouble in distinguishing the ideas of mathematics from principles of science, principles that are ultimately justified by observation of the world.

Where conflicts now sometimes arise between mathematicians and scientists, it is generally over the issue of mathematical rigor. Since the early nineteenth century, researchers in pure mathematics have regarded rigor as essential; definitions and assumptions must be precise, and deductions must follow with absolute certainty. Physicists are more opportunistic, demanding only enough precision and certainty to give them a good chance of avoiding serious mistakes. In the preface of my own treatise on the quantum theory of fields, I admit that "there are parts of this book that will bring tears to the eyes of the mathematically inclined reader."

This leads to problems in communication. Mathematicians have told me that they often find the literature of physics infuriatingly vague. Physicists like myself who need advanced mathematical tools often find that the mathematicians' search for rigor makes their writings complicated in ways that are of little physical interest.

There has been a noble effort by mathematically inclined physicists to put the formalism of modern elementary particle physics—the quantum theory of fields—on a mathematically rigorous basis, and some interesting progress has been made. But nothing in the development over the past half century of the Standard Model of elementary particles has depended on reaching a higher level of mathematical rigor.

Greek mathematics continued to thrive after Euclid. In Chapter 4 we will come to the great achievements of the later Hellenistic mathematicians Archimedes and Apollonius.

3

Motion and Philosophy

After Plato, the Greeks' speculations about nature took a turn toward a style that was less poetic and more argumentative. This change appears above all in the work of Aristotle. Neither a native Athenian nor an Ionian, Aristotle was born in 384 BC at Stagira in Macedon. He moved to Athens in 367 BC to study at the school founded by Plato, the Academy. After the death of Plato in 347 BC, Aristotle left Athens and lived for a while on the Aegean island of Lesbos and at the coastal town of Assos. In 343 BC Aristotle was called back to Macedon by Philip II to tutor his son Alexander, later Alexander the Great.

Macedon came to dominate the Greek world after Philip's army defeated Athens and Thebes at the battle of Chaeronea in 338 BC. After Philip's death in 336 BC Aristotle returned to Athens, where he founded his own school, the Lyceum. This was one of the four great schools of Athens, the others being Plato's Academy, the Garden of Epicurus, and the Colonnade (or Stoa) of the Stoics. The Lyceum continued for centuries, probably until it was closed in the sack of Athens by Roman soldiers under Sulla in 86 BC. It was outlasted, though, by Plato's Academy, which continued in one form or another until AD 529, enduring longer than any European university has lasted so far.

The works of Aristotle that survive appear to be chiefly notes for his lectures at the Lyceum. They treat an amazing variety of subjects: astronomy, zoology, dreams, metaphysics, logic, ethics, rhetoric, politics, aesthetics, and what is usually translated as "physics." According to one modern translator,[1] Aristotle's Greek is "terse, compact, abrupt, his arguments condensed, his thought dense," very unlike the poetic style of Plato. I confess that I find Aristotle frequently tedious, in a way that Plato is not, but although often wrong Aristotle is not silly, in the way that Plato sometimes is.

Plato and Aristotle were both realists, but in quite different senses. Plato was a realist in the medieval sense of the word: he believed in the reality of abstract ideas, in particular of ideal forms of things. It is the ideal form of a pine tree that is real, not the individual pine trees that only imperfectly realize this form. It is the forms that are changeless, in the way demanded by Parmenides and Zeno. Aristotle was a realist in a common modern sense: for him, though categories were deeply interesting, it was individual things, like individual pine trees, that were real, not Plato's forms.

Aristotle was careful to use reason rather than inspiration to justify his conclusions. We can agree with the classical scholar R. J. Hankinson that "we must not lose sight of the fact that Aristotle was a man of his time—and for that time he was extraordinarily perspicacious, acute, and advanced."[2] Nevertheless, there were principles running all through Aristotle's thought that had to be unlearned in the discovery of modern science.

For one thing, Aristotle's work was suffused with teleology: things are what they are because of the purpose they serve. In *Physics*,[3] we read, "But the nature is the end or that for the sake of which. For if a thing undergoes a continuous change toward some end, that last stage is actually that for the sake of which."

This emphasis on teleology was natural for someone like Aristotle, who was much concerned with biology. At Assos and Lesbos Aristotle had studied marine biology, and his father, Nicomachus, had been a physician at the court of Macedon. Friends who know more about biology than I do tell me that Aristotle's writing on animals is admirable. Teleology is natural for anyone who, like Aristotle in *Parts of Animals*, studies the heart or stomach of an animal—he can hardly help asking what purpose it serves.

Indeed, not until the work of Darwin and Wallace in the nineteenth century did naturalists came to understand that although bodily organs serve various purposes, there is no purpose underlying their evolution. They are what they are because they have been naturally selected over millions of years of undirected inheritable variations. And of course, long before Darwin, physicists had learned to study matter and force without asking about the purpose they serve.

Aristotle's early concern with zoology may also have inspired his strong emphasis on taxonomy, on sorting things out in categories. We still use some of this, for instance the Aristotelian classification of governments into monarchies, aristocracies, and not democracies but constitutional governments. But much of it seems pointless. I can imagine how Aristotle might have classified fruits: *All fruits come in three varieties—there are apples, and oranges, and fruits that are neither apples nor oranges.*

One of Aristotle's classifications was pervasive in his work, and became an obstacle for the future of science. He insisted on the distinction between the natural and the artificial. He begins Book II of *Physics*[4] with "Of things that exist, some exist by nature, some from other causes." It was only the natural that was worthy of his attention. Perhaps it was this distinction between the natural and the artificial that kept Aristotle and his followers from being interested in experimentation. What is the good of creating an artificial situation when what are really interesting are natural phenomena?

It is not that Aristotle neglected the observation of natural phenomena. From the delay between seeing lightning and hearing thunder, or seeing oars on a distant trireme striking the water and hearing the sound they make, he concluded that sound travels at a finite speed.[5] We will see that he also made good use of observation in reaching conclusions about the shape of the Earth and about the cause of rainbows. But this was all casual observation of natural phenomena, not the creation of artificial circumstances for the purpose of experimentation.

The distinction between the natural and artificial played a large role in Aristotle's thought about a problem of great importance in the history of science—the motion of falling bodies. Aristotle taught that solid bodies fall down because the natural place of the element earth is downward, toward the center of the cosmos, and sparks fly upward because the natural place of fire is in the heavens. The Earth is nearly a sphere, with its center at the center of the cosmos, because this allows the greatest proportion of earth to approach that center. Also, allowed to fall naturally, a falling body has a speed proportional to its weight. As we read in *On the Heavens*,[6] according to Aristotle, "A given weight moves a given distance in a given time; a weight which is as great and more moves the same distance in a less time, the times being in inverse proportion to the weights. For instance, if one weight is twice another, it will take half as long over a given movement."

Aristotle can't be accused of entirely ignoring the observation of falling bodies. Though he did not know the reason, the resistance of air or any other medium surrounding a falling body has the effect that the speed eventually approaches a constant value, the terminal velocity, which does increase with the falling body's weight. (See Technical Note 6.) Probably more important to Aristotle, the observation that the speed of a falling body increases with its weight fitted in well with his notion that the body falls because the natural place of its material is toward the center of the world.

For Aristotle, the presence of air or some other medium was essential in understanding motion. He thought that without any resistance, bodies would move at infinite speed, an absurdity that led him to deny the possibility of empty space. In *Physics,* he argues, "Let us explain that there is no void existing separately, as some maintain."[7] But in fact it is only the terminal velocity of a falling body that is inversely proportional to the resistance. The terminal velocity would indeed be infinite in the absence of all resistance, but in that case a falling body would never reach terminal velocity.

In the same chapter Aristotle gives a more sophisticated argument, that in a void there would be nothing to which motion could be relative: "in the void things must be at rest; for there is no place to which things can move more or less than to another; since the void in so far as it is void admits no difference."[8] But this is an argument against only an infinite void; otherwise motion in a void can be relative to whatever is outside the void.

Because Aristotle was acquainted with motion only in the presence of resistance, he believed that all motion has a cause.* (Aristotle distinguished four kinds of cause: material, formal, efficient, and final, of which the final cause is teleological—it is the purpose of the change.) That cause must itself be caused by something else, and so on, but the sequence of causes cannot go on forever. We read in *Physics*,[9] "Since everything that is in motion must be moved by something, let us take the case in which a thing is in locomotion and is moved by something that is itself in motion, and that again is moved by something else that is in motion, and that by something else, and so on continually; then the series cannot go on to infinity, but there must be some first mover." The doctrine of a first mover later provided Christianity and Islam with an argument for the existence of God. But as we will see, in the Middle Ages the conclusion that God could not make a void raised troubles for followers of Aristotle in both Islam and Christianity.

Aristotle was not bothered by the fact that bodies do not always move toward their natural place. A stone held in the hand does not fall, but for Aristotle this just showed the effect of ar-

* The Greek word *kineson*, which is usually translated as "motion," actually has a more general significance, referring to any sort of change. Thus Aristotle's classification of types of cause applied not only to change of position, but to any change. The Greek word *fora* refers specifically to change of location, and is usually translated as "locomotion."

tificial interference with the natural order. But he was seriously worried over the fact that a stone thrown upward continues for a while to rise, away from the Earth, even after it has left the hand. His explanation, really no explanation, was that the stone continues upward for a while because of the motion given to it by the air. In Book III of *On the Heavens*, he explains that "the force transmits the movement to the body by first, as it were, tying it up in the air. That is why a body moved by constraint continues to move even when that which gave it the impulse ceases to accompany it."[10] As we will see, this notion was frequently discussed and rejected in ancient and medieval times.

Aristotle's writing on falling bodies is typical at least of his physics—elaborate though non-mathematical reasoning based on assumed first principles, which are themselves based on only the most casual observation of nature, with no effort to test them.

I don't mean to say that Aristotle's philosophy was seen by his followers and successors as an alternative to science. There was no conception in the ancient or medieval world of science as something distinct from philosophy. Thinking about the natural world *was* philosophy. As late as the nineteenth century, when German universities instituted a doctoral degree for scholars of the arts and sciences to give them equal status with doctors of theology, law, and medicine, they invented the title "doctor of philosophy." When philosophy had earlier been compared with some other way of thinking about nature, it was contrasted not with science, but with mathematics.

No one in the history of philosophy has been as influential as Aristotle. As we will see in Chapter 9, he was greatly admired by some Arab philosophers, even slavishly so by Averroes. Chapter 10 tells how Aristotle became influential in Christian Europe in the 1200s, when his thought was reconciled with Christianity by Thomas Aquinas. In the high Middle Ages Aristotle was known simply as "The Philosopher," and Averroes as "The Commentator." After Aquinas the study of Aristotle became the center of university education. In the Prologue to Chaucer's *Canterbury Tales*, we are introduced to an Oxford scholar:

A Clerk there was of Oxenford also . . .
For he would rather have at his bed's head
Twenty books, clad in black or red,
Of Aristotle, and his philosophy,
Than robes rich, or fiddle, or gay psaltery.

Of course, things are different now. It was essential in the discovery of science to separate science from what is now called philosophy. There is active and interesting work on the philosophy *of* science, but it has very little effect on scientific research.

The precocious scientific revolution that began in the fourteenth century and is described in Chapter 10 was largely a revolt against Aristotelianism. In recent years students of Aristotle have mounted something of a counterrevolution. The very influential historian Thomas Kuhn described how he was converted from disparagement to admiration of Aristotle:[11]

> About motion, in particular, his writings seemed to me full of egregious errors, both of logic and of observation. These conclusions were, I felt, unlikely. Aristotle, after all, had been the much-admired codifier of ancient logic. For almost two millennia after his death, his work played the same role in logic that Euclid's played in geometry. . . . How could his characteristic talent have deserted him so systematically when he turned to the study of motion and mechanics? Equally, why had his writings in physics been taken so seriously for so many centuries after his death? . . . Suddenly the fragments in my head sorted themselves out in a new way, and fell in place together. My jaw dropped with surprise, for all at once Aristotle seemed a very good physicist indeed, but of a sort I'd never dreamed possible. . . . I had suddenly found the way to read Aristotelian texts.

I heard Kuhn make these remarks when we both received honorary degrees from the University of Padua, and later asked him to explain. He replied, "What was altered by my own first reading of [Aristotle's writings on physics] was my understanding, not my evaluation, of what they achieved." I didn't understand this: "a very good physicist indeed" seemed to me like an evaluation.

Regarding Aristotle's lack of interest in experiment: the historian David Lindberg[12] remarked, "Aristotle's scientific practice is not to be explained, therefore, as a result of stupidity or deficiency on his part—failure to perceive an obvious procedural improvement—but as a method compatible with the world as he perceived it and well suited to the questions that interest him." On the larger issue of how to judge Aristotle's success, Lindberg added, "It would be unfair and pointless to judge Aristotle's success by the degree to which he anticipated modern science (as though his goal was to answer our questions, rather than his own)." And in a second edition of the same work:[13] "The proper measure of a philosophical system or a scientific theory is not the degree to which it anticipated modern thought, but its degree of success in treating the philosophical and scientific problems of its own day."

I don't buy it. What is important in science (I leave philosophy to others) is not the solution of some popular scientific problems of one's own day, but understanding the world. In the course of this work, one finds out what sort of explanations are possible, and what sort of problems can lead to those explanations. The progress of science has been largely a matter of discovering what questions should be asked.

Of course, one has to try to understand the historical context of scientific discoveries. Beyond that, the task of a historian depends on what he or she is trying to accomplish. If the historian's aim is only to re-create the past, to understand "how it actually was," then it may not be helpful to judge a past scientist's success by modern standards. But this sort of judgment is indispensable if what one wants is to understand how science progressed from its past to its present.

This progress has been something objective, not just an evolution of fashion. Is it possible to doubt that Newton understood more about motion than Aristotle, or that we understand more than Newton? It never was fruitful to ask what motions are natural, or what is the purpose of this or that physical phenomenon.

I agree with Lindberg that it would be unfair to conclude that Aristotle was stupid. My purpose here in judging the past by the standards of the present is to come to an understanding of how difficult it was for even very intelligent persons like Aristotle to learn how to learn about nature. Nothing about the practice of modern science is obvious to someone who has never seen it done.

Aristotle left Athens at the death of Alexander in 323 BC, and died shortly afterward, in 322 BC. According to Michael Matthews,[14] this was "a death that signaled the twilight of one of the brightest intellectual periods in human history." It was indeed the end of the Classical era, but as we shall see, it was also the dawn of an age far brighter scientifically: the era of the Hellenistic.

4

Hellenistic Physics and Technology

Following Alexander's death his empire split into several successor states. Of these, the most important for the history of science was Egypt. Egypt was ruled by a succession of Greek kings, starting with Ptolemy I, who had been one of Alexander's generals, and ending with Ptolemy XV, the son of Cleopatra and (perhaps) Julius Caesar. This last Ptolemy was murdered soon after the defeat of Antony and Cleopatra at Actium in 31 BC, when Egypt was absorbed into the Roman Empire.

This age, from Alexander to Actium,[1] is commonly known as the Hellenistic period, a term (in German, *Hellenismus*) coined in the 1830s by Johann Gustav Droysen. I don't know if this was intended by Droysen, but to my car there is something pejorative about the English suffix "istic." Just as "archaistic," for instance, is used to describe an imitation of the archaic, the suffix seems to imply that Hellenistic culture was not fully Hellenic, that it was a mere imitation of the achievements of the Classical age of the fifth and fourth centuries BC. Those achievements were very great, especially in geometry, drama, historiography, architecture, and sculpture, and perhaps in other arts whose Classical productions have not survived, such as music and painting. But in the Hellenistic age science was brought to heights that not only dwarfed the scientific accomplishments of the Classical era but were not matched until the scientific revolution of the sixteenth and seventeenth centuries.

The vital center of Hellenistic science was Alexandria, the capital city of the Ptolemies, laid out by Alexander at one mouth of the Nile. Alexandria became the greatest city in the Greek world; and later, in the Roman Empire, it was second only to Rome in size and wealth.

Around 300 BC Ptolemy I founded the Museum of Alexandria, as part of his royal palace. It was originally intended as a center of literary and philological studies, dedicated to the nine Muses. But after the accession of Ptolemy II in 285 BC the Museum also became a center of scientific research. Literary studies continued at the Museum and Library of Alexandria, but now at the Museum the eight artistic Muses were outshone by their one scientific sister—Urania, the Muse of astronomy. The Museum and Greek science outlasted the kingdom of the Ptolemies, and, as we shall see, some of the greatest achievements of ancient science occurred in the Greek half of the Roman Empire, and largely in Alexandria.

The intellectual relations between Egypt and the Greek homeland in Hellenistic times were something like the connections between America and Europe in the twentieth century.[2] The riches of Egypt and the generous support of at least the first three Ptolemies brought to Alexandria scholars who had made their names in Athens, just as European scholars flocked to America from the 1930s on. Starting around 300 BC, a former member of the Lyceum, Demetrius of Phaleron, became the first director of the Museum, bringing his library with him from Athens. At around the same time Strato of Lampsacus, another member of the Lyceum, was called to Alexandria by Ptolemy I to serve as tutor to his son, and may have been responsible for the turn of the Museum toward science when that son succeeded to the throne of Egypt.

The sailing time between Athens and Alexandria during the Hellenistic and Roman periods was similar to the time it took for a steamship to go between Liverpool and New York in the twentieth century, and there was a great deal of coming and going between Egypt and Greece. For instance, Strato did not stay in Egypt; he returned to Athens to become the third director of the Lyceum.

Strato was a perceptive observer. He was able to conclude that falling bodies accelerate downward, by observing how drops of water falling from a roof become farther apart as they fall, a continuous stream of water breaking up into separating drops. This is because the drops that have fallen farthest have also been falling longest, and since they are accelerating this means that they are traveling faster than drops following them, which have been falling for a shorter time. (See Technical Note 7.) Strato noted also that when a body falls a very short distance the impact on the ground is negligible, but if it falls from a great height it makes a powerful impact, showing that its speed increases as it falls.[3]

It is probably no coincidence that centers of Greek natural philosophy like Alexandria as well as Miletus and Athens were also centers of commerce. A lively market brings together people from different cultures, and relieves the monotony of agriculture. The commerce of Alexandria was far-ranging: seaborne cargoes being taken from India to the Mediterranean world would cross the Arabian Sea, go up the Red Sea, then go overland to the Nile and down the Nile to Alexandria.

But there were great differences in the intellectual climates of Alexandria and Athens. For one thing, the scholars of the Museum generally did not pursue the kind of all-embracing theories that had preoccupied the Greeks from Thales to Aristotle. As Floris Cohen has remarked,[4] "Athenian thought was comprehensive, Alexandrian piecemeal." The Alexandrians concentrated on understanding specific phenomena, where real progress could be made. These topics included optics and hydrostatics, and above all astronomy, the subject of Part II.

It was no failing of the Hellenistic Greeks that they retreated from the effort to formulate a general theory of everything. Again and again, it has been an essential feature of scientific progress to understand which problems are ripe for study and which are not. For instance, leading physicists at the turn of the twentieth century, including Hendrik Lorentz and Max Abraham, devoted themselves to understanding the structure of the recently discovered electron. It was hopeless; no one could have made progress in understanding the nature of the electron before the advent of quantum mechanics some two decades later. The development of the special theory of relativity by Albert Einstein was made possible by Einstein's refusal to worry about what electrons are. Instead he worried about how observations of anything (including electrons) depend on the motion of the observer. Then Einstein himself in his later years addressed the problem of the unification of the forces of nature, and made no progress because no one at the time knew enough about these forces.

Another important difference between Hellenistic scientists and their Classical predecessors is that the Hellenistic era was less afflicted by a snobbish distinction between knowledge for its own sake and knowledge for use—in Greek, *episteme* versus *techne* (or in Latin, *scientia* versus *ars*). Throughout history, many philosophers have viewed inventors in much the same way that the court chamberlain Philostrate in *A Midsummer Night's Dream* described Peter Quince and his actors: "Hard-handed men, who work now in Athens, and never yet labor'd with their minds." As a physicist whose research is on subjects like elementary particles and cosmology that have no immediate practical application, I am certainly not going to say anything against knowledge for its own sake, but doing scientific research to fill human needs has a wonderful way of forcing the scientist to stop versifying and to confront reality.[5]

Of course, people have been interested in technological improvement since early humans learned how to use fire to cook food and how to make simple tools by banging one stone on another. But the persistent intellectual snobbery of the Classical intelligentsia kept philosophers like Plato and Aristotle from directing their theories toward technological applications.

Though this prejudice did not disappear in Hellenistic times, it became less influential. Indeed, people, even those of ordinary birth, could become famous as inventors. A good example is Ctesibius of Alexandria, a barber's son, who around 250 BC invented suction and force pumps and a water clock that kept time more accurately than earlier water clocks by keeping a constant level of water in the vessel from which the water flowed. Ctesibius was famous enough to be remembered two centuries later by the Roman Vitruvius in his treatise *On Architecture*.

It is important that some technology in the Hellenistic age was developed by scholars who were also concerned with systematic scientific inquiries, inquiries that were sometimes themselves used in aid of technology. For instance, Philo of Byzantium, who spent time in Alexandria around 250 BC, was a military engineer who in *Mechanice syntaxism* wrote about harbors, fortifications, sieges, and catapults (work based in part on that of Ctesibius). But in *Pneumatics*, Philo also gave experimental arguments supporting the view of Anaximenes, Aristotle, and Strato that air is real. For instance, if an empty bottle is submerged in water with its mouth open but facing downward, no water will flow into it, because there is nowhere for the air in the bottle to go; but if a hole is opened so that air is allowed to leave the bottle, then water will flow in and fill the bottle.[6]

There was one scientific subject of practical importance to which Greek scientists returned again and again, even into the Roman period: the behavior of light. This concern dates to the beginning of the Hellenistic era, with the work of Euclid.

Little is known of the life of Euclid. He is believed to have lived in the time of Ptolemy I, and may have founded the study of mathematics at the Museum in Alexandria. His best-known work is the *Elements*,[7] which begins with a number of geometric definitions, axioms, and postulates, and moves on to more or less rigorous proofs of increasingly sophisticated theorems. But Euclid also wrote the *Optics*, which deals with perspective, and his name is associated with the *Catoptrics*, which studies reflection by mirrors, though modern historians do not believe that he was its author.

When one thinks of it, there is something peculiar about reflection. When you look at the reflection of a small object in a flat mirror, you see the image at a definite spot, not spread out over the mirror. Yet there are many paths one can draw from the object to various spots on the mirror and then to the eye.* Apparently there is just one path that is actually taken, so that the image appears at the one point where this path strikes the mirror. But what determines the location of that point on the mirror? In the *Catoptrics* there appears a fundamental principle that answers this question: the angles that a light ray makes with a flat mirror when it strikes the mirror and when it is reflected are equal. Only one light path can satisfy this condition.

We don't know who in the Hellenistic era actually discovered this principle. We do know, though, that sometime around AD 60 Hero of Alexandria in his own *Catoptrics* gave a mathematical proof of the equal-angles rule, based on the assumption that the path taken by a light ray in going from the object to the mirror and then to the eye of the observer is the path of shortest length. (See Technical Note 8.) By way of justification for this principle, Hero was content to say only, "It is agreed that Nature does nothing in vain, nor exerts herself needlessly."[8] Perhaps he was motivated by the teleology of Aristotle—everything happens for a purpose. But Hero was right; as we will see in Chapter 14, in the seventeenth century Huygens was able to deduce the principle of shortest distance (actually shortest time) from the wave nature of light. The same Hero who explored the fundamentals of optics used that knowledge to invent an instrument of practical surveying, the theodolite, and also explained the action of siphons and designed military catapults and a primitive steam engine.

The study of optics was carried further about AD 150 in Alex-

* It was generally supposed in the ancient world that when we see something the light travels from the eye to the object, as if vision were a sort of touching that requires us to reach out to what is seen. In the following discussion I will take for granted the modern understanding, that in vision light travels from the object to the eye. Fortunately, in analyzing reflection and refraction, it makes no difference which way the light is going.

andria by the great astronomer Claudius Ptolemy (no kin of the kings). His book *Optics* survives in a Latin translation of a lost Arabic version of the lost Greek original (or perhaps of a lost Syriac intermediary). In this book Ptolemy described measurements that verified the equal-angles rule of Euclid and Hero. He also applied this rule to reflection by curved mirrors, of the sort one finds today in amusement parks. He correctly understood that reflections in a curved mirror are just the same as if the mirror were a plane, tangent to the actual mirror at the point of reflection.

In the final book of *Optics* Ptolemy also studied refraction, the bending of light rays when they pass from one transparent medium such as air to another transparent medium such as water. He suspended a disk, marked with measures of angle around its edge, halfway in a vessel of water. By sighting a submerged object along a tube mounted on the disk, he could measure the angles that the incident and refracted rays make with the normal to the surface, a line perpendicular to the surface, with an accuracy ranging from a fraction of a degree to a few degrees.[9] As we will see in Chapter 13, the correct law relating these angles was worked out by Fermat in the seventeenth century by a simple extension of the principle that Hero had applied to reflection: in refraction, the path taken by a ray of light that goes from the object to the eye is not the shortest, but the one that takes the least time. The distinction between shortest distance and least time is irrelevant for reflection, where the reflected and incident ray are passing through the same medium, and distance is simply proportional to time; but it does matter for refraction, where the speed of light changes as the ray passes from one medium to another. This was not understood by Ptolemy; the correct law of refraction, known as Snell's law (or in France, Descartes' law), was not discovered experimentally until the early 1600s AD.

The most impressive of the scientist-technologists of the Hellenistic era (or perhaps any era) was Archimedes. Archimedes lived in the 200s BC in the Greek city of Syracuse in Sicily, but is believed to have made at least one visit to Alexandria. He is credited with inventing varieties of pulleys and screws, and various instruments of war, such as a "claw," based on his understanding of the lever, with which ships lying at anchor near shore could be seized and capsized. One invention used in agriculture for centuries was a large screw, by which water could be lifted from streams to irrigate fields. The story that Archimedes used curved mirrors that concentrated sunlight to defend Syracuse by setting Roman ships on fire is almost certainly a fable, but it illustrates his reputation for technological wizardry.

In *On the Equilibrium of Bodies*, Archimedes worked out the rule that governs balances: a bar with weights at both ends is in equilibrium if the distances from the fulcrum on which the bar rests to both ends are inversely proportional to the weights. For instance, a bar with five pounds at one end and one pound at the other end is in equilibrium if the distance from the fulcrum to the one-pound weight is five times larger than the distance from the fulcrum to the five-pound weight.

The greatest achievement of Archimedes in physics is contained in his book *On Floating Bodies*.[10] Archimedes reasoned that if some part of a fluid was pressed down harder than another part by the weight of fluid or floating or submerged bodies above it, then the fluid would move until all parts were pressed down by the same weight. As he put it,

> Let it be supposed that a fluid is of such a character that, the parts lying evenly and being continuous, that part which is thrust the less is driven along by that which is thrust the more; and that each of its parts is thrust by the fluid which is above it in a perpendicular direction if the fluid be sunk in anything and compressed by anything else.

From this Archimedes deduced that a floating body would sink to a level such that the weight of the water displaced would equal its own weight. (This is why the weight of a ship is called its "displacement.") Also, a solid body that is too heavy to float and is immersed in the fluid, suspended by a cord from the arm of a balance, "will be lighter than its true weight by the weight of the fluid displaced." (See Technical Note 9.) The ratio of the true weight of a body and the decrease in its weight when it is suspended in water thus gives the body's "specific gravity," the ratio of its weight to the weight of the same volume of water. Each material has a characteristic specific gravity: for gold it is 19.32, for lead 11.34, and so on. This method, deduced from a systematic theoretical study of fluid statics, allowed Archimedes to tell whether a crown was made of pure gold or gold alloyed with cheaper metals. It is not clear that Archimedes ever put this method into practice, but it was used for centuries to judge the composition of objects.

Even more impressive were Archimedes' achievements in mathematics. By a technique that anticipated the integral calculus, he was able to calculate the areas and volumes of various plane figures and solid bodies. For instance, the area of a circle is one-half the circumference times the radius (see Technical Note 10). Using geometric methods, he was able to show that what we call pi (Archimedes did not use this term), the ratio of the circumference of a circle to its diameter, is between $3\frac{1}{7}$ and $3\frac{10}{71}$. Cicero said that he had seen on the tombstone of Archimedes a cylinder circumscribed about a sphere, the surface of the sphere touching the sides and both bases of the cylinder, like a single tennis ball just fitting into a tin can. Apparently Archimedes was most proud of having proved that in this case the volume of the sphere is two-thirds the volume of the cylinder.

There is an anecdote about the death of Archimedes, related by the Roman historian Livy. Archimedes died in 212 BC during the sack of Syracuse by Roman soldiers under Marcus Claudius Marcellus. (Syracuse had been taken over by a pro-Carthaginian faction during the Second Punic War.) As Roman soldiers swarmed over Syracuse, Archimedes was supposedly found by the soldier who killed him, while he was working out a problem in geometry.

Aside from the incomparable Archimedes, the greatest Hellenistic mathematician was his younger contemporary Apollonius. Apollonius was born around 262 BC in Perga, a city on the southeast coast of Asia Minor, then under the control of the rising kingdom of Pergamon, but he visited Alexandria in the times of both Ptolemy III and Ptolemy IV, who between them ruled from 247 to 203 BC. His great work was on conic sections: the ellipse, parabola, and hyperbola. These are curves that can be formed by a plane slicing through a cone at various angles. Much later, the theory of conic sections was crucially important to Kepler and Newton, but it found no physical applications in the ancient world.

Brilliant work, but with its emphasis on geometry, there were techniques missing from Greek mathematics that are essential in modern physical science. The Greeks never learned to write and manipulate algebraic formulas. Formulas like $E = mc^2$ and $F = ma$ are at the heart of modern physics. (Formulas were used in purely mathematical work by Diophantus, who flourished in Alexandria around AD 250, but the symbols in his equations were restricted to standing for whole or rational numbers, quite unlike the symbols in the formulas of physics.) Even where geometry is important, the modern physicist tends to derive what is needed by expressing geometric facts algebraically, using the techniques of analytic geometry invented in the seventeenth century by René Descartes and others, and described in Chapter 13. Perhaps because of the deserved prestige of Greek mathematics, the geometric style persisted until well into the scientific revolution of the seventeenth century. When Galileo in his 1623 book *The Assayer* wanted to sing the praises of mathematics, he spoke of geometry:* "Philosophy is written in this all-encompassing book that is constantly open to our eyes, that is the universe; but it cannot be understood unless one first learns to understand the language and knows the characters in which it is written. It is written in

* *The Assayer* is a polemic against Galileo's Jesuit adversaries, taking the form of a letter to the papal chamberlain Virginio Cesarini. As we will see in Chapter 11, Galileo in *The Assayer* was attacking the correct view of Tycho Brahe and the Jesuits that comets are farther from Earth than the Moon is. (The quotation here is taken from the translation by Maurice A. Finocchiaro, in *The Essential Galileo*, Hackett, Indianapolis, Ind., 2008, p. 183.)

mathematical language, and its characters are triangles, circles, and other geometrical figures; without these it is humanly impossible to understand a word of it, and one wanders in a dark labyrinth." Galileo was somewhat behind the times in emphasizing geometry over algebra. His own writing uses some algebra, but is more geometric than that of some of his contemporaries, and far more geometric than what one finds today in physics journals.

In modern times a place has been made for pure science, science pursued for its own sake without regard to practical applications. In the ancient world, before scientists learned the necessity of verifying their theories, the technological applications of science had a special importance, for when one is going to use a scientific theory rather than just talk about it, there is a large premium on getting it right. If Archimedes by his measurements of specific gravity had identified a gilded lead crown as being made of solid gold, he would have become unpopular in Syracuse.

I don't want to exaggerate the extent to which science-based technology was important in Hellenistic or Roman times. Many of the devices of Ctesibius and Hero seem to have been no more than toys, or theatrical props. Historians have speculated that in an economy based on slavery there was no demand for labor-saving devices, such as might have been developed from Hero's toy steam engine. Military and civil engineering *were* important in the ancient world, and the kings in Alexandria supported the study of catapults and other artillery, perhaps at the Museum, but this work does not seem to have gained much from the science of the time.

The one area of Greek science that did have great practical value was also the one that was most highly developed. It was astronomy, to which we will turn in Part II.

There is a large exception to the remark above that the existence of practical applications of science provided a strong incentive to get the science right. It is the practice of medicine. Until modern times the most highly regarded physicians persisted in practices, like bleeding, whose value had never been established experimentally, and that in fact did more harm than good. When in the nineteenth century the really useful technique of antisepsis was introduced, a technique for which there *was* a scientific basis, it was at first actively resisted by most physicians. Not until well into the twentieth century were clinical trials required before medicines could be approved for use. Physicians did learn early on to recognize various diseases, and for some they had effective remedies, such as Peruvian bark—which contains quinine—for malaria. They knew how to prepare analgesics, opiates, emetics, laxatives, soporifics, and poisons. But it is often remarked that until sometime around the beginning of the twentieth century the average sick person would do better avoiding the care of physicians.

It is not that there was no theory behind the practice of medicine. There was humorism, the theory of the four humors—blood, phlegm, black bile, and yellow bile, which (respectively) make us sanguine, phlegmatic, melancholy, or choleric. Humorism was introduced in classical Greek times by Hippocrates, or by colleagues of his whose writings were ascribed to him. As briefly stated much later by John Donne in "The Good Morrow," the theory held that "whatever dies was not mixed equally." The theory of humorism was adopted in Roman times by Galen of Pergamon, whose writings became enormously influential among the Arabs and then in Europe after about AD 1000. I am not aware of any effort while humorism was generally accepted ever to test its effectiveness experimentally. (Humorism survives today in Ayurveda, traditional Indian medicine, but with just three humors: phlegm, bile, and wind.)

In addition to humorism, physicians in Europe until modern times were expected to understand another theory with supposed medical applications: astrology. Ironically, the opportunity for physicians to study these theories at universities gave medical doctors much higher prestige than surgeons, who knew how to do really useful things like setting broken bones but until modern times were not usually trained in universities.

So why did the doctrines and practices of medicine continue

so long without correction by empirical science? Of course, progress is harder in biology than in astronomy. As we will discuss in Chapter 8, the apparent motions of the Sun, Moon, and planets are so regular that it was not difficult to see that an early theory was not working very well; and this perception led, after a few centuries, to a better theory. But if a patient dies despite the best efforts of a learned physician, who can say what is the cause? Perhaps the patient waited too long to see the doctor. Perhaps he did not follow the doctor's orders with sufficient care.

At least humorism and astrology had an air of being scientific. What was the alternative? Going back to sacrificing animals to Aesculapius?

Another factor may have been the extreme importance to patients of recovery from illness. This gave physicians authority over them, an authority that physicians had to maintain in order to impose their supposed remedies. It is not only in medicine that persons in authority will resist any investigation that might reduce their authority.

5

Ancient Science and Religion

The pre-Socratic Greeks took a great step toward modern science when they began to seek explanations of natural phenomena without reference to religion. This break with the past was at best tentative and incomplete. As we saw in Chapter 1, Diogenes Laertius described the doctrine of Thales as not only that "water is the universal primary substance," but also that "the world is animate and full of divinities." Still, if only in the teachings of Leucippus and Democritus, a beginning had been made. Nowhere in their surviving writings on the nature of matter is there any mention of the gods.

It was essential for the discovery of science that religious ideas be divorced from the study of nature. This divorce took many centuries, not being largely complete in physical science until the eighteenth century, nor in biology even then.

It is not that the modern scientist makes a decision from the start that there are no supernatural persons. That happens to be my view, but there are good scientists who are seriously religious. Rather, the idea is to see how far one can go without supposing supernatural intervention. Only in this way can we do science, because once one invokes the supernatural, anything can be explained, and no explanation can be verified. This is why the "intelligent design" ideology being promoted today is not science—it is rather the abdication of science.

Plato's speculations were suffused with religion. In *Timaeus* he described how a god placed the planets in their orbits, and he may have thought that the planets were deities themselves. Even when Hellenic philosophers dispensed with the gods, some of them described nature in terms of human values and emotions, which generally interested them more than the inanimate world. As we have seen, in discussing changes in matter, Anaximander spoke of justice, and Empedocles of strife. Plato thought that the elements and other aspects of nature were worth studying not for their own sake, but because for him they exemplified a kind of goodness, present in the natural world as well as in human affairs. His religion was informed by this sense, as shown by a passage from the *Timaeus*: "For God desired that, so far as possible, all things should be good and nothing evil; wherefore, when He took over all that was visible, seeing that it was not in a state of rest but in a state of discordant and disorderly motion, He brought it into order out of disorder, deeming that the former state was in all ways better than the latter."[1]

Today, we continue to seek order in nature, but we do not think it is an order rooted in human values. Not everyone has been happy about this. The great twentieth-century physicist Erwin Schrödinger argued for a return to the example of antiquity,[2] with its fusion of science and human values. In the same spirit, the historian Alexandre Koyré considered the present divorce of science and what we now call philosophy "disastrous."[3] My own view is that this yearning for a holistic approach to nature is precisely what scientists have had to outgrow. We simply do not find anything in the laws of nature that in any way corresponds to ideas of goodness, justice, love, or strife, and we cannot rely on philosophy as a reliable guide to scientific explanation.

It is not easy to understand in just what sense the pagans actually believed in their own religion. Those Greeks who had traveled or read widely knew that a great variety of gods and goddesses were worshipped in the countries of Europe, Asia, and Africa. Some of the Greeks tried to see these as the same deities under different names. For instance, the pious historian Herodotus reported, not that the native Egyptians worshipped a goddess named Bubastus who resembled the Greek goddess Artemis, but rather that they worshipped Artemis under the name of Bubastus. Others supposed that these deities were all different and all real, and even included foreign gods in their own worship. Some of the Olympian gods, such as Dionysus and Aphrodite, were imports from Asia.

Among other Greeks, however, the multiplicity of gods and goddesses promoted disbelief. The pre-Socratic Xenophanes famously commented, "Ethiopians have gods with snub noses and black hair, Thracians gods with gray eyes and red hair," and remarked, "But if oxen (and horses) and lions had hands or could draw with hands and create works of art like those made by men, horses would draw pictures of gods like horses, and oxen of gods like oxen, and they would make the bodies [of their gods] in accordance with the form that each species itself possesses."[4] In contrast to Herodotus, the historian Thucydides showed no signs of religious belief. He criticized the Athenian general Nicias for a disastrous decision to suspend an evacuation of his troops from the campaign against Syracuse because of a lunar eclipse. Thucydides explained that Nicias was "over-inclined to divination and such things."[5]

Skepticism became especially common among Greeks who concerned themselves with understanding nature. As we have seen, the speculations of Democritus about atoms were entirely naturalistic. The ideas of Democritus were adopted as an antidote to religion, first by Epicurus of Samos, who settled in Athens and at the beginning of the Hellenistic era founded the Athenian school known as the Garden. Epicurus in turn inspired the Roman poet Lucretius. Lucretius' poem *On the Nature of Things* moldered in monastic libraries until its rediscovery in 1417, after which it had a large influence in Renaissance Europe. Stephen Greenblatt[6] has traced the impact of Lucretius on Machiavelli, More, Shakespeare, Montaigne, Gassendi,* Newton, and Jef-

* Pierre Gassendi was a French priest and philosopher who tried to reconcile the atomism of Epicurus and Lucretius with Christianity.

ferson. Even where paganism was not abandoned, there was a growing tendency among the Greeks to take it allegorically, as a clue to hidden truths. As Gibbon said, "The extravagance of the Grecian mythology proclaimed, with a clear and audible voice, that the pious inquirer, instead of being scandalized or satisfied with the literal sense, should diligently explore the occult wisdom, which had been disguised, by the prudence of antiquity, under the mask of folly and of fable."[7] The search for hidden wisdom led in Roman times to the emergence of the school known to moderns as Neoplatonism, founded in the third century AD by Plotinus and his student Porphyry. Though not scientifically creative, the Neoplatonists retained Plato's regard for mathematics; for instance, Porphyry wrote a life of Pythagoras and a commentary on Euclid's *Elements*. Looking for hidden meanings beneath surface appearances is a large part of the task of science, so it is not surprising that the Neoplatonists maintained at least an interest in scientific matters.

Pagans were not much concerned to police each other's private beliefs. There were no authoritative written sources of pagan religious doctrine analogous to the Bible or the Koran. The *Iliad* and *Odyssey* and Hesiod's *Theogony* were understood as literature, not theology. Paganism had plenty of poets and priests, but it had no theologians. Still, open expressions of atheism were dangerous. At least in Athens an accusation of atheism was occasionally used as a weapon in political debate, and philosophers who expressed disbelief in the pagan pantheon could feel the wrath of the state. The pre-Socratic philosopher Anaxagoras was forced to flee Athens for teaching that the Sun is not a god but a hot stone, larger than the Peloponnesus.

Plato in particular was anxious to preserve the role of religion in the study of nature. He was so appalled by the nontheistic teaching of Democritus that he decreed in Book 10 of the *Laws* that in his ideal society anyone who denied that the gods were real and that they intervened in human affairs would be condemned to five years of solitary confinement, with death to follow if the prisoner did not repent.

In this as in much else, the spirit of Alexandria was different from that of Athens. I do not know of any Hellenistic scientists whose writings expressed any interest in religion, nor do I know of any who suffered for their disbelief.

Religious persecution was not unknown under the Roman Empire. Not that there was any objection to foreign gods. The pantheon of the later Roman Empire expanded to include the Phrygian Cybele, the Egyptian Isis, and the Persian Mithras. But whatever else one believed, it was necessary as a pledge of loyalty to the state also to publicly honor the official Roman religion. According to Gibbon, the religions of the Roman Empire "were all considered by the people, as equally true, by the philosopher, as equally false, and by the magistrate, as equally useful."[8] Christians were persecuted not because they believed in Jehovah or Jesus, but because they publicly denied the Roman religion; they would generally be exonerated if they put a pinch of incense on the altar of the Roman gods.

None of this led to interference with the work of Greek scientists under the empire. Hipparchus and Ptolemy were never persecuted for their nontheistic theories of the planets. The pious pagan emperor Julian criticized the followers of Epicurus, but did nothing to persecute them.

Though illegal because of its rejection of the state religion, Christianity spread widely through the empire in the second and third centuries. It was made legal in the year 313 by Constantine I, and was made the sole legal religion of the empire by Theodosius I in 380. During those years, the great achievements of Greek science were coming to an end. This has naturally led historians to ask whether the rise of Christianity had something to do with the decline of original work in science.

In the past attention centered on possible conflicts between the teachings of religion and the discoveries of science. For instance, Copernicus dedicated his masterpiece *On the Revolutions of the Heavenly Bodies* to Pope Paul III, and in the dedication warned against using passages of Scripture to contradict the work of science. He cited as a horrible example the views of Lactantius, the Christian tutor of Constantine's eldest son:

> But if perchance there are certain "idle talkers" who take it on themselves to pronounce judgment, though wholly ignorant of mathematics, and if by shamelessly distorting the sense of some passage in Holy Writ to suit their purpose, they dare to reprehend and to attack my work; they worry me so little that I shall even scorn their judgments as foolhardy. For it is not unknown that Lactantius, otherwise a distinguished writer but hardly a mathematician, speaks in an utterly childish fashion concerning the shape of the Earth, when he laughs at those who said that the Earth has the form of a globe.[9]

This was not quite fair. Lactantius did say that it was impossible for sky to be under the Earth.[10] He argued that if the world were a sphere then there would have to be people and animals living at the antipodes. This is absurd; there is no reason why people and animals would have to inhabit every part of a spherical Earth. And what would be wrong if there were people and animals at the antipodes? Lactantius suggests that they would tumble into "the bottom part of the sky." He then acknowledges the contrary view of Aristotle (not quoting him by name) that "it is the nature of things for weight to be drawn to the center," only to accuse those who hold this view of "defending nonsense with nonsense." Of course it is Lactantius who was guilty of nonsense, but contrary to what Copernicus suggested, Lactantius was relying not on Scripture, but only on some extremely shallow reasoning about natural phenomena. All in all, I don't think that the direct conflict between Scripture and scientific knowledge was an important source of tension between Christianity and science.

Much more important, it seems to me, was the widespread view among the early Christians that pagan science is a distraction from the things of the spirit that ought to concern us. This goes back to the very beginnings of Christianity, to Saint Paul, who warned: "Beware lest any man spoil you through philosophy and vain deceit, after the tradition of men, after the rudiments of the world, and not after Christ." [11] The most famous statement along these lines is due to the church father Tertullian, who around the year 200 asked, "What does Athens have to do with Jerusalem, or the Academy with the Church?" (Tertullian chose Athens and the Academy to symbolize Hellenic philosophy, with which he presumably was more familiar than he was with the science of Alexandria.) We find a sense of disillusion with pagan learning in the most important of the church fathers, Augustine of Hippo. Augustine studied Greek philosophy when young (though only in Latin translations) and boasted of his grasp of Aristotle, but he later asked, "And what did it profit me that I could read and understand all the books I could get in the so-called 'liberal arts,' when I was actually a slave of wicked lust?" [12] Augustine was also concerned with conflicts between Christianity and pagan philosophy. Toward the end of his life, in 426, he looked back at his past writing, and commented, "I have been rightly displeased, too, with the praise with which I extolled Plato or the Platonists or the Academic philosophers beyond what was proper for such irreligious men, especially those against whose great errors Christian teaching must be defended." [13]

Another factor: Christianity offered opportunities for advancement in the church to intelligent young men, some of whom might otherwise have become mathematicians or scientists. Bishops and presbyters were generally exempt from the jurisdiction of the ordinary civil courts, and from taxation. A bishop such as Cyril of Alexandria or Ambrose of Milan could exercise considerable political power, much more than a scholar at the Museum in Alexandria or the Academy in Athens. This was something new. Under paganism religious offices had gone to men of wealth or political power, rather than wealth and power going to men of religion. For instance, Julius Caesar and his successors won the office of supreme pontiff, not as a recognition of piety or learning, but as a consequence of their political power.

Greek science survived for a while after the adoption of Christianity, though mostly in the form of commentaries on earlier work. The philosopher Proclus, working in the fifth century at the Neoplatonic successor to Plato's Academy in Athens, wrote a commentary on Euclid's *Elements*, with some original contributions. In Chapter 8 I will have occasion to quote a later member of the Academy, Simplicius, for his remarks, in a commentary on Aristotle, about Plato's views on planetary orbits. In the late 300s there was Theon of Alexandria, who wrote a commentary on Ptolemy's great work of astronomy, the *Almagest*, and prepared an improved edition of Euclid. His famous daughter Hypatia became head of the city's Neoplatonic school. A century later in Alexandria the Christian John of Philoponus wrote commentaries on Aristotle, in which he took issue with Aristotle's doctrines concerning motion. John argued that the reason bodies thrown upward do not immediately fall down is not that they are carried by the air, as Aristotle had thought, but rather that when they are thrown bodies are given some quality that keeps them moving, an anticipation of later ideas of impetus or momentum. But there were no more creative scientists or mathematicians of the caliber of Eudoxus, Aristarchus, Hipparchus, Euclid, Eratosthenes, Archimedes, Apollonius, Hero, or Ptolemy.

Whether or not because of the rise of Christianity, soon even the commentators disappeared. Hypatia was killed in 415 by a mob, egged on by Bishop Cyril of Alexandria, though it is difficult to say whether this was for religious or political reasons. In 529 the emperor Justinian (who presided over the reconquest of Italy and Africa, the codification of Roman law, and the building of the great church of Santa Sophia in Constantinople) ordered the closing of the Neoplatonic Academy of Athens. On this event, though Gibbon is predisposed against Christianity, he is too eloquent not to be quoted:

> The Gothic arms were less fatal to the schools of Athens than the establishment of a new religion, whose ministers superseded the exercise of reason, resolved every question by an article of faith, and condemned the infidel or skeptic to eternal

> flames. In many a volume of laborious controversy they espoused the weakness of the understanding and the corruption of the heart, insulted human nature in the sages of antiquity, and proscribed the spirit of philosophical inquiry, so repugnant to the doctrine, or at least to the temper, of a humble believer.[14]

The Greek half of the Roman Empire survived until AD 1453, but as we shall see in Chapter 9, long before then the vital center of scientific research had moved east, to Baghdad.

PART II

GREEK ASTRONOMY

The science that in the ancient world saw the greatest progress was astronomy. One reason is that astronomical phenomena are simpler than those on the Earth's surface. Though the ancients did not know it, then as now the Earth and the other planets moved around the Sun on nearly circular orbits, at nearly constant velocities, under the influence of a single force—gravitation—and they spun on their axes at essentially constant rates. The same applied to the Moon in its motion around the Earth. In consequence the Sun, Moon, and planets appeared from Earth to move in a regular and predictable way that could be and was studied with considerable precision.

The other special feature of ancient astronomy is that it was useful, in a way that ancient physics generally was not. The uses of astronomy are discussed in Chapter 6.

Chapter 7 discusses what, flawed as it was, can be considered a triumph of Hellenistic science: the measurement of the sizes of the Sun, Moon, and Earth, and the distances to the Sun and Moon. Chapter 8 treats the problem posed by the apparent motion of the planets, a problem that continued to concern astronomers through the Middle Ages, and that eventually led to the birth of modern science.

6

The Uses of Astronomy[1]

Even before the start of history, the sky must have been commonly used as a compass, a clock, and a calendar. It could not have been difficult to notice that the Sun rises every morning in more or less the same direction, that during the day one can tell how much time there is before night from the height of the Sun in the sky, and that hot weather will follow the time of year when the day lasts longest.

We know that the stars were used for similar purposes very early in history. Around 3000 BC the Egyptians knew that the crucial event in their agriculture, the flooding of the Nile in June, coincided with the heliacal rising of the star Sirius. (This is the day in the year when Sirius first becomes visible just before dawn; earlier in the year it is not visible at night, and later it is visible well before dawn.) Homer, writing before 700 BC, compares Achilles to Sirius, which is high in the sky at the end of summer: "that star, which comes on in the autumn and whose conspicuous brightness far outshines the stars that are numbered in the night's darkening, the star they give the name of Orion's Dog, which is brightest among the stars, and yet is wrought as a sign of evil and brings on the great fever for unfortunate mortals."[2] A little later, the poet Hesiod in *Works and Days* told farmers that grapes are best cut at the heliacal rising of Arcturus, and that plowing should be done at the cosmical setting of the Pleiades constellation. (This is the day in the year when these stars first are seen to set just before sunrise; earlier in the year they do not set at all before the Sun comes up, and later they set well before dawn.) Following Hesiod, calendars known as *parameg-mata*, which gave the risings and settings of conspicuous stars for each day, became widely used by Greeks whose city-states had no other shared way of identifying dates.

By watching the stars at night, not obscured by the light of modern cities, observers in many early civilizations could see clearly that, with a few exceptions (about which more later), the stars always remain in the same places relative to one another. This is why constellations do not change from night to night or from year to year. But the whole firmament of these "fixed" stars seems to revolve each night from east to west around a point in the sky that is always due north, and hence is known as the north celestial pole. In modern terms, this is the point toward which the axis of the Earth extends if it is continued from the Earth's north pole out into the sky.

This observation made it possible for stars to be used very early by mariners for finding directions at night. Homer tells how Odysseus on his way home to Ithaca is trapped by the nymph Calypso on her island in the western Mediterranean, until Zeus orders Calypso to send Odysseus on his way. She tells Odysseus to keep the "Great Bear, that some have called the Wain . . . on his left hand as he crossed the main."[3] The Bear, of course, is Ursa Major, the constellation also known as the Wagon (or Wain), and in modern times as the Big Dipper. Ursa Major is near the north celestial pole. Thus in the latitude of the Mediterranean Ursa Major never sets ("would never bathe in or dip in the Ocean stream," as Homer puts it) and is always more or less in the north. With the Bear on his left, Odysseus would keep sailing east, toward Ithaca.

Some Greeks learned to do better with other constellations. According to the biography of Alexander the Great by Arrian, although most sailors in his time used Ursa Major to tell north, the Phoenicians, the ace sailors of the ancient world, used Ursa

Minor, a constellation that is not as conspicuous as Ursa Major but is closer to the north celestial pole. The poet Callimachus, as quoted by Diogenes Laertius,[4] claimed that the use of Ursa Minor goes back to Thales.

The Sun also seems during the day to revolve from east to west around the north celestial pole. Of course, we cannot usually see stars during the day, but Heraclitus[5] and perhaps others before him seem to have realized that the stars are always there, though with their light blotted out during the day by the light of the Sun. Some stars can be seen just before dawn or just after sunset, when the position of the Sun in the sky is known, and from this it became clear that the Sun does not keep a fixed position relative to the stars. Rather, as was well known very early in Babylon and India, in addition to seeming to revolve from east to west every day along with the stars, the Sun also moves each year around the sky from west to east through a path known as the zodiac, marked in order by the traditional constellations Aries, Taurus, Gemini, Cancer, Leo, Virgo, Libra, Scorpio, Sagittarius, Capricorn, Aquarius, and Pisces. As we will see, the Moon and planets also travel through the zodiac, though not on precisely the same paths. The particular path through these constellations followed by the Sun is known as the "ecliptic."

Once the zodiac was understood, it was easy to locate the Sun in the background of stars. Just notice what constellation of the zodiac is highest in the sky at midnight; the Sun is in the constellation of the zodiac that is directly opposite. Thales is supposed to have given 365 days as the time it takes for the Sun to make one complete circuit of the zodiac.

One can think of the firmament of stars as a revolving sphere surrounding the Earth, with the north celestial pole above the Earth's north pole. But the zodiac is not the equator of this sphere. Rather, as Anaximander is supposed to have discovered, the zodiac is tilted by 23½° with respect to the celestial equator, with Cancer and Gemini closest to the north celestial pole, and Capricorn and Sagittarius farthest from it. In modern terms, this tilt, which is responsible for the seasons, is due to the fact that the axis of the Earth's rotation is not perpendicular to the plane of its orbit, which is pretty close to the plane in which almost all objects in the solar system move, but is tilted from the perpendicular by an angle of 23½°; in the northern summer or winter the Sun is respectively in the direction toward which or away from which the Earth's north pole is tilted.

Astronomy began to be a precise science with the introduction of a device known as a gnomon, which allowed accurate measurements of the Sun's apparent motions. The gnomon, credited by the fourth-century bishop Eusebius of Caesarea to Anaximander but by Herodotus to the Babylonians, is simply a vertical pole, placed in a level patch of ground open to the Sun's rays. With the gnomon, one can accurately tell when it is noon; it is the moment in the day when Sun is highest, so that the shadow of the gnomon is shortest. At noon anywhere north of the tropics the Sun is due south, and the shadow of the gnomon therefore points due north, so one can permanently mark out on the ground the points of the compass. The gnomon also provides a calendar. During the spring and summer the Sun rises somewhat north of east, whereas during the autumn and winter it comes up south of east. When the shadow of the gnomon at dawn points due west, the Sun is rising due east, and the date must be either the vernal equinox, when winter gives way to spring; or the autumnal equinox, when summer ends and autumn begins. The summer and winter solstices are the days in the year when the shadow of the gnomon at noon is respectively shortest or longest. (A sundial is different from a gnomon; its pole is parallel to the Earth's axis rather than to the vertical direction, so that its shadow at a given hour is in the same direction every day. This makes a sundial more useful as a clock, but useless as a calendar.)

The gnomon provides a nice example of an important link between science and technology: an item of technology invented for practical purposes can open the way to scientific discoveries. With the gnomon, it was possible to make a precise count of the days in each season, such as the period from one equinox to the next solstice, or from then to the following equinox. In this way,

Euctemon, an Athenian contemporary of Socrates, discovered that the lengths of the seasons are not precisely equal. This was not what one would expect if the Sun goes around the Earth (or the Earth around the Sun) in a circle at constant speed, with the Earth (or the Sun) at the center, in which case the seasons would be of equal length. Astronomers tried for centuries to understand the inequality of the seasons, but the correct explanation of this and other anomalies was not found until the seventeenth century, when Johannes Kepler realized that the Earth moves around the Sun on an orbit that is elliptical rather than circular, with the Sun not at the center of the orbit but off to one side at a point called a focus, and moves at a speed that increases and decreases as the Earth approaches closer to and recedes farther from the Sun.

The Moon also seems to revolve like the stars each night from east to west around the north celestial pole; and over longer times it moves, like the Sun, through the zodiac from west to east, but taking a little more than 27 days instead of a year to make a full circle against the background of stars. Because the Sun appears to move through the zodiac in the same direction, though more slowly, the Moon takes about 29½ days to return to the same position relative to the Sun. (Actually 29 days, 12 hours, 44 minutes, and 3 seconds.) Since the phases of the Moon depend on the relative position of it and the Sun, this interval of about 29½ days is the lunar month,* the time from one new moon to the next. It was noticed early that eclipses of the Moon occur at full moon about every 18 years, when the Moon's path against the background of stars crosses that of the Sun.†

In some respects the Moon provides a more convenient cal-

* To be more precise, this is known as the "synodic" lunar month. The 27-day period for the Moon to return to the same position relative to the fixed stars is known as the "sidereal" lunar month.

† This does not happen every month, because the plane of the orbit of the Moon around the Earth is slightly tilted with respect to the plane of the orbit of the Earth around the Sun. The Moon crosses the plane of the Earth's orbit twice every sidereal month, but this happens at full moon, when the Earth is between the Sun and the Moon, only about once every 18 years.

endar than the Sun. Observing the phase of the Moon on any given night, one can easily tell approximately how many days have passed since the last new moon—much more easily than one can judge the time of year just by looking at the Sun. So lunar calendars were common in the ancient world, and still survive, for example for religious purposes in Islam. But of course, for purposes of agriculture or sailing or war, one needs to anticipate the changes of seasons, and these are governed by the Sun. Unfortunately, there is not a whole number of lunar months in the year—the year is approximately 11 days longer than 12 lunar months—so the date of any solstice or equinox would not remain fixed in a calendar based on the phases of the Moon.

Another familiar complication is that the year itself is not a whole number of days. This led to the introduction, in the time of Julius Caesar, of a leap year every fourth year. But this created further problems, because the year is not precisely 365¼ days, but 11 minutes longer.

Countless efforts, far too many to go into here, have been made throughout history to construct calendars that take account of these complications. A fundamental contribution was made around 432 BC by Meton of Athens, possibly a partner of Euctemon. Perhaps by the use of Babylonian records, Meton noticed that 19 years is almost precisely 235 lunar months. They differ by only 2 hours. So one can make a calendar covering 19 years rather than one year, in which both the time of year and the phase of the Moon are correctly identified for each day. The calendar then repeats itself for every successive 19-year period. But though 19 years is nearly exactly 235 lunar months, it is about a third of a day less than 6,940 days, so Meton had to prescribe that after every few 19-year cycles a day would be dropped from the calendar.

The effort of astronomers to reconcile calendars based on the Sun and Moon is illustrated by the definition of Easter. The Council of Nicaea in AD 325 decreed that Easter should be celebrated on the first Sunday following the first full moon following the vernal equinox. In the reign of Theodosius I it was declared a capital crime to celebrate Easter on the wrong day. Unfortunately the precise date when the vernal equinox is actually observed varies from place to place on the surface of the Earth.* To avoid the horror of Easter being celebrated on different days in different places, it was necessary to prescribe a definite date for the vernal equinox, and also for the first full moon following it. The Roman church in late antiquity adopted the Metonic cycle for this purpose, but the monastic communities of Ireland adopted an older Jewish 84-year cycle. The struggle in the seventh century between Roman missionaries and Irish monks for control over the English church was largely a conflict over the date of Easter.

Until modern times the construction of calendars has been a major occupation of astronomers, leading up to the adoption of our modern calendar in 1582 under the auspices of Pope Gregory XIII. For purposes of calculating the date of Easter, the date of the vernal equinox is now fixed to be March 21, but it is March 21 as given by the Gregorian calendar in the West and by the Julian calendar in the Orthodox churches of the East. So Easter is still celebrated on different days in different parts of the world.

Though scientific astronomy found useful applications in the Hellenic era, this did not impress Plato. There is a revealing exchange in the *Republic* between Socrates and his foil Glaucon.[6] Socrates suggests that astronomy should be included in the education of philosopher kings, and Glaucon readily agrees: "I mean, it's not only farmers and sailors who need to be sensitive to the seasons, months, and phases of the year; it's just as important for military purposes as well." Socrates calls this naive. For

* The equinox is the moment when the Sun in its motion against the background of stars crosses the celestial equator. (In modern terms, it is the moment when the line between the Earth and the Sun becomes perpendicular to the Earth's axis.) At points on the Earth with different longitude, this moment occurs at different times of day, so there may be a one-day difference in the date that different observers report the equinox. Similar remarks apply to the phases of the Moon.

him, the point of astronomy is that "studying this kind of subject cleans and re-ignites a particular mental organ . . . and this organ is a thousand times more worth preserving than any eye, since it is the only organ which can see truth." This intellectual snobbery was less common in Alexandria than in Athens, but it appears for instance in the first century AD, in the writing of the philosopher Philo of Alexandria, who remarks that "that which is appreciable by the intellect is at all times superior to that which is visible to the outward senses."[7] Fortunately, perhaps under the pressure of practical needs, astronomers learned not to rely on intellect alone.

7

Measuring the Sun, Moon, and Earth

One of the most remarkable achievements of Greek astronomy was the measurement of the sizes of the Earth, Sun, and Moon, and the distances of the Sun and Moon from the Earth. It is not that the results obtained were numerically accurate. The observations on which these calculations were based were too crude to yield accurate sizes and distances. But for the first time mathematics was being correctly used to draw quantitative conclusions about the nature of the world.

In this work, it was essential first to understand the nature of eclipses of the Sun and Moon, and to discover that the Earth is a sphere. Both the Christian martyr Hippolytus and Aëtius, a much-quoted philosopher of uncertain date, credit the earliest understanding of eclipses to Anaxagoras, an Ionian Greek born around 500 BC at Clazomenae (near Smyrna), who taught in Athens.[1] Perhaps relying on the observation of Parmenides that the bright side of the Moon always faces the Sun, Anaxagoras concluded, "It is the Sun that endows the Moon with its brilliance."[2] From this, it was natural to infer that eclipses of the Moon occur when the Moon passes through the Earth's shadow. He is also supposed to have understood that eclipses of the Sun occur when the Moon's shadow falls on the Earth.

On the shape of the Earth, the combination of reason and observation served Aristotle very well. Diogenes Laertius and the Greek geographer Strabo credit Parmenides with knowing, long before Aristotle, that the Earth is a sphere, but we have no idea how (if at all) Parmenides reached this conclusion. In *On the Heavens* Aristotle gave both theoretical and empirical arguments for the spherical shape of the Earth. As we saw in Chapter 3, according to Aristotle's a priori theory of matter the heavy elements earth and (less so) water seek to approach the center of the cosmos, while air and (more so) fire tend to recede from it. The Earth is a sphere, whose center coincides with the center of the cosmos, because this allows the greatest amount of the element earth to approach this center. Aristotle did not rest on this theoretical argument, but added empirical evidence for the spherical shape of the Earth. The Earth's shadow on the Moon during a lunar eclipse is curved,* and the position of stars in the sky seems to change as we travel north or south:

> In eclipses the outline is always curved, and, since it is the interposition of the Earth that makes the eclipse, the form of the line will be caused by the form of the Earth's surface, which is therefore spherical. Again, our observation of the stars make[s] it evident, not only that the Earth is circular, but also that it is a circle of no great size. For quite a small change of position on our part to south or north causes a manifest alteration of the horizon. There is much change, I mean, in the stars which are overhead, and the stars seen are different, as one moves northward or southward. Indeed there are some stars seen in Egypt and in the neighborhood of Cyprus that are not seen in the northerly regions; and stars, which in the north are never beyond the range of observation, in those regions rise and set.[3]

* It has been argued (in O. Neugebauer, *A History of Ancient Mathematical Astronomy*, Springer-Verlag, New York, 1975, pp. 1093–94), that Aristotle's reasoning about the shape of the Earth's shadow on the Moon is inconclusive, since an infinite variety of terrestrial and lunar shapes would give the same curved shadow.

It is characteristic of Aristotle's attitude toward mathematics that he made no attempt to use these observations of stars to give a quantitative estimate of the size of the Earth. Apart from this, I find it puzzling that Aristotle did not also cite a phenomenon that must have been familiar to every sailor. When a ship at sea is first seen on a clear day at a great distance it is "hull down on the horizon"—the curve of the Earth hides all but the tops of its masts—but then, as it approaches, the rest of the ship becomes visible.*

Aristotle's understanding of the spherical shape of the Earth was no small achievement. Anaximander had thought that the Earth is a cylinder, on whose flat face we live. According to Anaximenes, the Earth is flat, while the Sun, Moon, and stars float on the air, being hidden from us when they go behind high parts of the Earth. Xenophanes had written, "This is the upper limit of the Earth that we see at our feet; but the part beneath goes down to infinity."[4] Later, both Democritus and Anaxagoras had thought like Anaximenes that the Earth is flat.

I suspect that the persistent belief in the flatness of the Earth may have been due to an obvious problem with a spherical Earth: if the Earth is a sphere, then why do travelers not fall off? This was nicely answered by Aristotle's theory of matter. Aristotle understood that there is no universal direction "down," along which objects placed anywhere tend to fall. Rather, everywhere on Earth things made of the heavy elements earth and water tend to fall toward the center of the world, in agreement with observation.

* Samuel Eliot Morison cited this argument in his biography of Columbus (*Admiral of the Ocean Sea*, Little Brown, Boston, Mass., 1942) to show, contrary to a widespread supposition, that it was well understood before Columbus set sail that the Earth is a sphere. The debate in the court of Castile over whether to support the proposed expedition of Columbus concerned not the shape of the Earth, but its *size*. Columbus thought the Earth was small enough so that he could sail from Spain to the east coast of Asia without running out of food and water. He was wrong about the size of the Earth, but of course was saved by the unexpected appearance of America between Europe and Asia.

In this respect, Aristotle's theory that the natural place of the heavier elements is in the center of the cosmos worked much like the modern theory of gravitation, with the important difference that for Aristotle there was just one center of the cosmos, while today we understand that any large mass will tend to contract to a sphere under the influence of its own gravitation, and then will attract other bodies toward its own center. Aristotle's theory did not explain why any body other than the Earth should be a sphere, and yet he knew that at least the Moon is a sphere, reasoning from the gradual change of its phases, from full to new and back again.[5]

After Aristotle, the overwhelming consensus among astronomers and philosophers (aside from a few like Lactantius) was that the Earth is a sphere. With the mind's eye, Archimedes even saw the spherical shape of the Earth in a glass of water; in Proposition 2 of *On Floating Bodies*, he demonstrates, "The surface of any fluid at rest is the surface of a sphere whose center is the Earth."[6] (This would be true only in the absence of surface tension, which Archimedes neglected.)

Now I come to what in some respects is the most impressive example of the application of mathematics to natural science in the ancient world: the work of Aristarchus of Samos. Aristarchus was born around 310 BC on the Ionian island of Samos; studied as a pupil of Strato of Lampsacus, the third head of the Lyceum in Athens; and then worked at Alexandria until his death around 230 BC. Fortunately, his masterwork *On the Sizes and Distances of the Sun and Moon* has survived.[7] In it, Aristarchus takes four astronomical observations as postulates:

1. "At the time of Half Moon, the Moon's distance from the Sun is less than a quadrant by one-thirtieth of a quadrant." (That is, when the Moon is just half full, the angle between the lines of sight to the Moon and to the Sun is less than 90° by 3°, so it is 87°.)
2. The Moon just covers the visible disk of the Sun during a solar eclipse.

3. "The breadth of the Earth's shadow is that of two Moons." (The simplest interpretation is that at the position of the Moon, a sphere with twice the diameter of the Moon would just fill the Earth's shadow during a lunar eclipse. This was presumably found by measuring the time from when one edge of the Moon began to be obscured by the Earth's shadow to when it became entirely obscured, the time during which it was entirely obscured, and the time from then until the eclipse was completely over.)
4. "The Moon subtends one fifteenth part of the zodiac." (The complete zodiac is a full 360° circle, but Aristarchus here evidently means one sign of the zodiac; the zodiac consists of 12 constellations, so one sign occupies an angle of 360°/12 = 30°, and one fifteenth part of that is 2°.)

From these assumptions, Aristarchus deduced in turn that:

1. The distance from the Earth to the Sun is between 19 and 20 times larger than the distance of the Earth to the Moon.
2. The diameter of the Sun is between 19 and 20 times larger than the diameter of the Moon.
3. The diameter of the Earth is between $^{108}/_{43}$ and $^{60}/_{19}$ times larger than the diameter of the Moon.
4. The distance from the Earth to the Moon is between 30 and $^{45}/_{2}$ times larger than the diameter of the Moon.

At the time of his work, trigonometry was not known, so Aristarchus had to go through elaborate geometric constructions to get these upper and lower limits. Today, using trigonometry, we would get more precise results; for instance, we would conclude from point 1 that the distance from the Earth to the Sun is larger than the distance from the Earth to the Moon by the secant (the reciprocal of the cosine) of 87°, or 19.1, which is indeed between 19 and 20. (This and the other conclusions of Aristarchus are re-derived in modern terms in Technical Note 11.)

From these conclusions, Aristarchus could work out the sizes of the Sun and Moon and their distances from the Earth, all in terms of the diameter of the Earth. In particular, by combining points 2 and 3, Aristarchus could conclude that the diameter of the Sun is between $^{361}/_{60}$ and $^{215}/_{27}$ times larger than the diameter of the Earth.

The reasoning of Aristarchus was mathematically impeccable, but his results were quantitatively way off, because points 1 and 4 in the data he used as a starting point were badly in error. When the Moon is half full, the actual angle between the lines of sight to the Sun and to the Moon is not 87° but 89.853°, which makes the Sun 390 times farther away from the Earth than the Moon, and hence much larger than Aristarchus thought. This measurement could not possibly have been made by naked-eye astronomy, though Aristarchus could have correctly reported that when the Moon is half full the angle between the lines of sight to the Sun and Moon *is not less* than 87°. Also, the visible disk of the Moon subtends an angle of 0.519°, not 2°, which makes the distance of the Earth to the Moon more like 111 times the diameter the Moon. Aristarchus certainly could have done better than this, and there is a hint in Archimedes' *The Sand Reckoner* that in later work he did so.*

* There is a fascinating remark by Archimedes in *The Sand Reckoner*, that Aristarchus had found that the "Sun appeared to be about $^{1}/_{720}$ part of the zodiac" (*The Works of Archimedes*, trans. T. L. Heath, Cambridge University Press, Cambridge, 1897, p. 223). That is, the angle subtended on Earth by the disk of the Sun is $^{1}/_{720}$ times 360°, or 0.5°, not far from the correct value 0.519°. Archimedes even claimed that he had verified this by his own observations. But as we have seen, in his surviving work Aristarchus had given the angle subtended by the disk of the Moon the value 2°, and he had noted that the disks of the Sun and Moon have the same apparent size. Was Archimedes quoting a later measurement by Aristarchus, of which no report has survived? Was he quoting his own measurement, and attributing it to Aristarchus? I have heard scholars suggest that the source of the discrepancy is a copying error or a misinterpretation of the text, but this seems very unlikely. As already noted, Aristarchus had concluded from his measurement of the angular size of the Moon that its distance from the Earth must be between 30 and $^{45}/_{2}$ times greater than the Moon's diameter, a result quite incompatible with an apparent size of around 0.5°. Modern trigonometry tells us on the other hand that if the Moon's apparent size were 2°, then its

It is not the errors in his observations that mark the distance between the science of Aristarchus and our own. Occasional serious errors continue to plague observational astronomy and experimental physics. For instance, in the 1930s the rate at which the universe is expanding was thought to be about seven times faster than we now know it actually is. The real difference between Aristarchus and today's astronomers and physicists is not that his observational data were in error, but that he never tried to judge the uncertainty in them, or even acknowledged that they might be imperfect.

Physicists and astronomers today are trained to take experimental uncertainty very seriously. Even though as an undergraduate I knew that I wanted to be a theoretical physicist who would never do experiments, I was required along with all other physics students at Cornell to take a laboratory course. Most of our time in the course was spent estimating the uncertainty in the measurements we made. But historically, this attention to uncertainty was a long time in coming. As far as I know, no one in ancient or medieval times ever tried seriously to estimate the uncertainty in a measurement, and as we will see in Chapter 14, even Newton could be cavalier about experimental uncertainties.

We see in Aristarchus a pernicious effect of the prestige of mathematics. His book reads like Euclid's *Elements*: the data in points 1 through 4 are taken as postulates, from which his results are deduced with mathematical rigor. The observational error in his results was very much greater than the narrow ranges that he rigorously demonstrated for the various sizes and distances. Perhaps Aristarchus did not mean to say that the angle between the lines of sight to the Sun and the Moon when half full is really 87°, but only took that as an example, to illustrate what could be deduced. Not for nothing was Aristarchus known to his con-

---

distance from the Earth would be 28.6 times its diameter, a number that is indeed between 30 and $45/2$. (*The Sand Reckoner* is not a serious work of astronomy, but a demonstration by Archimedes that he could calculate very large numbers, such as the number of grains of sand needed to fill the sphere of the fixed stars.)

temporaries as "the Mathematician," in contrast to his teacher Strato, who was known as "the Physicist."

But Aristarchus did get one important point qualitatively correct: the Sun is much bigger than the Earth. To emphasize the point, Aristarchus noted that the volume of the Sun is at least $(361/60)^3$ (about 218) times larger than the volume of the Earth. Of course, we now know that it is much bigger than that.

There are tantalizing statements by both Archimedes and Plutarch that Aristarchus had concluded from the great size of the Sun that it is not the Sun that goes around the Earth, but the Earth that goes around the Sun. According to Archimedes in *The Sand Reckoner*,[8] Aristarchus had concluded not only that the Earth goes around the Sun, but also that the Earth's orbit is tiny compared with the distance to the fixed stars. It is likely that Aristarchus was dealing with a problem raised by any theory of the Earth's motion. Just as objects on the ground seem to be moving back and forth when viewed from a carousel, so the stars ought to seem to move back and forth during the year when viewed from the moving Earth. Aristotle had seemed to realize this, when he commented[9] that if the Earth moved, then "there would have to be passings and turnings of the fixed stars. Yet no such thing is observed. The same stars always rise and set in the same parts of the Earth." To be specific, if the Earth goes around the Sun, then each star should seem to trace out in the sky a closed curve, whose size would depend on the ratio of the diameter of the Earth's orbit around the Sun to the distance to the star.

So if the Earth goes around the Sun, why didn't ancient astronomers see this apparent annual motion of the stars, known as annual parallax? To make the parallax small enough to have escaped observation, it was necessary to assume that the stars are at least a certain distance away. Unfortunately, Archimedes in *The Sand Reckoner* made no explicit mention of parallax, and we don't know if anyone in the ancient world used this argument to put a lower bound on the distance to the stars.

Aristotle had given other arguments against a moving Earth.

Some were based on his theory of natural motion toward the center of the universe, mentioned in Chapter 3, but one other argument was based on observation. Aristotle reasoned that if the Earth were moving, then bodies thrown straight upward would be left behind by the moving Earth, and hence would fall to a place different from where they were thrown. Instead, as he remarks,[10] "heavy bodies forcibly thrown quite straight upward return to the point from which they started, even if they are thrown to an unlimited distance." This argument was repeated many times, for instance by Claudius Ptolemy (whom we met in Chapter 4) around AD 150, and by Jean Buridan in the Middle Ages, until (as we will see in Chapter 10) an answer to this argument was given by Nicole Oresme.

It might be possible to judge how far the idea of a moving Earth spread in the ancient world if we had a good description of an ancient orrery, a mechanical model of the solar system.* Cicero in *On the Republic* tells of a conversation about an orrery in 129 BC, twenty-three years before he himself was born. In this conversation, one Lucius Furius Philus was supposed to have told about an orrery made by Archimedes that had been taken after the fall of Syracuse by its conqueror Marcellus, and that was later seen in the house of Marcellus' grandson. It is not easy to tell from this thirdhand account how the orrery worked (and some pages from this part of *De Re Publica* are missing), but at one point in the story Cicero quotes Philus as saying that on this orrery "were delineated the motion of the Sun and Moon and of those five stars that are called wanderers [planets]," which

* There is a famous ancient device known as the Antikythera Mechanism, discovered in 1901 by sponge divers off the island of Antikythera, in the Mediterranean between Crete and mainland Greece. It is believed to have been lost in a shipwreck sometime around 150 to 100 BC. Though the Antikythera Mechanism is now a corroded mass of bronze, scholars have been able to deduce its workings by X-ray studies of its interior. Apparently it is not an orrery but a calendrical device, which tells the apparent position of the Sun and planets in the zodiac on any date. The most important thing about it is that its intricate gearwork provides evidence of the high competence of Hellenistic technology.

certainly suggests that the orrery had a moving Sun, rather than a moving Earth.[11]

As we will see in Chapter 8, long before Aristarchus the Pythagoreans had the idea that both the Earth and the Sun move around a central fire. For this they had no evidence, but somehow their speculations were remembered, while that of Aristarchus was almost forgotten. Just one ancient astronomer is known to have adopted the heliocentric ideas of Aristarchus: the obscure Seleucus of Seleucia, who flourished around 150 BC. In the time of Copernicus and Galileo, when astronomers and churchmen wanted to refer to the idea that the Earth moves, they called it Pythagorean, not Aristarchean. When I visited the island of Samos in 2005, I found plenty of bars and restaurants named for Pythagoras, but none for Aristarchus of Samos.

It is easy to see why the idea of the Earth's motion did not take hold in the ancient world. We do not feel this motion, and no one before the fourteenth century understood that there is no reason why we *should* feel it. Also, neither Archimedes nor anyone else gave any indication that Aristarchus had worked out how the motion of the planets would appear from a moving Earth.

The measurement of the distance from the Earth to the Moon was much improved by Hipparchus, generally regarded as the greatest astronomical observer of the ancient world.[12] Hipparchus made astronomical observations in Alexandria from 161 BC to 146 BC, and then continued until 127 BC, perhaps on the island of Rhodes. Almost all his writings have been lost; we know about his astronomical work chiefly from the testimony of Claudius Ptolemy, three centuries later. One of his calculations was based on the observation of an eclipse of the Sun, now known to have occurred on March 14, 189 BC. In this eclipse the disk of the Sun was totally hidden at Alexandria, but only four-fifths hidden on the Hellespont (the modern Dardanelles, between Asia and Europe). Since the apparent diameters of the Moon and Sun are very nearly equal, and were measured by Hipparchus to be about 33' (minutes of arc) or 0.55°, Hipparchus could conclude that the direction to the Moon as seen from the Hellespont and from Alexandria differed by one-fifth of 0.55°, or 0.11°. From observations of the Sun Hipparchus knew the latitudes of the Hellespont and Alexandria, and he knew the location of the Moon in the sky at the time of the eclipse, so he was able to work out the distance to the Moon as a multiple of the radius of the Earth. Considering the changes during a lunar month of the apparent size of the Moon, Hipparchus concluded that the distance from the Earth to the Moon varies from 71 to 83 Earth radii. The average distance is actually about 60 Earth radii.

I should pause to say something about another great achievement of Hipparchus, even though it is not directly relevant to the measurement of sizes and distances. Hipparchus prepared a star catalog, a list of about 800 stars, with the celestial position given for each star. It is fitting that our best modern star catalog, which gives the positions of 118,000 stars, was made by observations from an artificial satellite named in honor of Hipparchus.

The measurements of star positions by Hipparchus led him to the discovery of a remarkable phenomenon, which was not understood until the work of Newton. To explain this discovery, it is necessary to say something about how celestial positions are described. The catalog of Hipparchus has not survived, and we don't know just how he described these positions. There are two possibilities commonly used from Roman times on. One method, used later in the star catalog of Ptolemy,[13] pictures the fixed stars as points on a sphere, whose equator is the ecliptic, the path through the stars apparently traced in a year by the Sun. Celestial latitude and longitude locate stars on this sphere in the same way that ordinary latitude and longitude give the location of points on the Earth's surface.* In a different method, which may have been used by Hipparchus,[14] the stars are again taken as points on a sphere, but this sphere is oriented with the Earth's

* The celestial latitude is the angular separation between the star and the ecliptic. While on Earth we measure longitude from the Greenwich meridian, the celestial longitude is the angular separation, on a circle of fixed celestial latitude, between the star and the celestial meridian on which lies the position of the Sun at the vernal equinox.

axis rather than the ecliptic; the north pole of this sphere is the north celestial pole, about which the stars seem to revolve every night. Instead of latitude and longitude, the coordinates on this sphere are known as declination and right ascension.

According to Ptolemy,[15] the measurements of Hipparchus were sufficiently accurate for him to notice that the celestial longitude (or right ascension) of the star Spica had changed by 2° from what had been observed long before at Alexandria by the astronomer Timocharis. It was not that Spica had changed its position relative to the other stars; rather, the location of the Sun on the celestial sphere at the autumnal equinox, the point from which celestial longitude was then measured, had changed.

It is difficult to be precise about how long this change took. Timocharis was born around 320 BC, about 130 years before Hipparchus; but it is believed that he died young around 280 BC, about 160 years before Hipparchus. If we guess that about 150 years separated their observations of Spica, then these observations indicate that the position of the Sun at the autumnal equinox changes by about 1° every 75 years.* At that rate, this equinoctal point would precess through the whole 360° circle of the zodiac in 360 times 75 years, or 27,000 years.

Today we understand that the precession of the equinoxes is caused by a wobble of the Earth's axis (like the wobble of the axis of a spinning top) around a direction perpendicular to the plane of its orbit, with the angle between this direction and the Earth's axis remaining nearly fixed at 23.5°. The equinoxes are the dates when the line separating the Earth and the Sun is perpendicular to the Earth's axis, so a wobble of the Earth's axis causes the equinoxes to precess. We will see in Chapter 14 that this wobble was first explained by Isaac Newton, as an effect of the gravitational attraction of the Sun and Moon for the equatorial bulge of the Earth. It actually takes 25,727 years for the Earth's axis to wobble by a full 360°. It is remarkable how accurately the work

* On the basis of his own observations of the star Regulus, Ptolemy in *Almagest* gave a figure of 1° in approximately 100 years.

of Hipparchus predicted this great span of time. (By the way, it is the precession of the equinoxes that explains why ancient navigators had to judge the direction of north from the position in the sky of constellations near the north celestial pole, rather from the position of the North Star, Polaris. Polaris has not moved relative to the other stars, but in ancient times the Earth's axis did not point at Polaris as it does now, and in the future Polaris will again not be at the north celestial pole.)

Returning now to celestial measurement, all of the estimates by Aristarchus and Hipparchus expressed the size and distances of the Moon and Sun as multiples of the size of the Earth. The size of the Earth was measured a few decades after the work of Aristarchus by Eratosthenes. Eratosthenes was born in 273 BC at Cyrene, a Greek city on the Mediterranean coast of today's Libya, founded around 630 BC, that had become part of the kingdom of the Ptolemies. He was educated in Athens, partly at the Lyceum, and then around 245 BC was called by Ptolemy III to Alexandria, where he became a fellow of the Museum and tutor to the future Ptolemy IV. He was made the fifth head of the Library around 234 BC. His main works—*On the Measurement of the Earth, Geographic Memoirs, and Hermes*—have all unfortunately disappeared, but were widely quoted in antiquity.

The measurement of the size of the Earth by Eratosthenes was described by the Stoic philosopher Cleomedes in *On the Heavens,*[16] sometime after 50 BC. Eratosthenes started with the observations that at noon at the summer solstice the Sun is directly overhead at Syene, an Egyptian city that Eratosthenes supposed to be due south of Alexandria, while measurements with a gnomon at Alexandria showed the noon Sun at the solstice to be one-fiftieth of a full circle, or 7.2°, away from the vertical. From this he could conclude that the Earth's circumference is 50 times the distance from Alexandria to Syene. (See Technical Note 12.) The distance from Alexandria to Syene had been measured (probably by walkers, trained to make each step the same length) as 5,000 stadia, so the circumference of the Earth must be 250,000 stadia.

How good was this estimate? We don't know the length of the

stadion as used by Eratosthenes, and Cleomedes probably didn't know it either, since (unlike our mile or kilometer) it had never been given a standard definition. But without knowing the length of the stadion, we *can* judge the accuracy of Eratosthenes' use of astronomy. The Earth's circumference is actually 47.9 times the distance from Alexandria to Syene (modern Aswan), so the conclusion of Eratosthenes that the Earth's circumference is 50 times the distance from Alexandria to Syene was actually quite accurate, whatever the length of the stadion.* In his use of astronomy, if not of geography, Eratosthenes had done quite well.

* Eratosthenes was lucky. Syene is not precisely due south of Alexandria (its longitude is 32.9° E, while that of Alexandria is 29.9° E) and the noon Sun at the summer solstice is not precisely overhead at Syene, but about 0.4° from the vertical. The two errors partly cancel. What Eratosthenes had really measured was the ratio of the circumference of the Earth to the distance from Alexandria to the Tropic of Cancer (called the summer tropical circle by Cleomedes), the circle on the Earth's surface where the noon Sun at the summer solstice really is directly overhead. Alexandria is at a latitude of 31.2°, while the latitude of the Tropic of Cancer is 23.5°, which is less than the latitude of Alexandria by 7.7°, so the circumference of the Earth is in fact $360°/7.7° = 46.75$ times greater than the distance between Alexandria and the Tropic of Cancer, just a little less than the ratio 50 given by Eratosthenes.

8

The Problem of the Planets

The Sun and Moon are not alone in moving from west to east through the zodiac while they share the quicker daily revolution of the stars from east to west around the north celestial pole. In several ancient civilizations it was noticed that over many days five "stars" travel from west to east through the fixed stars along pretty much the same path as the Sun and Moon. The Greeks called them wandering stars, or planets, and gave them the names of gods: Hermes, Aphrodite, Ares, Zeus, and Cronos, translated by the Romans into Mercury, Venus, Mars, Jupiter, and Saturn. Following the lead of the Babylonians, they also included the Sun and Moon as planets,* making seven in all, and on this based the week of seven days.†

The planets move through the sky at different speeds: Mercury and Venus take 1 year to complete one circuit of the zodiac; Mars takes 1 year and 322 days; Jupiter 11 years and 315 days;

* For the sake of clarity, when I refer to planets in this chapter, I will mean just the five: Mercury, Venus, Mars, Jupiter, and Saturn.

† We can see the correspondence of days of the week with planets and the associated gods in the names of the days of the week in English. Saturday, Sunday, and Monday are obviously associated with Saturn, the Sun, and the Moon; Tuesday, Wednesday, Thursday, and Friday are based on an association of Germanic gods with supposed Latin equivalents: Tyr with Mars, Wotan with Mercury, Thor with Jupiter, and Frigga with Venus.

and Saturn 29 years and 166 days. All these are average periods, because the planets do not move at constant speed through the zodiac—they even occasionally reverse the direction of their motion for a while, before resuming their eastward motion. Much of the story of the emergence of modern science deals with the effort, extending over two millennia, to explain the peculiar motions of the planets.

An early attempt at a theory of the planets and Sun and Moon was made by the Pythagoreans. They imagined that the five planets, together with the Sun and Moon and also the Earth, all revolve around a central fire. To explain why we on Earth do not see the central fire, the Pythagoreans supposed that we live on the side of the Earth that faces outward, away from the fire. (Like almost all the pre-Socratics, the Pythagoreans believed the Earth to be flat; they thought of it as a disk always presenting the same side to the central fire, with us on the other side. The daily motion of the Earth around the central fire was supposed to explain the apparent daily motion of the more slowly moving Sun, Moon, planets, and stars around the Earth.)[1] According to Aristotle and Aëtius, the Pythagorean Philolaus of the fifth century BC invented a counter-Earth, orbiting where on our side of the Earth we can't see it, either between the Earth and the central fire or on the other side of the central fire from the Earth. Aristotle explained the introduction of the counter-Earth as a result of the Pythagoreans' obsession with numbers. The Earth, Sun, Moon, and five planets together with the sphere of the fixed stars made nine objects about the central fire, but the Pythagoreans supposed that the number of these objects must be 10, a perfect number in the sense that 10 = 1 + 2 + 3 + 4. As described somewhat scornfully by Aristotle,[2] the Pythagoreans

> supposed the elements of numbers to be the elements of all things, and the whole heaven to be a musical scale and a number. And all the properties of numbers and scales which they could show to agree with the attributes and parts and the whole arrangement of the heavens, they collected and fitted into their

> scheme, and if there was a gap anywhere, they readily made additions so as to make their whole theory coherent. For example, as the number 10 is thought to be perfect and to comprise the whole nature of numbers, they say that the bodies which move through the heavens are ten, but as the visible bodies are only nine, to meet this they invent a tenth—the "counter-Earth."

Apparently the Pythagoreans never tried to show that their theory explained in detail the apparent motions in the sky of the Sun, Moon, and planets against the background of fixed stars. The explanation of these apparent motions was a task for the following centuries, not completed until the time of Kepler.

This work was aided by the introduction of devices like the gnomon, for studying the motions of the Sun, and other instruments that allowed the measurement of angles between the lines of sight to various stars and planets, or between such astronomical objects and the horizon. Of course, all this was naked-eye astronomy. It is ironic that Claudius Ptolemy, who had deeply studied the phenomena of refraction and reflection (including the effects of refraction in the atmosphere on the apparent positions of stars) and who as we will see played a crucial role in the history of astronomy, never realized that lenses and curved mirrors could be used to magnify the images of astronomical bodies, as in Galileo Galilei's refracting telescope and the reflecting telescope invented by Isaac Newton.

It was not just physical instruments that furthered the great advances of scientific astronomy among the Greeks. These advances were made possible also by improvements in the discipline of mathematics. As matters worked out, the great debate in ancient and medieval astronomy was not between those who thought that the Earth or the Sun was in motion, but between two different conceptions of how the Sun and Moon and planets revolve around a stationary Earth. As we will see, much of this debate had to do with different conceptions of the role of mathematics in the natural sciences.

This story begins with what I like to call Plato's homework

problem. According to the Neoplatonist Simplicius, writing around AD 530 in his commentary on Aristotle's *On the Heavens*,

> Plato lays down the principle that the heavenly bodies' motion is circular, uniform, and constantly regular. Therefore he sets the mathematicians the following problem: What circular motions, uniform and perfectly regular, are to be admitted as hypotheses so that it might be possible to save the appearances presented by the planets?[3]

"Save (or preserve) the appearances" is the traditional translation; Plato is asking what combinations of motion of the planets (here including the Sun and Moon) in circles at constant speed, always in the same direction, would present an appearance just like what we actually observe.

This question was first addressed by Plato's contemporary, the mathematician Eudoxus of Cnidus.[4] He constructed a mathematical model, described in a lost book, *On Speeds*, whose contents are known to us from descriptions by Aristotle[5] and Simplicius.[6] According to this model, the stars are carried around the Earth on a sphere that revolves once a day from east to west, while the Sun and Moon and planets are carried around the Earth on spheres that are themselves carried by other spheres. The simplest model would have two spheres for the Sun. The outer sphere revolves around the Earth once a day from east to west, with the same axis and speed of rotation as the sphere of the stars; but the Sun is on the equator of an inner sphere, which shares the rotation of the outer sphere as if it were attached to it, but that also revolves around its own axis from west to east once a year. The axis of the inner sphere is tilted by 23½° to the axis of the outer sphere. This would account both for the Sun's daily apparent motion, and for its annual apparent motion through the zodiac. Likewise the Moon could be supposed to be carried around the Earth by two other counter-rotating spheres, with the difference that the inner sphere on which the Moon rides makes a full rotation from west to east once a month, rather than once a year. For reasons that are not clear, Eudoxus is supposed to have added a third sphere each for the Sun and Moon. Such theories are called "homocentric," because the spheres associated with the planets as well as the Sun and the Moon all have the same center, the center of the Earth.

The irregular motions of the planets posed a more difficult problem. Eudoxus gave each planet four spheres: the outer sphere rotating once a day around the Earth from east to west, with the same axis of rotation as the sphere of the fixed stars and the outer spheres of the Sun and Moon; the next sphere like the inner spheres of the Sun and Moon revolving more slowly at various speeds from west to east around an axis tilted by about 23½° to the axis of the outer sphere; and the two innermost spheres rotating, at exactly the same rates, in opposite directions around two nearly parallel axes tilted at large angles to the axes of the two outer spheres. The planet is attached to the innermost sphere. The two outer spheres give each planet its daily revolution following the stars around the Earth and its *average* motion over longer periods through the zodiac. The effects of the two oppositely rotating inner spheres would cancel if their axes were precisely parallel, but because these axes are supposed to be not quite parallel, they superimpose a figure eight motion on the average motion of each planet through the zodiac, accounting for the occasional reversals of direction of the planet. The Greeks called this path a *hippopede* because it resembled the tethers used to keep horses from straying.

The model of Eudoxus did not quite agree with observations of the Sun, Moon, and planets. For instance, its picture of the Sun's motion did not account for the differences in the lengths of the seasons that, as we saw in Chapter 6, had been found with the use of the gnomon by Euctemon. It quite failed for Mercury, and did not do well for Venus or Mars. To improve things, a new model was proposed by Callippus of Cyzicus. He added two more spheres to the Sun and Moon, and one more each to Mercury, Venus, and Mars. The model of Callippus generally worked better than that of Eudoxus, though it introduced some new fictitious peculiarities to the apparent motions of the planets.

In the homocentric models of Eudoxus and Callippus, the Sun, Moon, and planets were each given a separate suite of spheres, all with outer spheres rotating in perfect unison with a separate sphere carrying the fixed stars. This is an early example of what modern physicists call "fine-tuning." We criticize a proposed theory as fine-tuned when its features are adjusted to make some things equal, without any understanding of why they should be equal. The appearance of fine-tuning in a scientific theory is like a cry of distress from nature, complaining that something needs to be better explained.

A distaste for fine-tuning led modern physicists to make a discovery of fundamental importance. In the late 1950s two types of unstable particle called tau and theta had been identified that decay in different ways—the theta into two lighter particles called pions, and the tau into three pions. Not only did the tau and theta particles have the same mass—they had the same average lifetime, even though their decay modes were entirely different! Physicists assumed that the tau and the theta could not be the same particle, because for complicated reasons the symmetry of nature between right and left (which dictates that the laws of nature must appear the same when the world is viewed in a mirror as when it is viewed directly) would forbid the same particle from decaying sometimes into two pions and sometimes into three. With what we knew at the time, it would have been possible to adjust the constants in our theories to make the masses and lifetimes of the tau and theta equal, but one could hardly stomach such a theory—it seemed hopelessly fine-tuned. In the end, it was found that no fine-tuning was necessary, because the two particles are in fact the same particle. The symmetry between right and left, though obeyed by the forces that hold atoms and their nuclei together, is simply not obeyed in various decay processes, including the decay of the tau and theta.[7] The physicists who realized this were right to distrust the idea that the tau and the theta particles just happened to have the same mass and lifetime—that would take too much fine-tuning.

Today we face an even more distressing sort of fine-tuning. In 1998 astronomers discovered that the expansion of the universe is not slowing down, as would be expected from the gravitational attraction of galaxies for each other, but is instead speeding up. This acceleration is attributed to an energy associated with space itself, known as dark energy. Theory indicates that there are several different contributions to dark energy. Some contributions we can calculate, and others we can't. The contributions to dark energy that we can calculate turn out to be larger than the value of the dark energy observed by astronomers by about 56 orders of magnitude—that is, 1 followed by 56 zeroes. It's not a paradox, because we can suppose that these calculable contributions to dark energy are nearly canceled by contributions we can't calculate, but the cancellation would have to be precise to 56 decimal places. This level of fine-tuning is intolerable, and theorists have been working hard to find a better way to explain why the amount of dark energy is so much smaller than that suggested by our calculations. One possible explanation is mentioned in Chapter 11.

At the same time, it must be acknowledged that some apparent examples of fine-tuning are just accidents. For instance, the distances of the Sun and Moon from the Earth are in just about the same ratio as their diameters, so that seen from Earth, the Sun and Moon appear about the same size, as shown by the fact that the Moon just covers the Sun during a total solar eclipse. There is no reason to suppose that this is anything but a coincidence.

Aristotle took a step to reduce the fine-tuning of the models of Eudoxus and Callippus. In *Metaphysics*[8] he proposed to tie all the spheres together in a single connected system. Instead of giving the outermost planet, Saturn, four spheres like Eudoxus and Callippus, he gave it only their three inner spheres; the daily motion of Saturn from east to west was explained by tying these three spheres to the sphere of the fixed stars. Aristotle also added, inside the three of Saturn, three extra spheres that rotated in opposite directions, so as to cancel the effect of the motion of the three spheres of Saturn on the spheres of the next planet, Jupiter, whose outer sphere was attached to the innermost of the three extra spheres between Jupiter and Saturn.

At the cost of adding these three extra counter-rotating spheres, by tying the outer sphere of Saturn to the sphere of the fixed stars Aristotle had accomplished something rather nice. It was no longer necessary to wonder why the daily motion of Saturn should precisely follow that of the stars—Saturn was physically tied to the sphere of the stars. But then Aristotle spoiled it all: he gave Jupiter all *four* spheres that had been given to it by Eudoxus and Callippus. The trouble with this was that Jupiter would then get a daily motion from that of Saturn and also from the outermost of its own four spheres, so that *it would go around the Earth twice a day.* Did he forget that the three counter-rotating spheres inside the spheres of Saturn would cancel only the special motions of Saturn, not its daily revolution around the Earth?

Worse yet, Aristotle added only three counter-rotating spheres inside the four spheres of Jupiter, to cancel its own special motions but not its daily motion, and then gave Mars, the next planet, the full five spheres given to it by Callippus, so that Mars would go around the Earth three times a day. Continuing in this way, in Aristotle's scheme Venus, Mercury, the Sun, and the Moon would in a day respectively go around the Earth four, five, six, and seven times.

I was struck by this apparent failure when I read Aristotle's *Metaphysics*, and then I learned that it had already been noticed by several authors, including J. L. E. Dreyer, Thomas Heath, and W. D. Ross.[9] Some of them blamed it on a corrupt text. But if Aristotle really did present the scheme described in the standard version of *Metaphysics*, then this cannot be explained as a matter of his thinking in different terms from ours, or being interested in different problems from ours. We would have to conclude that on his own terms, in working on a problem that interested him, he was being careless or stupid.

Even if Aristotle had put in the right number of counter-rotating spheres, so that each planet would follow the stars around the Earth just once each day, his scheme still relied on a great deal of fine-tuning. The counter-rotating spheres introduced inside the spheres of Saturn to cancel the effect of Saturn's special motions on the motions of Jupiter would have to revolve at precisely the same speed as the three spheres of Saturn for the cancellation to work, and likewise for the planets closer to the Earth. And, just as for Eudoxus and Callippus, in Aristotle's scheme the second spheres of Mercury and Venus would have to revolve at precisely the same speed as the second sphere of the Sun, in order to account for the fact that Mercury, Venus, and the Sun move together through the zodiac, so that the inner planets are never seen far in the sky from the Sun. Venus, for instance, is always the morning star or the evening star, never seen high in the sky at midnight.

At least one ancient astronomer seems to have taken the problem of fine-tuning very seriously. This was Heraclides of Pontus. He was a student at Plato's Academy in the fourth century BC, and may have been left in charge of it when Plato went to Sicily. Both Simplicius[10] and Aëtius say that Heraclides taught that the Earth rotates on its axis,* eliminating at one blow the supposed simultaneous daily revolution of the stars, planets, Sun, and Moon around the Earth. This proposal of Heraclides was occasionally mentioned by writers in late antiquity and the Middle Ages, but it did not became popular until the time of Copernicus, again presumably because we do not feel the Earth's rotation. There is no indication that Aristarchus, writing a century after

* In a year of 365¼ days, the Earth actually rotates on its axis 366¼ times. The Sun seems to go around the Earth only 365¼ times in this period, because at the same time that the Earth is rotating 366¼ times on its axis, it is going around the Sun once in the same direction, giving 365¼ apparent revolutions of the Sun around the Earth. Since it takes 365.25 days of 24 hours for the Earth to spin 366.25 times relative to the stars, the time it takes the Earth to spin once is (365.25 × 24 hours)/366.25, or 23 hours, 56 minutes, and 4 seconds. This is known as the sidereal day.

Heraclides, suspected that the Earth not only moves around the Sun but also rotates on its own axis.

According to Chalcidius (or Calcidius), a Christian who translated the *Timaeus* from Greek to Latin in the fourth century, Heraclides also proposed that since Mercury and Venus are never seen far in the sky from the Sun, they revolve about the Sun rather than about the Earth, thus removing another bit of fine-tuning from the schemes of Eudoxus, Callippus, and Aristotle: the artificial coordination of the revolutions of the second spheres of the Sun and inner planets. But the Sun and Moon and three outer planets were still supposed to revolve about a stationary, though rotating, Earth. This theory works very well for the inner planets, because it gives them precisely the same apparent motions as the simplest version of the Copernican theory, in which Mercury, Venus, and the Earth all go at constant speed on circles around the Sun. As far as the inner planets are concerned, the only difference between Heraclides and Copernicus is point of view—either based on the Earth or based on the Sun.

Besides the fine-tuning inherent in the schemes of Eudoxus, Callippus, and Aristotle, there was another problem: these homocentric schemes did not agree very well with observation. It was believed then that the planets shine by their own light, and since in these schemes the spheres on which the planets ride always remain at the same distance from the Earth's surface, the planets' brightness should never change. It was obvious however that their brightness changed very much. As quoted by Simplicius,[11] around AD 200 the philosopher Sosigenes the Peripatetic had commented:

> However the [hypotheses] of the associates of Eudoxus do not preserve the phenomena, and just those which had been known previously and were accepted by themselves. And what necessity is there to speak about other things, some of which Callippus of Cyzicus also tried to preserve when Eudoxus had not been able to do so, whether or not Callippus did preserve

> them? . . . What I mean is that there are many times when the planets appear near, and there are times when they appear to have moved away from us. And in the case of some [planets] this is apparent to sight. For the star which is called Venus and also the one which is called Mars appear many times larger when they are in the middle of their retrogressions so that in moonless nights Venus causes shadows to be cast by bodies.

Where Simplicius or Sosigenes refers to the size of planets, we presumably should understand their apparent luminosity; with the naked eye we can't actually see the disk of any planet, but the brighter a point of light is, the larger it *seems* to be.

Actually, this argument is not as conclusive as Simplicius thought. The planets (like the Moon) shine by the reflected light of the Sun, so their brightness would change even in the schemes of Eudoxus et al. as they go through different phases (like the phases of the Moon). This was not understood until the work of Galileo. But even if the phases of the planets had been taken into account, the changes in brightness that would be expected in homocentric theories would not have agreed with what is actually seen.

For professional astronomers (if not for philosophers) the homocentric theory of Eudoxus, Callippus, and Aristotle was supplanted in the Hellenistic and Roman eras by a theory that did much better at accounting for the apparent motions of the Sun and planets. This theory is based on three mathematical devices—the epicycle, the eccentric, and the equant—to be described below. We do not know who invented the epicycle and eccentric, but they were definitely known to the Hellenistic mathematician Apollonius of Perga and to the astronomer Hipparchus of Nicaea, whom we met in Chapters 6 and 7.[12] We know about the theory of epicycles and eccentrics through the writings of Claudius Ptolemy, who invented the equant, and with whose name the theory has ever after been associated.

Ptolemy flourished around AD 150, in the age of the Antonine emperors at the height of the Roman Empire. He worked at the Museum of Alexandria, and died sometime after AD 161. We have already discussed his study of reflection and refraction in Chapter 4. His astronomical work is described in *Megale Syntaxis*, a title transformed by the Arabs to *Almagest*, by which name it became generally known in Europe. The *Almagest* was so successful that scribes stopped copying the works of earlier astronomers like Hipparchus; as a result, it is difficult now to distinguish Ptolemy's own work from theirs.

The *Almagest* improved on the star catalog of Hipparchus, listing 1,028 stars, hundreds more than Hipparchus, and giving some indication of their brightness as well as their position in the sky.* Ptolemy's theory of the planets and the Sun and Moon was much more important for the future of science. In one respect the work on this theory described in the *Almagest* is strikingly modern in its methods. Mathematical models are proposed for planetary motions containing various free numerical parameters, which are then found by constraining the predictions of the models to agree with observation. We will see an example of this below, in connection with the eccentric and equant.

In its simplest version, the Ptolemaic theory has each planet revolving in a circle known as an "epicycle," not about the Earth, but about a moving point that goes around the Earth on another circle known as a "deferent." The inner planets, Mercury and Venus, go around the epicycle in 88 and 225 days, respectively, while the model is fine-tuned so that the center of the epicycle goes around the Earth on the deferent in precisely one year, always remaining on the line between the Earth and the Sun.

* The apparent luminosity of stars in catalogs from Ptolemy's time to the present is described in terms of their "magnitude." Magnitude increases with *decreasing* luminosity. The brightest star, Sirius, has magnitude –1.4, the bright star Vega has magnitude zero, and stars that are just barely visible to the naked eye are of sixth magnitude. In 1856 the astronomer Norman Pogson compared the measured apparent luminosity of a number of stars with the magnitudes that had historically been attributed to them, and on that basis decreed that if one star has a magnitude greater than another by 5 units, it is 100 times dimmer.

We can see why this works. Nothing in the apparent motion of planets tells us how far away they are. Hence in the theory of Ptolemy, the apparent motion of any planet in the sky does not depend on the absolute sizes of the epicycle and deferent; it depends only on the *ratio* of their sizes. If Ptolemy had wanted to he could have expanded the sizes of both the epicycle and the deferent of Venus, keeping their ratio fixed, and likewise of Mercury, so that both planets would have the same deferent, namely, the orbit of the Sun. The Sun would then be the point on the deferent about which the inner planets travel on their epicycles. This is not the theory proposed by Hipparchus or Ptolemy, but it gives the motion of the inner planets the same appearance, because it differs only in the overall scale of the orbits, which does not affect apparent motions. This special case of the epicycle theory is just the same as the theory attributed to Heraclides discussed above, in which Mercury and Venus go around the Sun while the Sun goes around the Earth. As already mentioned, Heraclides' theory works well because it is equivalent to one in which the Earth and inner planets go around the Sun, the two theories differing only in the point of view of the astronomer. So it is no accident that the epicycle theory of Ptolemy, which gives Mercury and Venus the same apparent motions as the theory of Heraclides, also works pretty well in comparison with observation.

Ptolemy could have applied the same theory of epicycles and deferents to the outer planets—Mars, Jupiter, and Saturn—but to make the theory work it would have been necessary to make the planets' motion around the epicycles much slower than the motion of the epicycles' centers around the deferents. I don't know what would have been wrong with this, but for one reason or another Ptolemy chose a different path. In the simplest version of his scheme, each outer planet goes on its epicycle around a point on the deferent once a year, and that point on the deferent goes around the Earth in a longer time: 1.88 years for Mars, 11.9 years for Jupiter, and 29.5 years for Saturn. Here there is a different sort of fine-tuning—the line from the center of the epicycle to the planet is always parallel to the line from the Earth to the

Sun. This scheme agrees fairly well with the observed apparent motions of the outer planets because here, as for the inner planets, the different special cases of this theory that differ only in the scale of the epicycle and deferent (keeping their ratio fixed) all give the same apparent motions, and there is one special value of this scale that makes this model the same as the simplest Copernican theory, differing only in point of view: Earth or Sun. For the outer planets, this special choice of scale is the one for which the radius of the epicycle equals the distance of the Sun from the Earth. (See Technical Note 13.)

Ptolemy's theory nicely accounted for the apparent reversal in direction of planetary motions. For instance, Mars seems to go backward in its motion through the zodiac when it is on a point in its epicycle closest to the Earth, because then its supposed motion around the epicycle is in the direction opposite to the supposed motion of the epicycle around the deferent, and faster. This is just a transcription into a frame of reference based on the Earth of the modern statement that Mars seems to go backward in the zodiac when the Earth is passing it as they both go around the Sun. This is also the time when it is brightest (as noted in the above quotation from Simplicius), because at this time it is closest to the Earth, and the side of Mars that we see faces the Sun.

The theory developed by Hipparchus, Apollonius, and Ptolemy was not a fantasy that, by good luck, just happened to agree fairly well with observation but had no relation to reality. As far as the apparent motions of the Sun and planets are concerned, in its simplest version, with just one epicycle for each planet and no other complications, this theory gives *precisely* the same predictions as the simplest version of the theory of Copernicus—that is, a theory in which the Earth and the other planets go in circles at constant speed with the Sun at the center. As already explained in connection with Mercury and Venus (and further explained in Technical Note 13), this is because the Ptolemaic theory is in a class of theories that all give the same apparent motions of the Sun and planets, and one member of that class (though not the one adopted by Ptolemy) gives *precisely* the same actual motions of the Sun and planets relative to one another as given by the simplest version of the Copernican theory.

It would be nice to end the story of Greek astronomy here. Unfortunately, as Copernicus himself well understood, the predictions of the simplest version of the Copernican theory for the apparent motions of the planets do not quite agree with observation, and so neither do the predictions of the simplest version of the Ptolemaic theory, which are identical. We have known since the time of Kepler and Newton that the orbits of the Earth and the other planets are not exactly circular, the Sun is not exactly at the center of these orbits, and the Earth and the other planets do not travel around their orbits at exactly constant speed. Of course, none of that was understood in modern terms by the Greek astronomers. Much of the history of astronomy until Kepler was taken up with trying to accommodate the small inaccuracies in the simplest versions of both the Ptolemaic and the Copernican theories.

Plato had called for circles and uniform motion, and as far as is known no one in antiquity conceived that astronomical bodies could have any motion other than one compounded of circular motions, though Ptolemy was willing to compromise on the issue of *uniform* motion. Working under the limitation to orbits composed of circles, Ptolemy and his forerunners invented various complications to make their theories agree more accurately with observation, for the Sun and Moon as well as for the planets.*

One complication was just to add more epicycles. The only planet for which Ptolemy found this necessary was Mercury, whose orbit differs from a circle more than that of any other planet. Another complication was the "eccentric"; the Earth was taken to be, not at the center of the deferent for each planet, but

* In one of the few hints to the origin of the use of epicycles, Ptolemy at the beginning of Book XII of the *Almagest* credits Apollonius of Perga with proving a theorem relating the use of epicycles and eccentrics in accounting for the apparent motion of the Sun.

at some distance from it. For instance, in Ptolemy's theory the center of the deferent of Venus was displaced from the Earth by 2 percent of the radius of the deferent.*

The eccentric could be combined with another mathematical device introduced by Ptolemy, the "equant." This is a prescription for giving a planet a varying speed in its orbit, apart from the variation due to the planet's epicycle. One might imagine that, sitting on the Earth, we should see each planet, or more precisely the center of each planet's epicycle, going around us at a constant rate (say, in degrees of arc per day), but Ptolemy knew that this did not quite agree with actual observation. Once an eccentric was introduced, one might instead imagine that we should see the centers of the planets' epicycles go at a constant rate, not around the Earth, but around the centers of the planets' deferents. Alas, that didn't work either. Instead, for each planet Ptolemy introduced what came to be called an equant,† a point on the *opposite* side of the center of the deferent from the Earth, but at an equal distance from this center; and he supposed that the centers of the planets' epicycles go at a constant angular rate about the equant. The fact that the Earth and the equant are at an equal distance from the center of the deferent was not assumed on the basis of philosophical preconceptions, but found by leaving these distances as free parameters, and finding the values of the distances for which the predictions of the theory would agree with observation.

There were still sizable discrepancies between Ptolemy's model and observation. As we will see when we come to Kepler in Chapter 11, if consistently used the combination of a single epicycle for each planet and an eccentric and equant for the Sun and

* In the theory of the Sun's motion an eccentric can be regarded as a sort of epicycle, on which the line from the center of the epicycle to the Sun is always parallel to the line between the Earth and the center of the Sun's deferent, thus shifting the center of the Sun's orbit away from the Earth. Similar remarks apply to the Moon and planets.

† The term "equant" was not used by Ptolemy. He referred instead to a "bisected eccentric," indicating that the center of the deferent is taken to be in the middle of the line connecting the equant and the Earth.

for each planet can do a good job of imitating the actual motion of planets including the Earth in elliptical orbits—good enough to agree with almost any observation that could be made without telescopes. But Ptolemy was not consistent. He did not use the equant in describing the supposed motion of the Sun around the Earth; and this omission—since the locations of planets are referred to the position of the Sun—also messed up the predictions of planetary motions. As George Smith has emphasized,[13] it is a sign of the distance between ancient or medieval astronomy and modern science that no one after Ptolemy appears to have taken these discrepancies seriously as a guide to a better theory.

The Moon presented special difficulties: the sort of theory that worked pretty well for the apparent motions of the Sun and planets did not work well for the Moon. It was not understood until the work of Isaac Newton that this is because the Moon's motion is significantly affected by the gravitation of two bodies—the Sun as well as the Earth—while the planets' motion is almost entirely governed by the gravitation of a single body: the Sun. Hipparchus had proposed a theory of the Moon's motion with a single epicycle, which was adjusted to account for the length of time between eclipses; but as Ptolemy recognized, this model did not do well in predicting the location of the Moon on the zodiac between eclipses. Ptolemy was able to fix this flaw with a more complicated model, but his theory had its own problems: the distance between the Moon and the Earth would vary a good deal, leading to a much larger change in the apparent size of the Moon than is observed.

As already mentioned, in the system of Ptolemy and his predecessors there is no way that observation of the planets could have indicated the sizes of their deferents and epicycles; observation could have fixed only the ratio of these sizes for each planet.*

* The same is true when eccentrics and equants are added; observation could have fixed only the *ratios* of the distances of the Earth and the equant from the center of the deferent and the radii of the deferent and epicycle, separately for each planet.

Ptolemy filled this gap in *Planetary Hypotheses*, a follow-up to the *Almagest*. In this work he invoked an a priori principle, perhaps taken from Aristotle, that there should be no gaps in the system of the world. Each planet as well as the Sun and Moon was supposed to occupy a shell, extending from the minimum to the maximum distance of the planet or Sun or Moon from the Earth, and these shells were supposed to fit together with no gaps. In this scheme the *relative* sizes of the orbits of the planets and Sun and Moon were all fixed, once one decided on their order going outward from the Earth. Also, the Moon is close enough to the Earth so that its absolute distance (in units of the radius of the Earth) could be estimated in various ways, including the method of Hipparchus discussed in Chapter 7. Ptolemy himself developed the method of parallax: the ratio of the distance to the Moon and the radius of the Earth can be calculated from the observed angle between the zenith and the direction to the Moon and the calculated value that this angle would have if the Moon were observed from the center of the Earth.[14] (See Technical Note 14.) Hence, according to Ptolemy's assumptions, to find the distances of the Sun and planets all that was necessary was to know the order of their orbits around the Earth.

The innermost orbit was always taken to be that of the Moon, because the Sun and the planets are each occasionally eclipsed by the Moon. Also, it was natural to suppose that the farthest planets are those that appear to take the longest to go around the Earth, so Mars, Jupiter, and Saturn were generally taken in that order going away from the Earth. But the Sun, Venus, and Mercury all on average appear to take a year to go around the Earth, so their order remained a subject of controversy. Ptolemy guessed that the order going out from the Earth is the Moon, Mercury, Venus, the Sun, and then Mars, Jupiter, and Saturn. Ptolemy's results for the distances of the Sun, Moon, and planets as multiples of the diameter of the Earth were much smaller than their actual values, and for the Sun and Moon similar (perhaps not coincidentally) to the results of Aristarchus discussed in Chapter 7.

The complications of epicycles, equants, and eccentrics have given Ptolemaic astronomy a bad name. But it should not be thought that Ptolemy was stubbornly introducing these complications in order to make up for the mistake of taking the Earth as the unmoving center of the solar system. The complications, beyond just a single epicycle for each planet (and none for the Sun), had nothing to do with whether the Earth goes around the Sun or the Sun around the Earth. They were made necessary by the fact, not understood until Kepler's time, that the orbits are not circles, the Sun is not at the center of the orbits, and the velocities are not constant. The same complications also affected the original theory of Copernicus, who assumed that the orbits of planets and the Earth had to be circles and the speeds constant. Fortunately, this is a pretty good approximation, and the simplest version of the epicycle theory, with just one epicycle for each planet and none for the Sun, worked far better than the homocentric spheres of Eudoxus, Callippus, and Aristotle. If Ptolemy had included an equant along with an eccentric for the Sun as well as for each planet, the discrepancies between theory and observation would have been too small to be detected with the methods then available.

But this did not settle the issue between the Ptolemaic and Aristotelian theories of planetary motions. The Ptolemaic theory agreed better with observation, but it did violence to the assumption of Aristotelian physics that all celestial motions must be composed of circles whose center is the center of the Earth. Indeed, the queer looping motion of planets moving on epicycles would have been hard to swallow even for someone who had no stake in any other theory.

For fifteen hundred years the debate continued between the defenders of Aristotle, often called physicists or philosophers, and the supporters of Ptolemy, generally referred to as astronomers or mathematicians. The Aristotelians often acknowledged that the model of Ptolemy fitted the data better, but they regarded this as just the sort of thing that might interest mathematicians, not relevant for understanding reality. Their attitude was expressed in a statement by Geminus of Rhodes, who flourished around 70 BC, quoted about three centuries later by Alexander of Aphrodisias, who in turn was quoted by Simplicius,[15] in a commentary on Aristotle's *Physics*. This statement lays out the great debate between natural scientists (sometimes translated "physicists") and astronomers:

> It is the concern of physical inquiry to enquire into the substance of the heavens and the heavenly bodies, their powers and the nature of their coming-to-be and passing away; by Zeus, it can reveal the truth about their size, shape, and positioning. Astronomy does not attempt to pronounce on any of these questions, but reveals the ordered nature of the phenomena in the heavens, showing that the heavens are indeed an ordered cosmos, and it also discusses the shapes, sizes, and relative distances of the Earth, Sun, and Moon, as well as eclipses, the conjunctions of the heavenly bodies, and qualities and quantities inherent in their paths. Since astronomy touches on the study of the quantity, magnitude, and quality of their shapes, it understandably has recourse to arithmetic and geometry in this respect. And about these questions, which are the only ones it promised to give an account of, it has the power to reach results through the use of arithmetic and geometry. The astronomer and the natural scientist will accordingly on many occasions set out to achieve the same objective, for example, that the Sun is a sizeable body, that the Earth is spherical, but they do not use the same methodology. For the natural scientist will prove each of his points from the substance of the heavenly bodies, either from their powers, or from the fact that they are better as they are, or from their coming-to-be and change, while the astronomer argues from the properties of their shapes and sizes, or from quantity of movement and the time that corresponds to it. . . . In general it is not the concern of the astronomer to know what by nature is at rest and what by nature is in motion; he must rather make assumptions about what stays at rest and what moves, and consider with which assumptions the appearances in the heavens are consistent. He must get his first basic

> principles from the natural scientist, namely that the dance of the heavenly bodies is simple, regular, and ordered; from these principles he will be able to show that the movement of all the heavenly bodies is circular, both those that revolve in parallel courses and those that wind along oblique circles.

The "natural scientists" of Geminus share some characteristics of today's theoretical physicists, but with huge differences. Following Aristotle, Geminus sees the natural scientists as relying on first principles, including teleological principles: the natural scientist supposes that the heavenly bodies "are better as they are." For Geminus it is only the astronomer who uses mathematics, as an adjunct to his observations. What Geminus does not imagine is the give-and-take that has developed between theory and observation. The modern theoretical physicist does make deductions from basic principles, but he uses mathematics in this work, and the principles themselves are expressed mathematically and are learned from observation, certainly not by considering what is "better."

In the reference by Geminus to the motions of planets "that revolve in parallel courses and those that wind along oblique circles" one can recognize the homocentric spheres rotating on tilted axes of the schemes of Eudoxus, Callippus, and Aristotle, to which Geminus as a good Aristotelian would naturally be loyal. On the other hand, Adrastus of Aphrodisias, who around AD 100 wrote a commentary on the *Timaeus*, and a generation later the mathematician Theon of Smyrna were sufficiently convinced by the theory of Apollonius and Hipparchus that they tried to make it respectable, by interpreting the epicycles and deferents as solid transparent spheres, like the homocentric spheres of Aristotle, but now not homocentric.

Some writers, facing the conflict between the rival theories of the planets, threw up their hands, and declared that humans were not meant to understand celestial phenomena. Thus, in the mid–fifth century AD, in his commentary on the *Timaeus*, the Neoplatonist pagan Proclus proclaimed:[16]

> When we are dealing with sublunary things, we are content, because of the instability of the material which goes to constitute them, to grasp what happens in most instances. But when we want to know heavenly things, we use sensibility and call upon all sorts of contrivances quite removed from likelihood. . . . That this is the way things stand is plainly shown by the discoveries made about these heavenly things—from different hypotheses we draw the same conclusions relative to the same objects. Among these hypotheses are some which save the phenomena by means of epicycles, others which do so by means of eccentrics, still others which save the phenomena by means of counterturning spheres devoid of planets. Surely the god's judgement is more certain. But as for us, we must be satisfied to "come close" to those things, for we are men, who speak according to what is likely, and whose lectures resemble fables.

Proclus was wrong on three counts. He missed the point that the Ptolemaic theories that used epicycles and eccentrics did a far better job of "saving the phenomena" than the Aristotelian theory using the hypothesis of homocentric "counterturning spheres." There is also a minor technical point: in referring to hypotheses "which save the phenomena by means of epicycles, others which do so by means of eccentrics" Proclus seems not to realize that in the case where an epicycle can play the role of an eccentric (discussed in footnote * on page 92), these are not different hypotheses but different ways of describing what is mathematically the same hypothesis. Above all, Proclus was wrong in supposing that it is harder to understand heavenly motions than those here on Earth, below the orbit of the Moon. Just the reverse is true. We know how to calculate the motions of bodies in the solar system with exquisite precision, but we still can't predict earthquakes or hurricanes. But Proclus was not alone. We will see his unwarranted pessimism regarding the possibility of understanding the motion of the planets repeated centuries later, by Moses Maimonides.

Writing in the first decade of the twentieth century, the physicist turned philosopher Pierre Duhem[17] took the side of the Ptolemaics because their model fitted the data better, but he disapproved of Theon and Adrastus for trying to lend reality to the model. Perhaps because he was deeply religious, Duhem sought to restrict the role of science merely to the construction of mathematical theories that agree with observation, rather than encompassing efforts to explain anything. I am not sympathetic to this view, because the work of my generation of physicists certainly feels like explanation as we ordinarily use the word, not like mere description.[18] True, it is not so easy to draw a precise distinction between description and explanation. I would say that we explain some generalization about the world by showing how it follows from some more fundamental generalization, but what do we mean by fundamental? Still, I think we know what we mean when we say that Newton's laws of gravitation and motion are more fundamental than Kepler's three laws of planetary motion. The great success of Newton was in *explaining* the motions of the planets, not merely describing them. Newton did not explain gravitation, and he knew that he had not, but that is the way it always is with explanation—something is always left for future explanation.

Because of their odd motions, the planets were useless as clocks or calendars or compasses. They were put to a different sort of use in Hellenistic times and afterward for purposes of astrology, a false science learned from the Babylonians.* The sharp modern distinction between astronomy and astrology was less clear in the ancient and medieval worlds, because the lesson had not

* The association of astrology with the Babylonians is illustrated in Ode XI of Book 1 of Horace: "Do not inquire (we are not allowed to know) what ends the gods have assigned to you and me, Leoconoe, and do not meddle with Babylonian horoscopes. How much better to endure whatever it proves to be." (Horace, *Odes and Epodes*, ed. and trans. Niall Rudd, Loeb Classical Library, Harvard University Press, Cambridge, Mass., 2004, pp. 44–45). It sounds better in Latin: "Tu ne quaesieris—scire nefas—quem mihi, quem tibi, finem di dederint, Leuconoë, nec Babylonios temptaris numerous, ut melius, quidquid erit, pati."

yet been learned that human concerns were irrelevant to the laws governing the stars and planets. Governments from the Ptolemies on supported the study of astronomy largely in the hope that it would reveal the future, and so naturally astronomers spent much of their time on astrology. Indeed, Claudius Ptolemy was the author not only of the greatest astronomical work of antiquity, the *Almagest*, but also of a textbook of astrology, the *Tetrabiblos*.

But I can't leave Greek astronomy on this sour note. For a happier ending to Part II of this book, I'll quote Ptolemy on his pleasure in astronomy:[19]

> I know that I am mortal and the creature of a day; but when I search out the massed wheeling circles of the stars, my feet no longer touch the Earth, but, side by side with Zeus himself, I take my fill of ambrosia, the food of the gods.

PART III

THE MIDDLE AGES

Science reached heights in the Greek part of the ancient world that were not regained until the scientific revolution of the sixteenth and seventeenth centuries. The Greeks made the great discovery that some aspects of nature, especially in optics and astronomy, could be described by precise mathematical naturalistic theories that agree with observation. What was learned about light and the heavens was important, but even more important was what was learned about the sort of thing that *could* be learned, and how to learn it.

Nothing during the Middle Ages, either in the Islamic world or in Christian Europe, compares with this. But the millennium intervening between the fall of Rome and the scientific revolution was not an intellectual desert. The achievements of Greek science were preserved and in some cases improved in the institutions of Islam and then in the universities of Europe. In this way, the ground was prepared for the scientific revolution.

It was not only the achievements of Greek science that were preserved in the Middle Ages. We will see in medieval Islam and Christendom a continuation of the ancient debates over the role in science of philosophy, of mathematics, and of religion.

9

The Arabs

After the collapse in the fifth century of the western Roman Empire, the Greek-speaking eastern half continued as the Byzantine Empire, and even increased in extent. The Byzantine Empire achieved a climactic military success during the reign of the emperor Heraclius, whose army in AD 627 in the battle of Nineveh destroyed the army of the Persian Empire, the ancient enemy of Rome. But within a decade the Byzantines had to confront a more formidable adversary.

The Arabs were known in antiquity as a barbarian people, living in the borderland of the Roman and Persian empires, "that just divides the desert and the sown." They were pagans, with a religion centered at the city of Mecca, in the settled portion of western Arabia known as the Hejaz. Starting at the end of the 500s, Muhammad, an inhabitant of Mecca, set out to convert his fellow citizens to monotheism. Meeting opposition, he and his acolytes fled in 622 to Medina, which they then used as a military base for the conquest of Mecca and of most of the Arabian Peninsula.

After Muhammad's death in 632, a majority of Muslims followed the authority of four successive leaders headquartered at first at Medina: his companions and relatives Abu Bakr, Omar, Othman, and Ali. They are recognized today by Sunni Muslims as the "four rightly guided caliphs." The Muslims conquered the

Byzantine province of Syria in 636, just nine years after the battle of Nineveh, and then went on to seize Persia, Mesopotamia, and Egypt.

Their conquests introduced the Arabs to a more cosmopolitan world. For instance, the Arab general Amrou, who conquered Alexandria, reported to the caliph Omar, "I have taken a city, of which I can only say that it contains 6000 palaces, 4000 baths, 400 theatres, 12,000 greengrocers, and 40,000 Jews."[1]

A minority, the forerunners of today's Shiites, accepted only the authority of Ali, the fourth caliph and the husband of Muhammad's daughter Fatima. The split in the world of Islam became permanent after a revolt against Ali, in which Ali as well as his son Hussein was killed. A new dynasty, the Sunni Ummayad caliphate, was established at Damascus in 661.

Under the Ummayads Arab conquests expanded to include the territories of modern Afghanistan, Pakistan, Libya, Tunisia, Algeria, and Morocco, most of Spain, and much of central Asia beyond the Oxus River. From the formerly Byzantine lands that they now ruled they began to absorb Greek science. Some Greek learning also came from Persia, whose rulers had welcomed Greek scholars (including Simplicius) before the rise of Islam, when the Neoplatonic Academy was closed by the emperor Justinian. Christendom's loss became Islam's gain.

It was in the time of the next Sunni dynasty, the caliphate of the Abbasids, that Arab science entered its golden age. Baghdad, the capital city of the Abbasids, was built on both sides of the Tigris River in Mesopotamia by al-Mansur, caliph from 754 to 775. It became the largest city in the world, or at least the largest outside China. Its best-known ruler was Harun al-Rashid, caliph from 786 to 809, famous from *A Thousand and One Nights*. It was under al-Rashid and his son al-Mamun, caliph from 813 to 833, that translation from Greece, Persia, and India reached its greatest scope. Al-Mamun sent a mission to Constantinople that brought back manuscripts in Greek. The delegation probably included the physician Hunayn ibn Ishaq, the greatest of the ninth-century translators, who founded a dynasty of translators, training his son and nephew to carry on the work. Hunayn translated works of Plato and Aristotle, as well as medical texts of Dioscorides, Galen, and Hippocrates. Mathematical works of Euclid, Ptolemy, and others were also translated into Arabic at Baghdad, some through Syriac intermediaries. The historian Philip Hitti has nicely contrasted the state of learning at this time at Baghdad with the illiteracy of Europe in the early Middle Ages: "For while in the East al-Rashid and al-Mamun were delving into Greek and Persian philosophy, their contemporaries in the West, Charlemagne and his lords, were dabbling in the art of writing their names."[2]

It is sometimes said that the greatest contribution to science of the Abbasid caliphs was the foundation of an institute for translation and original research, the Bayt al-Hikmah, or House of Wisdom. This institute is supposed to have served for the Arabs somewhat the same function that the Museum and Library of Alexandria served for the Greeks. This view has been challenged by a scholar of Arabic language and literature, Dimitri Gutas.[3] He points out that Bayt al-Hikmah is a translation of a Persian term that had long been used in pre-Islamic Persia for storehouses of books, mostly of Persian history and poetry rather than of Greek science. There are only a few known examples of works that were translated at the Bayt al-Hikmah in the time of al-Mamun, and those are from Persian rather than Greek. Some astronomical research, as we shall see, was going on at the Bayt al-Hikmah, but little is known of its scope. What is not in dispute is that, whether or not at the Bayt al-Hikmah, the city of Baghdad itself in the time of al-Mamun and al-Rashid was a great center of translation and research.

Arab science was not limited to Baghdad, but spread west to Egypt, Spain, and Morocco, and east to Persia and central Asia. Participating in this work were not only Arabs but also Persians, Jews, and Turks. They were very much a part of Arab civilization and wrote in Arabic (or at least in Arabic script). Arabic then had something like the status in science that English has today. In some cases it is difficult to decide on the ethnic background of these figures. I will consider them all together, under the heading "Arabs."

As a rough approximation, we can identify two different scientific traditions that divided the Arab savants. On one hand, there were real mathematicians and astronomers who were not much concerned with what today we would call philosophy. Then there were philosophers and physicians, not very active in mathematics, and strongly influenced by Aristotle. Their interest in astronomy was chiefly astrological. Where they were concerned at all with the theory of the planets, the philosopher/physicians favored the Aristotelian theory of spheres centered on the Earth, while the astronomer/mathematicians generally followed the Ptolemaic theory of epicycles and deferents discussed in Chapter 8. This was an intellectual schism that, as we shall see, would persist in Europe until the time of Copernicus.

The achievements of Arab science were the work of many individuals, none of them clearly standing out from the rest as, say, Galileo and Newton stand out in the scientific revolution. What follows is a brief gallery of medieval Arab scientists that I hope may give some idea of their accomplishments and variety.

The first of the important astronomer/mathematicians at Baghdad was al-Khwarizmi,* a Persian born around 780 in what is now Uzbekistan. Al-Khwarizmi worked at the Bayt al-Hikmah and prepared widely used astronomical tables based in part on Hindu observations. His famous book on mathematics was *Hisab al-Jabr w-al-Muqabalah*, dedicated to the caliph al-Mamun (who was half Persian himself). From its title we derive the word "algebra." But this was not really a book on what is today called algebra. Formulas like the one for the solution of quadratic equations were given in words, not in the symbols that are an essential element of algebra. (In this respect, al-Khwarizmi's mathematics

* His full name is Abū Abdallāh Muhammad ibn Mūsā al-Khwārizmī. Full Arab names tend to be long, so I will usually just give the abbreviated name by which these persons are generally known. I will also dispense with diacritical marks such as bars over vowels, as in ā, which have no significance for readers (like myself) ignorant of Arabic.

was less advanced than that of Diophantus.) From al-Khwarizmi we also get our name for a rule for solving problems, "algorithm." The text of *Hisab al-Jabr w-al-Muqabalah* contains a confusing mixture of Roman numerals; Babylonian numbers based on powers of 60; and a new system of numbers learned from India, based on powers of 10. Perhaps the most important mathematical contribution of al-Khwarizmi was his explanation to the Arabs of these Hindu numbers, which in turn became known in Europe as Arabic numbers.

In addition to the senior figure of al-Khwarizmi, there were collected in Baghdad a productive group of other ninth-century astronomers, including al-Farghani (Alfraganus),* who wrote a popular summary of Ptolemy's *Almagest* and developed his own version of the planetary scheme described in Ptolemy's *Planetary Hypotheses*.

It was a major occupation of this Baghdad group to improve on Eratosthenes' measurement of the size of the Earth. Al-Farghani in particular reported a smaller circumference, which centuries later encouraged Columbus (as mentioned in the footnote on page 65) to think that he could survive an ocean voyage westward from Spain to Japan, perhaps the luckiest miscalculation in history.

The Arab who was most influential among European astronomers was al-Battani (Albatenius), born around 858 BC in northern Mesopotamia. He used and corrected Ptolemy's *Almagest*, making more accurate measurements of the ~23½° angle between the Sun's path through the zodiac and the celestial equator, of the lengths of the year and the seasons, of the precession of the equinoxes, and of the positions of stars. He introduced a trigonometric quantity, the sine, from India, in place of the closely related chord used and calculated by Hipparchus. (See Technical Note 15.) His work was frequently quoted by Copernicus and Tycho Brahe.

* Alfraganus is the latinized name by which al-Farghani became known in medieval Europe. In what follows, the latinized names of other Arabs will be given, as here, in parentheses.

The Persian astronomer al-Sufi (Azophi) made a discovery whose cosmological significance was not recognized until the twentieth century. In 964, in his *Book of the Fixed Stars*, he described a "little cloud" always present in the constellation Andromeda. This was the earliest known observation of what are now called galaxies, in this case the large spiral galaxy M31. Working at Isfahan, al-Sufi also participated in translating works of Greek astronomy into Arabic.

Perhaps the most impressive astronomer of the Abbasid era was al-Biruni. His work was unknown in medieval Europe, so there is no latinized version of his name. Al-Biruni lived in central Asia, and in 1017 visited India, where he lectured on Greek philosophy. He considered the possibility that the Earth rotates, gave accurate values for the latitude and longitude of various cities, prepared a table of the trigonometric quantity known as the tangent, and measured specific gravities of various solids and liquids. He scoffed at the pretensions of astrology. In India, al-Biruni invented a new method for measuring the circumference of the Earth. As he described it:[4]

> When I happened to be living in the fort of Nandana in the land of India, I observed from a high mountain standing to the west of the fort, a large plain lying south of the mountain. It occurred to me that I should examine this method [a method described previously] there. So, from the top of the mountain, I made an empirical measurement of the contact between the Earth and the blue sky. I found that the line of sight [to the horizon] had dipped below the reference line [that is, the horizontal direction] by the amount 34 minutes of arc. Then I measured the perpendicular of the mountain [that is, its height] and found it to be 652.055 cubits, where the cubit is a standard of length used in that region for measuring cloth.*

* Al-Biruni actually used a mixed decimal and sexigesimal system of numbers. He gave the height of the mountain in cubits as 652;3;18, that is, 652 plus 3/60 plus 18/3,600, which equals 652.055 in modern decimal notation.

From these data, al-Biruni concluded that the radius of the Earth is 12,803,337.0358 cubits. Something went wrong with his calculation; from the data he quoted, he should have calculated the radius as about 13.3 million cubits. (See Technical Note 16.) Of course, he could not possibly have known the height of the mountain to the accuracy he stated, so there was no practical difference between 12.8 million cubits and 13.3 million cubits. In giving the radius of the Earth to 12 significant figures, al-Biruni was guilty of misplaced precision, the same error that we saw in Aristarchus: carrying out calculations and quoting results to a much greater degree of precision than is warranted by the accuracy of the measurements on which the calculation is based.

I once got into trouble in this way. I had a summer job long ago, calculating the path of atoms through a series of magnets in an atomic beam apparatus. This was before desktop computers or pocket electronic calculators, but I had an electromechanical calculating machine that could add, subtract, multiply, and divide to eight significant figures. Out of laziness, in my report I gave the results of the calculations to eight significant figures just as they came from the calculating machine, without bothering to round them off to a realistic precision. My boss complained to me that the magnetic field measurements on which my calculation was based were accurate to only a few percent, and that any precision beyond this was meaningless.

In any case, we can't now judge the accuracy of al-Biruni's result that the Earth's radius is about 13 million cubits, because no one now knows the length of his cubit. Al-Biruni said there are 4,000 cubits in a mile, but what did he mean by a mile?

Omar Khayyam, the poet and astronomer, was born in 1048 in Nishapur, in Persia, and died there around 1131. He headed the observatory at Isfahan, where he compiled astronomical tables and planned calendar reform. In Samarkand in central Asia he wrote about topics in algebra, such as the solution of cubic equations. He is best known to English-speaking readers as a poet, through the magnificent nineteenth-century translation by Edward Fitzgerald of 75 out of a larger number of quatrains written by Omar Khayyam in Persian, and known as *The Rubaiyat*. Unsurprisingly for the hardheaded realist who wrote these verses, he strongly opposed astrology.

The greatest Arab contributions to physics were made in optics, first at the end of the tenth century by Ibn Sahl, who may have worked out the rule giving the direction of refracted rays of light (about which more in Chapter 13), and then by the great al-Haitam (Alhazen). Al-Haitam was born in Basra, in southern Mesopotamia, around 965, but worked in Cairo. His extant books include *Optics, The Light of the Moon, The Halo and the Rainbow, On Paraboloidal Burning Mirrors, The Formation of Shadows, The Light of the Stars, Discourse on Light, The Burning Sphere*, and *The Shape of the Eclipse*. He correctly attributed the bending of light in refraction to the change in the speed of light when it passes from one medium to another, and found experimentally that the angle of refraction is proportional to the angle of incidence only for small angles. But he did not give the correct general formula. In astronomy, he followed Adrastus and Theon in attempting to give a physical explanation to the epicycles and deferents of Ptolemy.

An early chemist, Jabir ibn Hayyan, is now believed to have flourished in the late eighth or early ninth century. His life is obscure, and it is not clear whether the many Arabic works attributed to him are really the work of a single person. There is also a large body of Latin works that appeared in Europe in the thirteenth and fourteenth centuries attributed to a "Geber," but it is now thought that the author of these works is not the same as the author of the Arabic works attributed to Jabir ibn Hayyan. Jabir developed techniques of evaporation, sublimation, melting, and crystallization. He was concerned with transmuting base metals into gold, and hence is often called an alchemist, but the distinction between chemistry and alchemy as practiced in his time is artificial, for there was then no fundamental scientific theory to tell anyone that such transmutations are impossible. To my mind, a distinction more important for the future of science is between those chemists or alchemists who followed Democritus in viewing the workings of matter in a purely naturalistic way, whether their theories were right or wrong, and those like Plato (and, unless they were speaking metaphorically, Anaximander and Empedocles), who brought human or religious values into the study of matter. Jabir probably belongs to the latter class. For instance, he made much of the chemical significance of 28, the number of letters in the alphabet of Arabic, the language of the Koran. Somehow it was important that 28 is the product of 7, supposed to be the number of metals, and 4, the number of qualities: cold, warm, wet, and dry.

The earliest major figure in the Arab medical/philosophical tradition was al-Kindi (Alkindus), who was born in Basra of a noble family but worked in Baghdad in the ninth century. He was a follower of Aristotle, and tried to reconcile Aristotle's doctrines with those of Plato and of Islam. Al-Kindi was a polymath, very interested in mathematics, but like Jabir he followed the Pythagoreans in using it as a sort of number magic. He wrote about optics and medicine, and attacked alchemy, though he defended astrology. Al-Kindi also supervised some of the work of translation from Greek to Arabic.

More impressive was al-Razi (Rhazes), an Arabic-speaking Persian of the generation following al-Kindi. His works include *A Treatise on the Small Pox and Measles*. In *Doubts Concerning Galen*, he challenged the authority of the influential Roman physician and disputed the theory, going back to Hippocrates, that health is a matter of balance of the four humors (described in Chapter 4). He explained, "Medicine is a philosophy, and this is not compatible with renouncement of criticism with regard to the leading authors." In an exception to the typical views of Arab physicians, al-Razi also challenged Aristotle's teaching, such as the doctrine that space must be finite.

The most famous of the Islamic physicians was Ibn Sina (Avicenna), another Arabic-speaking Persian. He was born in 980 near Bokhara in central Asia, became court physician to the sultan of Bokhara, and was appointed governor of a province. Ibn Sina was an Aristotelian, who like al-Kindi tried to reconcile Aristotle with Islam. His *Al Qanum* was the most influential medical text of the Middle Ages.

At the same time, medicine began to flourish in Islamic Spain. Al-Zahrawi (Abulcasis) was born in 936 near Córdoba, the metropolis of Andalusia, and worked there until his death in 1013. He was the greatest surgeon of the Middle Ages, and highly influential in Christian Europe. Perhaps because surgery was less burdened than other branches of medicine by ill-founded theory, al-Zahrawi sought to keep medicine separate from philosophy and theology.

The divorce of medicine from philosophy did not last. In the following century the physician Ibn Bajjah (Avempace) was born in Saragossa, and worked there and in Fez, Seville, and Granada. He was an Aristotelian who criticized Ptolemy and rejected Ptolemaic astronomy, but he took exception to Aristotle's theory of motion.

Ibn Bajjah was succeeded by his student Ibn Tufayl (Abubacer), also born in Muslim Spain. He practiced medicine in Granada, Ceuta, and Tangier, and he became vizier and physician to the sultan of the Almohad dynasty. He argued that there is no contradiction between Aristotle and Islam, and like his teacher rejected the epicycles and eccentrics of Ptolemaic astronomy.

In turn, Ibn Tufayl had a distinguished student, al-Bitruji. He was an astronomer but inherited his teacher's commitment to Aristotle and rejection of Ptolemy. Al-Bitruji unsuccessfully attempted to reinterpret the motion of planets on epicycles in terms of homocentric spheres.

One physician of Muslim Spain became more famous as a philosopher. Ibn Rushd (Averroes) was born in 1126 at Córdoba, the grandson of the city's imam. He became *cadi* (judge) of Seville in 1169 and of Córdoba in 1171, and then on the recommendation of Ibn Tufayl became court physician in 1182. As a medical scientist, Ibn Rushd is best known for recognizing the function of the retina of the eye, but his fame rests chiefly on his work as a commentator on Aristotle. His praise of Aristotle is almost embarrassing to read:

> [Aristotle] founded and completed logic, physics, and metaphysics. I say that he founded them because the works written before him on these sciences are not worth talking about and are quite eclipsed by his own writings. And I say that he completed them because no one who has come after him up to our own time, that is, for nearly fifteen hundred years, has been able to add anything to his writings or to find any error of any importance in them.[5]

The father of the modern author Salman Rushdie chose the surname Rushdie to honor the secular rationalism of Ibn Rushd.

Naturally Ibn Rushd rejected Ptolemaic astronomy, as contrary to physics, meaning Aristotle's physics. He was aware that Aristotle's homocentric spheres did not "save the appearances," and he tried to reconcile Aristotle with observation but concluded that this was a task for the future:

> In my youth I hoped it would be possible for me to bring this research [in astronomy] to a successful conclusion. Now, in my old age, I have lost hope, for several obstacles have stood in my way. But what I say about it will perhaps attract the attention of future researchers. The astronomical science of our days surely offers nothing from which one can derive an existing reality. The model that has been developed in the times in which we live accords with the computations, not with existence.[6]

Of course, Ibn Rushd's hopes for future researchers were unfulfilled; no one ever was able to make the Aristotelian theory of the planets work.

There was also serious astronomy done in Muslim Spain. In Toledo al-Zarqali (Arzachel) in the eleventh century was the first to measure the precession of the apparent orbit of the Sun around the Earth (actually of course the precession of the orbit of the Earth around the Sun), which is now known to be mostly due to the gravitational attraction between the Earth and other planets. He gave a value for this precession of 12.9" (seconds of arc) per year, in fair agreement with the modern value, 11.6" per year.[7] A group of astronomers including al-Zarqali used the earlier work of al-Khwarizmi and al-Battani to construct the *Tables of Toledo*, a successor to the *Handy Tables* of Ptolemy. These astronomical tables and their successors described in detail the apparent motions of the Sun, Moon, and planets through the zodiac and were landmarks in the history of astronomy.

Under the Ummayad caliphate and its successor, the Berber Almoravid dynasty, Spain was a cosmopolitan center of learning, hospitable to Jews as well as Muslims. Moses ben Maimon (Maimonides), a Jew, was born in 1135 at Córdoba during this happy time. Jews and Christians were never more than second-class citizens under Islam, but during the Middle Ages the condition of Jews was generally far better under the Arabs than in Christian Europe. Unfortunately for ben Maimon, during his youth Spain came under the rule of the fanatical Islamist Almohad caliphate, and he had to flee, trying to find refuge in Almeira, Marrakesh, Caesarea, and Cairo, finally coming to rest in Fustat, a suburb of Cairo. There until his death in 1204 he worked both as a rabbi, with influence throughout the world of medieval Jewry, and as a highly prized physician to both Arabs and Jews. His best-known work is the *Guide to the Perplexed*, which takes the form of letters to a perplexed young man. In it he expressed his rejection of Ptolemaic astronomy as contrary to Aristotle:[8]

> You know of Astronomy as much as you have studied with me, and learned from the book *Almagest*; we had not sufficient time to go beyond this. The theory that the spheres move regularly, and that the assumed courses of the stars are in harmony with observation, depends, as you are aware, on two hypotheses: we must assume either epicycles, or eccentric spheres, or a combination of both. Now I will show that each of these two hypotheses is irregular, and totally contrary to the results of Natural Science.

He then went on to acknowledge that Ptolemy's scheme agrees with observation, while Aristotle's does not, and as Proclus did before him, ben Maimon despaired at the difficulty of understanding the heavens:

> But of the things in the heavens man knows nothing except a few mathematical calculations, and you see how far these go. I say in the words of the poet[9] "The heavens are the Lord's, but the Earth he has given to the sons of man"; that is to say, God alone has a perfect and true knowledge of the heavens, their nature, their essence, their form, their motion, and their causes; but He gave man power to know the things which are under the heavens.

Just the opposite turned out to be true; it was the motion of heavenly bodies that was first understood in the early days of modern science.

There is testimony to the influence of Arab science on Europe in a long list of words derived from Arabic originals: not only algebra and algorithm, but also names of stars like Aldebaran, Algol, Alphecca, Altair, Betelgeuse, Mizar, Rigel, Vega, and so on, and chemical terms like alkali, alembic, alcohol, alizarin, and of course alchemy.

This brief survey leaves us with a question: why was it specifically those who practiced medicine, such as Ibn Bajjah, Ibn Tufayl, Ibn Rushd, and ben Maimon, who held on so firmly to the teachings of Aristotle? I can think of three possible reasons. First, physicians would naturally be most interested in Aristotle's writings on biology, and in these Aristotle was at his best. Also, Arab physicians were powerfully influenced by the writings of Galen, who greatly admired Aristotle. Finally, medicine is a field in which the precise confrontation of theory and observation was very difficult (and still is), so that the failings of Aristotelian physics and astronomy to agree in detail with observation may not have seemed so important to physicians. In contrast, the work of astronomers was used for purposes where correct precise results are essential, such as constructing calendars; measuring distances on Earth; telling the correct times for daily prayers; and determining the *qibla*, the direction to Mecca, which should be faced during prayer. Even astronomers who applied their science to astrology had to be able to tell precisely in what sign of the zodiac the Sun and planets were on any given date; and they were not likely to tolerate a theory like Aristotle's that gave the wrong answers.

The Abbasid caliphate came to an end in 1258, when the Mongols under Hulegu Khan sacked Baghdad and killed the caliph. Abbasid rule had disintegrated well before that. Political and military power had passed from the caliphs to Turkish sultans, and even the caliph's religious authority was weakened by the founding of independent Islamic governments: a translated Ummayad caliphate in Spain, the Fatimid caliphate in Egypt, the Almoravid dynasty in Morocco and Spain, succeeded by the Almohad caliphate in North Africa and Spain. Parts of Syria and Palestine were temporarily reconquered by Christians, first by Byzantines and then by Frankish crusaders.

Arab science had already begun to decline before the end of the Abbasid caliphate, perhaps beginning about AD 1100. After that, there were no more scientists with the stature of al-Battani, al-Biruni, Ibn Sina, and al-Haitam. This is a controversial point, and the bitterness of the controversy is heightened by today's politics. Some scholars deny that there was any decline.[10]

It is certainly true that some science continued even after the end of the Abbasid era, under the Mongols in Persia and then in India, and later under the Ottoman Turks. For instance, the building of the Maragha observatory in Persia was ordered by Hulegu in 1259, just a year after his sacking of Baghdad, in gratitude for the help that he thought that astrologers had given him in his conquests. Its founding director, the astronomer al-Tusi, wrote about spherical geometry (the geometry obeyed by great circles on a spherical surface, like the notional sphere of the fixed stars), compiled astronomical tables, and suggested modifications to Ptolemy's epicycles. Al-Tusi founded a scientific dynasty: his student al-Shirazi was an astronomer and mathematician, and al-Shirazi's student al-Farisi did groundbreaking work on optics, explaining the rainbow and its colors as the result of the refraction of sunlight in raindrops.

More impressive, it seems to me, is Ibn al-Shatir, a fourteenth-century astronomer of Damascus. Following earlier work of the Maragha astronomers, he developed a theory of planetary motions in which Ptolemy's equant was replaced with a pair of epicycles, thus satisfying Plato's demand that the motion of planets must be compounded of motions at constant speed around circles. Ibn al-Shatir also gave a theory of the Moon's motion based on epicycles: it avoided the excessive variation in the distance of the Moon from the Earth that had afflicted Ptolemy's lunar theory. The early work of Copernicus reported in his *Commentariolus* presents a lunar theory that is identical to Ibn al-Shatir's, and a planetary theory that gives the same apparent motions as the theory of al-Shatir.[11] It is now thought that Copernicus learned of these results (if not of their source) as a young student in Italy.

Some authors have made much of the fact that a geometric construction, the "Tusi couple," which had been invented by al-Tusi in his work on planetary motion, was later used by Copernicus. (This was a way of mathematically converting rotary motion of two touching spheres into oscillation in a straight line.) It is a matter of some controversy whether Copernicus learned of the Tusi couple from Arab sources, or invented it himself.[12] He was not unwilling to give credit to Arabs, and quoted five of them, including al-Battani, al-Bitruji, and Ibn Rushd, but made no mention of al-Tusi.

It is revealing that, whatever the influence of al-Tusi and Ibn al-Shatir on Copernicus, their work was not followed up among Islamic astronomers. In any case, the Tusi couple and the planetary epicycles of Ibn al-Shatir were means of dealing with the complications that (though neither al-Tusi nor al-Shatir nor Copernicus knew it) are actually due to the elliptical orbits of planets and the off-center location of the Sun. These are complications that (as discussed in Chapters 8 and 11) equally affected Ptolemaic and Copernican theories, and had nothing to do with whether the Sun goes around the Earth or the Earth around the Sun. No Arab astronomer before modern times seriously proposed a heliocentric theory.

Observatories continued to be built in Islamic countries. The greatest may have been an observatory in Samarkand, built in the 1420s by the ruler Ulugh Beg of the Timurid dynasty founded by Timur Lenk (Tamburlaine). There more accurate values were calculated for the sidereal year (365 days, 5 hours, 49 minutes, and 15 seconds) and the precession of the equinoxes (70 rather than 75 years per degree of precession, as compared with the modern value of 71.46 years per degree).

An important advance in medicine was made just after the end of the Abbasid period. This was the discovery by the Arab physician Ibn al-Nafis of the pulmonary circulation, the circulation of blood from the right side of the heart through the lungs, where it mixes with air, and then flows back to the heart's left side. Ibn al-Nafis worked at hospitals in Damascus and Cairo, and also wrote on ophthalmology.

These examples notwithstanding, it is hard to avoid the impression that science in the Islamic world began to lose momentum toward the end of the Abbasid era, and then continued to decline. When the scientific revolution came, it took place only in Europe, not in the lands of Islam, and it was not joined by Islamic scientists. Even after telescopes became available in the seventeenth century, astronomical observatories in Islamic countries continued to be limited to naked-eye astronomy[13] (though aided by elaborate instruments), undertaken largely for calendrical and religious rather than scientific purposes.

This picture of decline inevitably raises the same question that was raised by the decline of science toward the end of the Roman Empire—do these declines have anything to do with the advance of religion? For Islam, as for Christianity, the case for a conflict between science and religion is complicated, and I won't attempt a definite answer. There are at least two questions here.

First, what was the general attitude of Islamic scientists toward religion? That is, was it only those who set aside the influence of their religion who were creative scientists? And second, what was the attitude toward science of Muslim society?

Religious skepticism was widespread among scientists of the Abbasid era. The clearest example is provided by the astronomer Omar Khayyam, generally regarded as an atheist. He reveals his skepticism in several verses of the *Rubaiyat*:[14]

Some for the Glories of the World, and some
Sigh for the Prophet's Paradise to come;
Ah, take the Cash, and let the Credit go,
Nor heed the rumble of a distant Drum!

Why all the Saints and Sages who discuss'd
Of the two Worlds so learnedly, are thrust
Like foolish Prophets forth; their Words to Scorn
Are scatter'd, and their mouths are stopt with Dust.

Myself when young did eagerly frequent
Doctor and Saint, and heard great argument
About it, and about: but evermore
Came out by the same door as in I went.

(The literal translation into English is of course less poetic, but expresses essentially the same attitude.) Not for nothing was Khayyam after his death called "a stinging serpent to the Shari'ah." Today in Iran, government censorship requires that published versions of the poetry of Khayyam must be edited to remove or revise his atheistic sentiments.

The Aristotelian Ibn Rushd was banished around 1195 on suspicion of heresy. Another physician, al-Razi, was an outspoken skeptic. In his *Tricks of the Prophets* he argued that miracles are mere tricks, that people do not need religious leaders, and that Euclid and Hippocrates are more useful to humanity than religious teachers. His contemporary, the astronomer al-Biruni, was sufficiently sympathetic to these views to write an admiring biography of al-Razi.

On the other hand, the physician Ibn Sina had a nasty correspondence with al-Biruni, and said that al-Razi should have stuck to things he understood, like boils and excrement. The astronomer al-Tusi was a devout Shiite, and wrote about theology. The name of the astronomer al-Sufi suggests that he was a Sufi mystic.

It is hard to balance these individual examples. Most Arab scientists have left no record of their religious leanings. My own guess is that silence is more likely an indication of skepticism and perhaps fear than of devotion.

Then there is the question of the attitude of Muslims in general toward science. The caliph al-Mamun who founded the House of Wisdom was certainly an important supporter of science, and it may be significant that he belonged to a Muslim sect, the Mutazalites, which sought a more rational interpretation of the Koran, and later came under attack for this. But the Mutazalites should not be regarded as religious skeptics. They had no doubt that the Koran is the word of God; they argued only that it was created by God, and had not always existed. Nor should they be confused with modern civil libertarians; they persecuted Muslims who thought that there was no need for God to have created the eternal Koran.

By the eleventh century, there were signs in Islam of outright hostility to science. The astronomer al-Biruni complained about antiscientific attitudes among Islamic extremists:[15]

> The extremist among them would stamp the sciences as atheistic, and would proclaim that they lead people astray in order to make ignoramuses, like him, hate the sciences. For this will help him to conceal his ignorance, and to open the door to the complete destruction of science and scientists.

There is a well-known anecdote, according to which al-Biruni was criticized by a religious legalist, because the astronomer was using an instrument that listed the months according to their names in Greek, the language of the Christian Byzantines. Al-Biruni replied, "The Byzantines also eat food."

The key figure in the growth of tension between science and Islam is often said to be al-Ghazali (Algazel). Born in 1058 in Persia, he moved to Syria and then to Baghdad. He also moved about a good deal intellectually, from orthodox Islam to skepticism and then back to orthodoxy, but combined with Sufi mysticism. After absorbing the works of Aristotle, and summarizing them in *Inventions of the Philosophers*, he later attacked rationalism in his best-known work, *The Incoherence of the Philosophers*.[16] (Ibn Rushd, the partisan of Aristotle, wrote a riposte, *The Incoherence of the Incoherence*.) Here is how al-Ghazali expressed his view of Greek philosophy:

> The heretics in our times have heard the awe-inspiring names of people like Socrates, Hippocrates, Plato, Aristotle, etc. They have been deceived by the exaggerations made by the followers of these philosophers—exaggerations to the effect that the ancient masters possessed extraordinary intellectual powers; that the mathematical, logical, physical and metaphysical sciences developed by them are the most profound; that their excellent intelligence justifies their bold attempts to discover the Hidden Things by deductive methods; and that with all the subtlety of their intelligence and the originality of their accomplishments they repudiated the authority of religious laws: denied the validity of the positive contents of historical religions, and believe that all such things are only sanctimonious lies and trivialities.

Al-Ghazali's attack on science took the form of "occasionalism"—the doctrine that whatever happens is a singular occasion, governed not by any laws of nature but directly by the will of God. (This doctrine was not new in Islam—it had been advanced a century earlier by al-Ashari, an opponent of the Mutazalites.) In al-Ghazali's Problem XVII, "Refutation of

Their Belief in the Impossibility of a Departure from the Natural Course of Events," one reads:

> In our view, the connection between what are believed to be the cause and the effect [is] not necessary. . . . [God] has the power to create the satisfaction of hunger without eating, or death without the severance of the head, or even the survival of life when the head has been cut off, or any other thing from among the connected things (independently of what is supposed to be its cause). The philosophers deny this possibility; indeed, they assert its impossibility. Since the inquiry concerning these things (which are innumerable) may go to an indefinite length, let us consider only one example—viz., the burning of a piece of cotton at the time of its contact with fire. We admit the possibility of a contact between the two which will not result in burning, as also we admit the possibility of a transformation of cotton into ashes without coming into contact with fire. And they reject this possibility. . . . We say that it is God who—through the intermediacy of angels, or directly—is the agent of the creation of blackness in cotton; or of the disintegration of its parts, and their transformation into a smouldering heap or ashes. Fire, which is an inanimate thing, has no action.

Other religions, such as Christianity and Judaism, also admit the possibility of miracles, departures from the natural order, but here we see that al-Ghazali denied the significance of any natural order whatsoever.

This is hard to understand, because we certainly observe some regularities in nature. I doubt that al-Ghazali was unaware that it was not safe to put one's hand into fire. He could have saved a place for science in the world of Islam, as a study of what God *usually* wills to happen, a position taken in the seventeenth century by Nicolas Malebranche. But al-Ghazali did not take this path. His reason is spelled out in another work, *The Beginning of Sciences*,[17] in which he compared science to wine. Wine strengthens the body, but is nevertheless forbidden to Muslims. In the same way, astronomy and mathematics strengthen the mind, but "we nevertheless fear that one might be attracted through them to doctrines that are dangerous."

It is not only the writings of al-Ghazali that bear witness to a growing Islamic hostility to science in the Middle Ages. In 1194 in Almohad Córdoba, at the other end of the Islamic world from Baghdad, the *Ulama* (the local religious scholars) burned all medical and scientific books. And in 1449 religious fanatics destroyed Ulugh Beg's observatory in Samarkand.

We see in Islam today signs of the same concerns that troubled al-Ghazali. My friend the late Abdus Salam, a Pakistani physicist who won the first Nobel Prize in science awarded to a Muslim (for work done in England and Italy), once told me that he had tried to persuade the rulers of the oil-rich Persian Gulf states to invest in scientific research. He found that they were enthusiastic about supporting technology, but they feared that pure science would be culturally corrosive. (Salam was himself a devout Muslim. He was loyal to a Muslim sect, the Ahmadiyya, which has been regarded as heretical in Pakistan, and for years he could not return to his home country.)

It is ironic that in the twentieth century Sayyid Qutb, a guiding spirit of modern radical Islamism, called for the replacement of Christianity, Judaism, and the Islam of his own day with a universal purified Islam, in part because he hoped in this way to create an Islamic science that would close the gap between science and religion. But Arab scientists in their golden age were not doing Islamic science. They were doing science.

10

Medieval Europe

As the Roman Empire decayed in the West, Europe outside the realm of Byzantium became poor, rural, and largely illiterate. Where some literacy did survive, it was concentrated in the church, and there only in Latin. In Western Europe in the early Middle Ages virtually no one could read Greek.

Some fragments of Greek learning had survived in monastery libraries as Latin translations, including parts of Plato's *Timaeus* and translations around AD 500 by the Roman aristocrat Boethius of Aristotle's work on logic and of a textbook of arithmetic. There were also works written in Latin by Romans, describing Greek science. Most notable was a fifth-century encyclopedia oddly titled *The Marriage of Mercury and Philology* by Martianus Capella, which treated (as handmaidens of philology) seven liberal arts: grammar, logic, rhetoric, geography, arithmetic, astronomy, and music. In his discussion of astronomy Martianus described the old theory of Heraclides that Mercury and Venus go around the Sun while the Sun goes around the Earth, a description praised a millennium later by Copernicus. But even with these shreds of ancient learning, Europeans in the early Middle Ages knew almost nothing of the great scientific achievements of the Greeks. Battered by repeated invasions of Goths, Vandals, Huns, Avars, Arabs, Magyars, and Northmen, the people of Western Europe had other concerns.

Europe began to revive in the tenth and eleventh centuries. The invasions were winding down, and new techniques improved the productivity of agriculture.[1] It was not until the late thirteenth century that significant scientific work would begin again, and not much would be accomplished until the sixteenth century, but in the interval an institutional and intellectual foundation was laid for the rebirth of science.

In the tenth and eleventh centuries—a religious age—much of the new wealth of Europe naturally went not to the peasantry but to the church. As wonderfully described around AD 1030 by the French chronicler Raoul (or Radulfus) Glaber, "It was as if the world, shaking itself and putting off the old things, were putting on the white robe of churches." For the future of learning, most important were the schools attached to cathedrals, such as those at Orléans, Reims, Laon, Cologne, Utrecht, Sens, Toledo, Chartres, and Paris.

These schools trained the clergy not only in religion but also in a secular liberal arts curriculum left over from Roman times, based in part on the writings of Boethius and Martianus: the trivium of grammar, logic, and rhetoric; and, especially at Chartres, the quadrivium of arithmetic, geometry, astronomy, and music. Some of these schools went back to the time of Charlemagne, but in the eleventh century they began to attract schoolmasters of intellectual distinction, and at some schools there was a renewed interest in reconciling Christianity with knowledge of the natural world. As remarked by the historian Peter Dear,[2] "Learning about God by learning what He had made, and understanding the whys and wherefores of its fabric, was seen by many as an eminently pious enterprise." For instance, Thierry of Chartres, who taught at Paris and Chartres and became chancellor of the school at Chartres in 1142, explained the origin of the world as described in Genesis in terms of the theory of the four elements he learned from the *Timaeus*.

Another development was even more important than the flowering of the cathedral schools, though not unrelated to it. This was a new wave of translations of the works of earlier scientists.

Translations were at first not so much directly from Greek as from Arabic: either the works of Arab scientists, or works that had earlier been translated from Greek to Arabic or Greek to Syriac to Arabic.

The enterprise of translation began early, in the middle of the tenth century, for instance at the monastery of Santa Maria de Ripoli in the Pyrenees, near the border between Christian Europe and Ummayad Spain. For an illustration of how this new knowledge could spread in medieval Europe, and its influence on the cathedral schools, consider the career of Gerbert d'Aurillac. Born in 945 in Aquitaine of obscure parents, he learned some Arab mathematics and astronomy in Catalonia; spent time in Rome; went to Reims, where he lectured on Arabic numbers and the abacus and reorganized the cathedral school; became abbot and then archbishop of Reims; assisted in the coronation of the founder of a new dynasty of French kings, Hugh Capet; followed the German emperor Otto III to Italy and Magdeburg; became archbishop of Ravenna; and in 999 was elected pope, as Sylvester II. His student Fulbert of Chartres studied at the cathedral school of Reims and then became bishop of Chartres in 1006, presiding over the rebuilding of its magnificent cathedral.

The pace of translation accelerated in the twelfth century. At the century's start, an Englishman, Adelard of Bath, traveled extensively in Arab countries; translated works of al-Khwarizmi; and, in *Natural Questions*, reported on Arab learning. Somehow Thierry of Chartres learned of the use of zero in Arab mathematics, and introduced it into Europe. Probably the most important twelfth-century translator was Gerard of Cremona. He worked in Toledo, which had been the capital of Christian Spain before the Arab conquests, and though reconquered by Castilians in 1085 remained a center of Arab and Jewish culture. His Latin translation from Arabic of Ptolemy's *Almagest* made Greek astronomy available to medieval Europe. Gerard also translated Euclid's *Elements* and works by Archimedes, al-Razi, al-Ferghani, Galen, Ibn Sina, and al-Khwarizmi. After Arab Sicily fell to the Normans in 1091, translations were also made directly from Greek to Latin, with no reliance on Arabic intermediaries.

The translations that had the greatest immediate impact were of Aristotle. It was in Toledo that the bulk of Aristotle's work was translated from Arabic sources; for instance, there Gerard translated *On the Heavens*, *Physics*, and *Meteorology*.

Aristotle's works were not universally welcomed in the church. Medieval Christianity had been far more influenced by Platonism and Neoplatonism, partly through the example of Saint Augustine. Aristotle's writings were naturalistic in a way that Plato's were not, and his vision of a cosmos governed by laws, even laws as ill-developed as his were, presented an image of God's hands in chains, the same image that had so disturbed al-Ghazali. The conflict over Aristotle was at least in part a conflict between two new mendicant orders: the Franciscans, or gray friars, founded in 1209, who opposed the teaching of Aristotle; and the Dominicans, or black friars, founded around 1216, who embraced "The Philosopher."

This conflict was chiefly carried out in new European institutions of higher learning, the universities. One of the cathedral schools, at Paris, received a royal charter as a university in 1200. (There was a slightly older university at Bologna, but it specialized in law and medicine, and did not play an important role in medieval physical science.) Almost immediately, in 1210, scholars at the University of Paris were forbidden to teach the books of Aristotle on natural philosophy. Pope Gregory IX in 1231 called for Aristotle's works to be expurgated, so that the useful parts could be safely taught.

The ban on Aristotle was not universal. His works were taught at the University of Toulouse from its founding in 1229. At Paris the total ban on Aristotle was lifted in 1234, and in subsequent decades the study of Aristotle became the center of education there. This was largely the work of two thirteenth-century clerics: Albertus Magnus and Thomas Aquinas. In the fashion of the times, they were given grand doctoral titles: Albertus was the "Universal Doctor," and Thomas the "Angelic Doctor."

Albertus Magnus studied in Padua and Cologne, became a Dominican friar, and in 1241 went to Paris, where from 1245 to 1248 he occupied a professorial chair for foreign savants. Later he moved to Cologne, where he founded its university. Albertus was a moderate Aristotelian who favored the Ptolemaic system over Aristotle's homocentric spheres but was concerned about its conflict with Aristotle's physics. He speculated that the Milky Way consists of many stars and (contrary to Aristotle) that the markings on the Moon are intrinsic imperfections. The example of Albertus was followed a little later by another German Dominican, Dietrich of Freiburg, who independently duplicated some of al-Farisi's work on the rainbow. In 1941 the Vatican declared Albertus the patron saint of all scientists.

Thomas Aquinas was born a member of the minor nobility of southern Italy. After his education at the monastery of Monte Cassino and the University of Naples, he disappointed his family's hopes that he would become the abbot of a rich monastery; instead, like Albertus Magnus, he became a Dominican friar. Thomas went to Paris and Cologne, where he studied under Albertus. He then returned to Paris, and served as professor at the university in 1256–1259 and 1269–1272.

The great work of Aquinas was the *Summa Theologica*, a comprehensive fusion of Aristotelian philosophy and Christian theology. In it, he took a middle ground between extreme Aristotelians, known as Averroists after Ibn Rushd; and the extreme anti-Aristotelians, such as members of the newly founded Augustinian order of friars. Aquinas strenuously opposed a doctrine that was widely (but probably unjustly) attributed to thirteenth-century Averroists like Siger of Brabant and Boethius of Dacia. According to this doctrine, it is possible to hold opinions true in philosophy, such as the eternity of matter or the impossibility of the resurrection of the dead, while acknowledging that they are false in religion. For Aquinas, there could be only one truth. In astronomy, Aquinas leaned toward Aristotle's homocentric theory of the planets, arguing that this theory was founded on reason while Ptolemaic theory merely agreed with observation, and another hypothesis might also fit the data. On the other hand, Aquinas disagreed with Aristotle on the theory of motion; he argued that even in a vacuum any motion would take a finite time. It is thought that Aquinas encouraged the Latin translation of Aristotle, Archimedes, and others directly from Greek sources by his contemporary, the English Dominican William of Moerbeke. By 1255 students at Paris were being examined on their knowledge of the works of Aquinas.

But Aristotle's troubles were not over. Starting in the 1250s, the opposition to Aristotle at Paris was forcefully led by the Franciscan Saint Bonaventure. Aristotle's works were banned at Toulouse in 1245 by Pope Innocent IV. In 1270 the bishop of Paris, Étienne Tempier, banned the teaching of 13 Aristotelian propositions. Pope John XXI ordered Tempier to look further into the matter, and in 1277 Tempier condemned 219 doctrines of Aristotle or Aquinas.[3] The condemnation was extended to England by Robert Kilwardy, the archbishop of Canterbury, and renewed in 1284 by his successor, John Pecham.

The propositions condemned in 1277 can be divided according to the reasons for their condemnation. Some presented a direct conflict with scripture—for instance, propositions that state the eternity of the world:

9. That there was no first man, nor will there be a last; on the contrary, there always was and always will be the generation of man from man.
87. That the world is eternal as to all the species contained in it; and that time is eternal, as are motion, matter, agent, and recipient.

Some of the condemned doctrines described methods of learning truth that challenged religious authority, for instance:

38. That nothing should be believed unless it is self-evident or could be asserted from things that are self-evident.

150. That on any question, a man ought not to be satisfied with certitude based upon authority.
153. That nothing is known better because of knowing theology.

Finally, some of the condemned propositions had raised the same issue that had concerned al-Ghazali, that philosophical and scientific reasoning seems to limit the freedom of God, for example:

34. That the first cause could not make several worlds.
49. That God could not move the heavens with rectilinear motion, and the reason is that a vacuum would remain.
141. That God cannot make an accident exist without a subject nor make more [than three] dimensions exist simultaneously.

The condemnation of propositions of Aristotle and Aquinas did not last. Under the authority of a new pope who had been educated by Dominicans, John XXII, Thomas Aquinas was canonized in 1323. In 1325 the condemnation was rescinded by the bishop of Paris, who decreed: "We wholly annul the aforementioned condemnation of articles and judgments of excommunication as they touch, or are said to touch, the teaching of blessed Thomas, mentioned above, and because of this we neither approve nor disapprove of these articles, but leave them for free scholastic discussion."[4] In 1341 masters of arts at the University of Paris were required to swear they would teach "the system of Aristotle and his commentator Averroes, and of the other ancient commentators and expositors of the said Aristotle, except in those cases that are contrary to the faith."[5]

Historians disagree about the importance for the future of science of this episode of condemnation and rehabilitation. There are two questions here: What would have been the effect on science if the condemnation had not been rescinded? And what would have been the effect on science if there had never been any condemnation of the teachings of Aristotle and Aquinas?

It seems to me that the effect on science of the condemnation if not rescinded would have been disastrous. This is not because of the importance of Aristotle's conclusions about nature. Most of them were wrong, anyway. Contrary to Aristotle, there was a time before there were any men; there certainly are many planetary systems, and there may be many big bangs; things in the heavens can and often do move in straight lines; there is nothing impossible about a vacuum; and in modern string theories there are more than three dimensions, with the extra dimensions unobserved because they are tightly curled up. The danger in the condemnation came from the *reasons* why propositions were condemned, not from the denial of the propositions themselves.

Even though Aristotle was wrong about the laws of nature, it was important to believe that there *are* laws of nature. If the condemnation of generalizations about nature like propositions 34, 49, and 141, on the ground that God can do anything, had been allowed to stand, then Christian Europe might have lapsed into the sort of occasionalism urged on Islam by al-Ghazali.

Also, the condemnation of articles that questioned religious authority (such as articles 38, 150, and 153 quoted above) was in part an episode in the conflict between the faculties of liberal arts and theology in medieval universities. Theology had a distinctly higher status; its study led to a degree of doctor of theology, while liberal arts faculties could confer no degree higher than master of arts. (Academic processions were headed by doctors of theology, law, and medicine in that order, followed by the masters of arts.) Lifting the condemnation did not give the liberal arts equal status with theology, but it helped to free the liberal arts faculties from intellectual control by their theological colleagues.

It is harder to judge what would have been the effect if the condemnations had never occurred. As we will see, the authority of Aristotle on matters of physics and astronomy was increasingly challenged at Paris and Oxford in the fourteenth century, though sometimes new ideas had to be camouflaged as being merely *secundum imaginationem*—that is, something imagined, rather than asserted. Would challenges to Aristotle have been possible if his authority had not been weakened by the condemnations of the thirteenth century? David Lindberg[6] cites the example of Nicole Oresme (about whom more later), who in 1377 argued that it is permissible to imagine that the Earth moves in a straight line through infinite space, because "To say the contrary is to maintain an article condemned in Paris."[7] Perhaps the course of events in the thirteenth century can be summarized by saying that the condemnation saved science from dogmatic Aristotelianism, while the lifting of the condemnation saved science from dogmatic Christianity.

After the era of translation and the conflict over the reception of Aristotle, creative scientific work began at last in Europe in the fourteenth century. The leading figure was Jean Buridan, a Frenchman born in 1296 near Arras, who spent much of his life in Paris. Buridan was a cleric, but secular—that is, not a member of any religious order. In philosophy he was a nominalist, who believed in the reality of individual things, not of classes of things. Twice Buridan was honored by election as rector of the University of Paris, in 1328 and 1340.

Buridan was an empiricist, who rejected the logical necessity of scientific principles: "These principles are not immediately evident; indeed, we may be in doubt concerning them for a long time. But they are called principles because they are indemonstrable, and cannot be deduced from other premises nor be proved by any formal procedure, but they are accepted because they have been observed to be true in many instances and to be false in none."[8]

Understanding this was essential for the future of science, and not so easy. The old impossible Platonic goal of a purely deductive natural science stood in the way of progress that could be based only on careful analysis of careful observation. Even today one sometimes encounters confusion about this. For instance, the psychologist Jean Piaget[9] thought he had detected signs that children have an innate understanding of relativity, which they lose later in life, as if relativity were somehow logically or philosophically necessary, rather than a conclusion ultimately based on observations of things that travel at or near the speed of light.

Though an empiricist, Buridan was not an experimentalist. Like Aristotle's, his reasoning was based on everyday observation, but he was more cautious than Aristotle in reaching broad conclusions. For instance, Buridan confronted an old problem of Aristotle: why a projectile thrown horizontally or upward does not immediately start what was supposed to be its natural motion, straight downward, when it leaves the hand. On several grounds, Buridan rejected Aristotle's explanation that the projectile continues for a while to be carried by the air. First, the air must resist rather than assist motion, since it must be divided apart for a solid body to penetrate it. Further, why does the air move, when the hand that threw the projectile stops moving? Also, a lance that is pointed in back moves through the air as well as or better than one that has a broad rear on which the air can push.

Rather than supposing that air keeps projectiles moving, Buridan supposed that this is an effect of something called "impetus," which the hand gives the projectile. As we have seen, a somewhat similar idea had been proposed by John of Philoponus, and Buridan's impetus was in turn a foreshadowing of what Newton was to call "quantity of motion," or in modern terms momentum, though it is not precisely the same. Buridan shared with Aristotle the assumption that something has to keep moving things in motion, and he conceived of impetus as playing this role, rather than as being only a property of motion, like momentum. He never identified the impetus carried by a body as its mass times its velocity, which is how momentum is defined in Newtonian physics. Nevertheless, he was onto something. The amount of force that is required to stop a moving body in a given time is proportional to its momentum, and in this sense momentum plays the same role as Buridan's impetus.

Buridan extended the idea of impetus to circular motion, supposing that planets are kept moving by their impetus, an impetus given to them by God. In this way, Buridan was seeking a compromise between science and religion of a sort that became popular centuries later: God sets the machinery of the cosmos in motion, after which what happens is governed by the laws of nature. But although the conservation of momentum does keep the planets moving, by itself it could not keep them moving on curved orbits as Buridan thought was done by impetus; that requires an additional force, eventually recognized as the force of gravitation.

Buridan also toyed with an idea due originally to Heraclides, that the Earth rotates once a day from west to east. He recognized that this would give the same appearance as if the heavens rotated around a stationary Earth once a day from east to west. He also acknowledged that this is a more natural theory, since the Earth is so much smaller that the firmament of Sun, Moon, planets, and stars. But he rejected the rotation of the Earth, reasoning that if the Earth rotated, then an arrow shot straight upward would fall to the west of the archer, since the Earth would have moved under the arrow while it was in flight. It is ironic that Buridan might have been saved from this error if he had realized that the Earth's rotation would give the arrow an impetus that would carry it to the east along with the rotating Earth. Instead, he was misled by the notion of impetus; he considered only the vertical impetus given to the arrow by the bow, not the horizontal impetus it takes from the rotation of the Earth.

Buridan's notion of impetus remained influential for centuries. It was being taught at the University of Padua when Copernicus studied medicine there in the early 1500s. Later in that century Galileo learned about it as a student at the University of Pisa.

Buridan sided with Aristotle on another issue, the impossibility of a vacuum. But he characteristically based his conclusion on observations: when air is sucked out of a drinking straw, a vacuum is prevented by liquid being pulled up into the straw; and when the handles of a bellows are pulled apart, a vacuum is prevented by air rushing into the bellows. It was natural to conclude that nature abhors a vacuum. As we will see in Chapter 12, the correct explanation for these phenomena in terms of air pressure was not understood until the 1600s.

Buridan's work was carried further by two of his students: Albert of Saxony and Nicole Oresme. Albert's writings on philosophy became widely circulated, but it was Oresme who made the greater contribution to science.

Oresme was born in 1325 in Normandy, and came to Paris to study with Buridan in the 1340s. He was a vigorous opponent of looking into the future by means of "astrology, geomancy, necromancy, or any such arts, if they can be called arts." In 1377 Oresme was appointed bishop of the city of Lisieux, in Normandy, where he died in 1382.

Oresme's book *On the Heavens and the Earth*[10] (written in the vernacular for the convenience of the king of France) has the form of an extended commentary on Aristotle, in which again and again he takes issue with The Philosopher. In this book Oresme reconsidered the idea that the heavens do not rotate about the Earth from east to west but, rather, the Earth rotates on its axis from west to east. Both Buridan and Oresme recognized that we observe only relative motion, so seeing the heavens move leaves open the possibility that it is instead the Earth that is moving. Oresme went through various objections to the idea, and picked them apart. Ptolemy in the *Almagest* had argued that if the Earth rotated, then clouds and thrown objects would be left behind; and as we have seen, Buridan had argued against the Earth's rotation by reasoning that if the Earth rotated from west to east, then an arrow shot straight upward would be left behind by the Earth's rotation, contrary to the observation that the arrow seems to fall straight down to the same spot on the Earth's surface from which it was shot vertically upward. Oresme replied that the Earth's rotation carries the arrow with it, along with the archer and the air and everything else on the Earth's surface, thus applying Buridan's theory of impetus in a way that its author had not understood.

Oresme answered another objection to the rotation of the Earth—an objection of a very different sort, that there are passages in Holy Scripture (such as in the Book of Joshua) that refer to the Sun going daily around the Earth. Oresme replied that this was just a concession to the customs of popular speech, as where it is written that God became angry or repented—things that could not be taken literally. In this, Oresme was following the lead of Thomas Aquinas, who had wrestled with the passage in Genesis where God is supposed to proclaim, "Let there be a firmament above the waters, and let it divide the waters from the waters." Aquinas had explained that Moses was adjusting his speech to the capacity of his audience, and should not be taken literally. Biblical literalism could have been a drag on the progress of science, if there had not been many inside the church like Aquinas and Oresme who took a more enlightened view.

Despite all his arguments, Oresme finally surrendered to the common idea of a stationary Earth, as follows:

> Afterward, it was demonstrated how it cannot be proved conclusively by argument that the heavens move. . . . However, everyone maintains, and I think myself, that the heavens do move and not the Earth: For God has established the world which shall not be moved, in spite of contrary reasons because they are clearly not conclusive persuasions. However, after considering all that has been said, one could then believe that the Earth moves and not the heavens, for the opposite is not self-evident. However, at first sight, this seems as much against natural reason than all or many of the articles of our faith. What I have said by way of diversion or intellectual exercise can in this manner serve as a valuable means of refuting and checking those who would like to impugn our faith by argument.[11]

We do not know if Oresme really was unwilling to take the final step toward acknowledging that the Earth rotates, or whether he was merely paying lip service to religious orthodoxy.

Oresme also anticipated one aspect of Newton's theory of gravitation. He argued that heavy things do not necessarily tend to fall toward the center of our Earth, if they are near some other world. The idea that there may be other worlds, more or less like the Earth, was theologically daring. Did God create humans on those other worlds? Did Christ come to those other worlds to redeem those humans? The questions are endless, and subversive.

Unlike Buridan, Oresme was a mathematician. His major mathematical contribution led to an improvement on work done earlier at Oxford, so we must now shift our scene from France to England, and back a little in time, though we will return soon to Oresme.

By the twelfth century Oxford had become a prosperous market town on the upper reaches of the Thames, and began to attract students and teachers. The informal cluster of schools at Oxford became recognized as a university in the early 1200s. Oxford conventionally lists its line of chancellors starting in 1224 with Robert Grosseteste, later bishop of Lincoln, who began the concern of medieval Oxford with natural philosophy. Grosseteste read Aristotle in Greek, and he wrote on optics and the calendar as well as on Aristotle. He was frequently cited by the scholars who succeeded him at Oxford.

In *Robert Grosseteste and the Origins of Experimental Science*,[12] A. C. Crombie went further, giving Grosseteste a pivotal role in the development of experimental methods leading to the advent of modern physics. This seems rather an exaggeration of Grosseteste's importance. As is clear from Crombie's account, "experiment" for Grosseteste was the passive observation of nature, not very different from the method of Aristotle. Neither Grosseteste nor any of his medieval successors sought to learn general principles by experiment in the modern sense, the aggressive manipulation of natural phenomena. Grosseteste's theorizing has also been praised,[13] but there is nothing in his work that bears comparison with the development of quantitatively successful theories of light by Hero, Ptolemy, and al-Haitam, or of planetary motion by Hipparchus, Ptolemy, and al-Biruni, among others.

Grosseteste had a great influence on Roger Bacon, who in his intellectual energy and scientific innocence was a true representative of the spirit of his times. After studying at Oxford,

Bacon lectured on Aristotle in Paris in the 1240s, went back and forth between Paris and Oxford, and became a Franciscan friar around 1257. Like Plato, he was enthusiastic about mathematics but made little use of it. He wrote extensively on optics and geography, but added nothing important to the earlier work of Greeks and Arabs. To an extent that was remarkable for the time, Bacon was also an optimist about technology:

> Also cars can be made so that without animals they will move with unbelievable rapidity. . . . Also flying machines can be constructed so that a man sits in the midst of the machine revolving some engine by which artificial wings are made to beat the air like a flying bird.[14]

Appropriately, Bacon became known as "Doctor Mirabilis."

In 1264 the first residential college was founded at Oxford by Walter de Merton, at one time the chancellor of England and later bishop of Rochester. It was at Merton College that serious mathematical work at Oxford began in the fourteenth century. The key figures were four fellows of the college: Thomas Bradwardine (c. 1295–1349), William Heytesbury (fl. 1335), Richard Swineshead (fl. 1340–1355), and John of Dumbleton (fl. 1338–1348). Their most notable achievement, known as the Merton College mean speed theorem, for the first time gives a mathematical description of nonuniform motion—that is, motion at a speed that does not remain constant.

The earliest surviving statement of this theorem is by William of Heytesbury (chancellor of the University of Oxford in 1371), in *Regulae solvendi sophismata*. He defined the velocity at any instant in nonuniform motion as the ratio of the distance traveled to the time that would have elapsed if the motion had been uniform at that velocity. As it stands, this definition is circular, and hence useless. A more modern definition, possibly what Heytesbury meant to say, is that the velocity at any instant in nonuniform motion is the ratio of the distance traveled to the time elapsed if the velocity were the same as it is in a very short interval of time around that instant, so short that during this interval the change in velocity is negligible. Heytesbury then defined uniform acceleration as nonuniform motion in which the velocity increases by the same increment in each equal time. He then went on to state the theorem:[15]

> When any mobile body is uniformly accelerated from rest to some given degree [of velocity], it will in that time traverse one-half the distance that it would traverse if, in that same time, it were moved uniformly at the degree of velocity terminating that increment of velocity. For that motion, as a whole, will correspond to the mean degree of that increment of velocity, which is precisely one-half that degree of velocity which is its terminal velocity.

That is, the distance traveled during an interval of time when a body is uniformly accelerated is the distance it would have traveled in uniform motion if its velocity in that interval equaled the average of the actual velocity. If something is uniformly accelerated from rest to some final velocity, then its average velocity during that interval is half the final velocity, so the distance traveled is half the final velocity times the time elapsed.

Various proofs of this theorem were offered by Heytesbury, by John of Dumbleton, and then by Nicole Oresme. Oresme's proof is the most interesting, because he introduced a technique of representing algebraic relations by graphs. In this way, he was able to reduce the problem of calculating the distance traveled when a body is uniformly accelerated from rest to some final velocity to the problem of calculating the area of a right triangle, whose sides that meet at the right angle have lengths equal respectively to the time elapsed and to the final velocity. (See Technical Note 17.) The mean speed theorem then follows immediately from an elementary fact of geometry, that the area of a right triangle is half the product of the two sides that meet at the right angle.

Neither any don of Merton College nor Nicole Oresme seems to have applied the mean speed theorem to the most important case where it is relevant, the motion of freely falling bodies. For the dons and Oresme the theorem was an intellectual exercise, undertaken to show that they were capable of dealing mathematically with nonuniform motion. If the mean speed theorem is evidence of an increasing ability to use mathematics, it also shows how uneasy the fit between mathematics and natural science still was.

It must be acknowledged that although it is obvious (as Strato had demonstrated) that falling bodies accelerate, it is not obvious that the speed of a falling body increases in proportion to the *time*, the characteristic of uniform acceleration, rather than to the *distance* fallen. If the rate of change of the distance fallen (that is, the speed) were proportional to the distance fallen, then the distance fallen once the body starts to fall would increase exponentially with time,* just as a bank account that receives interest proportional to the amount in the account increases exponentially with time (though if the interest rate is low it takes a long time to see this). The first person to guess that the increase in the speed of a falling body is proportional to the time elapsed seems to have been the sixteenth-century Dominican friar Domingo de Soto,[16] about two centuries after Oresme.

From the mid-fourteenth to the mid-fifteenth century Europe was harried by catastrophe. The Hundred Years' War between England and France drained England and devastated France. The church underwent a schism, with a pope in Rome and another in Avignon. The Black Death destroyed a large fraction of the population everywhere.

Perhaps as a result of the Hundred Years' War, the center of scientific work shifted eastward in this period, from France and England to Germany and Italy. The two regions were spanned in the career of Nicholas of Cusa. Born around 1401 in the town of Kues on the Moselle in Germany, he died in 1464 in the Umbrian province of Italy. Nicholas was educated at both Heidelberg and Padua, becoming a canon lawyer, a diplomat, and after 1448 a

* But see the second footnote on page 191.

cardinal. His writing shows the continuing medieval difficulty of separating natural science from theology and philosophy. Nicholas wrote in vague terms about a moving Earth and a world without limits, but with no use of mathematics. Though he was later cited by Kepler and Descartes, it is hard to see how they could have learned anything from him.

The late Middle Ages also show a continuation of the Arab separation of professional mathematical astronomers, who used the Ptolemaic system, and physician-philosophers, followers of Aristotle. Among the fifteenth-century astronomers, mostly in Germany, were Georg von Peurbach and his pupil Johann Müller of Königsberg (Regiomontanus), who together continued and extended the Ptolemaic theory of epicycles.* Copernicus later made much use of the *Epitome of the Almagest* of Regiomontanus. The physicians included Alessandro Achillini (1463–1512) of Bologna and Girolamo Fracastoro of Verona (1478–1553), both educated at Padua, at the time a stronghold of Aristotelianism.

Fracastoro gave an interestingly biased account of the conflict:[17]

> You are well aware that those who make profession of astronomy have always found it extremely difficult to account for the appearances presented by the planets. For there are two ways of accounting for them: the one proceeds by means of those spheres called homocentric, the other by means of so-called eccentric spheres [epicycles]. Each of these methods has its dangers, each its stumbling blocks. Those who employ homocentric spheres never manage to arrive at an explanation of phenomena. Those who use eccentric spheres do, it is true, seem to

* A later writer, Georg Hartmann (1489–1564), claimed that he had seen a letter by Regiomontanus containing the sentence "The motion of the stars must vary a tiny bit on account of the motion of the Earth" (*Dictonary of Scientific Biography*, Scribner, New York, 1975, Volume II, p. 351). If this is true, then Regiomontanus may have anticipated Copernicus, though the sentence is also consistent with the Pythagorean doctrine that the Earth and Sun both revolve around the center of the world.

> explain the phenomena more adequately, but their conception of these divine bodies is erroneous, one might almost say impious, for they ascribe positions and shapes to them that are not fit for the heavens. We know that, among the ancients, Eudoxus and Callippus were misled many times by these difficulties. Hipparchus was among the first who chose rather to admit eccentric spheres than to be found wanting by the phenomena. Ptolemy followed him, and soon practically all astronomers were won over by Ptolemy. But against these astronomers, or at least, against the hypothesis of eccentrics, the whole of philosophy has raised continuing protest. What am I saying? Philosophy? Nature and the celestial spheres themselves protest unceasingly. Until now, no philosopher has ever been found who would allow that these monstrous spheres exist among the divine and perfect bodies.

To be fair, observations were not all on the side of Ptolemy against Aristotle. One of the failings of the Aristotelian system of homocentric spheres, which as we have seen had been noted around AD 200 by Sosigenes, is that it puts the planets always at the same distance from the Earth, in contradiction with the fact that the brightness of planets increases and decreases as they appear to go around the Earth. But Ptolemy's theory seemed to go too far in the other direction. For instance, in Ptolemy's theory the maximum distance of Venus from the Earth is 6.5 times its minimum distance, so if Venus shines by its own light, then (since apparent brightness goes as the inverse square of the distance) its maximum brightness should be $6.5^2 = 42$ times greater than its minimum brightness, which is certainly not the case. Ptolemy's theory had been criticized on this ground at the University of Vienna by Henry of Hesse (1325–1397). The resolution of the problem is of course that planets shine not by their own light, but by the reflected light of the Sun, so their apparent brightness depends not only on their distance from the Earth but also, like the Moon's brightness, on their phase. When Venus is farthest from the Earth it is on the side of the Sun away from the Earth, so its face is fully illuminated; when it is closest to the Earth it is more or less between the Earth and the Sun and we mostly see its dark side. For Venus the effects of phase and distance therefore partly cancel, moderating its variations in brightness. None of this was understood until Galileo discovered the phases of Venus.

Soon the controversy between Ptolemaic and Aristotelian astronomy was swept away in a deeper conflict, between those who followed either Ptolemy or Aristotle, all of them accepting that the heavens revolve around a stationary Earth; and a new revival of the idea of Aristarchus, that it is the Sun that is at rest.

PART IV

THE SCIENTIFIC REVOLUTION

Historians used to take it for granted that physics and astronomy underwent revolutionary changes in the sixteenth and seventeenth centuries, after which these sciences took something like their modern form, providing a paradigm for the future development of all science. The importance of this revolution seemed obvious. Thus the historian Herbert Butterfield* declared that the scientific revolution "outshines everything since the rise of Christianity and reduces the Renaissance and Reformation to the rank of mere episodes, mere internal displacements, within the system of medieval Christendom."[1]

There is something about this sort of consensus that always attracts the skeptical attention of the next generation of historians. In the past few decades some historians have expressed doubt about the importance or even the existence of the scientific revolution.[2] For instance, Steven Shapin famously began a book with the sentence "There was no such thing as the scientific revolution, and this is a book about it."[3]

Criticisms of the notion of a scientific revolution take two

* Butterfield coined the phrase "the Whig interpretation of history," which he used to criticize historians who judge the past according to its contribution to our present enlightened practices. But when it came to the scientific revolution, Butterfield was thoroughly Whiggish, as am I.

anonymous work, later titled *De hypothesibus motuum coelestium a se constitutis commentariolus*, and generally known as the *Commentariolus*, or *Little Commentary*.[1] The *Commentariolus* was not published until long after its author's death, and so was not as influential as his later writings, but it gives a good account of the ideas that guided his work.

After a brief criticism of earlier theories of the planets, Copernicus in the *Commentariolus* states seven principles of his new theory. Here is a paraphrase, with some added comments:

1. There is no one center of the orbits of the celestial bodies. (There is disagreement among historians whether Copernicus thought that these bodies are carried on material spheres,[2] as supposed by Aristotle.)
2. The center of the Earth is not the center of the universe, but only the center of the Moon's orbit, and the center of gravity toward which bodies on Earth are attracted.
3. All the heavenly bodies except the Moon revolve about the Sun, which is therefore the center of the universe. (But as discussed below, Copernicus took the center of the orbits of the Earth and other planets to be, not the Sun, but rather a point near the Sun.)
4. The distance between the Earth and the Sun is negligible compared with the distance of the fixed stars. (Presumably Copernicus made this assumption to explain why we do not see annual parallax, the apparent annual motion of the stars caused by the Earth's motion around the Sun. But the problem of parallax is nowhere mentioned in the *Commentariolus*.)
5. The apparent daily motion of the stars around the Earth arises entirely from the Earth's rotation on its axis.
6. The apparent motion of the Sun arises jointly from the rotation of the Earth on its axis and the Earth's revolution (like that of the other planets) around the Sun.
7. The apparent retrograde motion of the planets arises from the Earth's motion, occurring when the Earth passes Mars, Jupiter, or Saturn, or is passed in its orbit by Mercury or Venus.

Copernicus could not claim in the *Commentariolus* that his scheme fitted observation better than that of Ptolemy. For one thing, it didn't. Indeed, it couldn't, since for the most part Copernicus based his theory on data he inferred from Ptolemy's *Almagest*, rather than on his own observations.[3] Instead of appealing to new observations, Copernicus pointed out a number of his theory's aesthetic advantages.

One advantage is that the motion of the Earth accounted for a wide variety of apparent motions of the Sun, stars, and the other planets. In this way, Copernicus was able to eliminate the "fine-tuning" assumed in the Ptolemaic theory, that the center of the epicycles of Mercury and Venus had to remain always on the line between the Earth and the Sun, and that the lines between Mars, Jupiter, and Saturn and the centers of their respective epicycles had to remain always parallel to the line between the Earth and the Sun. In consequence the revolution of the center of the epicycle of each inner planet around the Earth and the revolution of each outer planet by a full turn on its epicycle all had to be fine-tuned to take precisely one year. Copernicus saw that these unnatural requirements simply mirrored the fact that we view the solar system from a platform revolving about the Sun.

Another aesthetic advantage of the Copernican theory had to do with its greater definiteness regarding the sizes of planetary orbits. Recall that the apparent motion of the planets in Ptolemaic astronomy depends, not on the sizes of the epicycles and deferents, but only on the ratio of the radii of the epicycle and deferent for each planet. If one liked, one could even take the deferent of Mercury to be larger than the deferent of Saturn, as long as the size of Mercury's epicycle was adjusted accordingly. Following the lead of Ptolemy in *Planetary Hypotheses*, astronomers customarily assigned sizes to the orbits, on the assumption that the maximum distance of one planet from the Earth equals the minimum distance from the Earth of the next planet outward. This fixed the relative sizes of planetary orbits for any given choice of the order of the planets going out from the Earth, but that choice was still quite arbitrary. In any case, the assumptions of *Planetary Hypotheses* were neither based on observation nor confirmed by observation.

In contrast, for the scheme of Copernicus to agree with observation, the radius of every planet's orbit had to have a definite ratio to the radius of the Earth's orbit.* Specifically, because of the difference in the way that Ptolemy had introduced epicycles for the inner and outer planets (and leaving aside complications associated with the ellipticity of the orbits), the ratio of the radii of the epicycles and deferents must equal the ratio of the distances from the Sun of the planets and Earth for the inner planets, and equal the inverse of this ratio for the outer planets. (See Technical Note 13.) Copernicus did not present his results this way; he gave them in terms of a complicated "triangulation scheme," which conveyed a false impression that he was making new predictions confirmed by observation. But he did in fact get the right radii of planetary orbits. He found that going out from the Sun, the planets are in order Mercury, Venus, Earth, Mars, Jupiter, and Saturn; this is precisely the same as the order of their periods, which Copernicus estimated to be respectively 3 months, 9 months, 1 year, 2½ years, 12 years, and 30 years. Though there was as yet no theory that dictated the speeds of the planets in their orbits, it must have seemed to Copernicus evidence of cosmic order that the larger the orbit of a planet, the more slowly it moves around the Sun.[4]

The theory of Copernicus provides a classic example of how a theory can be selected on aesthetic criteria, with no experimental evidence that favors it over other theories. The case for the Co-

* As mentioned in Chapter 8, there is only one special case of the simplest version of Ptolemy's theory (with one epicycle for each planet, and none for the Sun) that is equivalent to the simplest version of the Copernican theory, differing only in point of view: it is the special case in which the deferents of the inner planets are each taken to coincide with the orbit of the Sun around the Earth, while the radii of the epicycles of the outer planets all equal the distance of the Sun from the Earth. The radii of the epicycles of the inner planets and the radii of the deferents of the outer planets in this special case of the Ptolemaic theory coincide with the radii of planetary orbits in the Copernican theory.

pernican theory in the *Commentariolus* was simply that a great deal of what was peculiar about the Ptolemaic theory was explained at one blow by the revolution and rotation of the Earth, and that the Copernican theory was much more definite than the Ptolemaic theory about the order of the planets and the sizes of their orbits. Copernicus acknowledged that the idea of a moving Earth had long before been proposed by the Pythagoreans, but he also noted (correctly) that this idea had been "gratuitously asserted" by them, without arguments of the sort he himself was able to advance.

There was something else about the Ptolemaic theory that Copernicus did not like, besides its fine-tuning and its uncertainty regarding the sizes and order of planetary orbits. True to Plato's dictum that planets move on circles at constant speed, Copernicus rejected Ptolemy's use of devices like the equant to deal with the actual departures from circular motion at fixed speed. As had been done by Ibn al-Shatir, Copernicus instead introduced more epicycles: six for Mercury; three for the Moon; and four each for Venus, Mars, Jupiter, and Saturn. Here he made no improvement over the *Almagest*.

This work of Copernicus illustrates another recurrent theme in the history of physical science: a simple and beautiful theory that agrees pretty well with observation is often closer to the truth than a complicated ugly theory that agrees better with observation. The simplest realization of the general ideas of Copernicus would have been to give each planet including the Earth a circular orbit at constant speed with the Sun at the center of all orbits, and no epicycles anywhere. This would have agreed with the simplest version of Ptolemaic astronomy, with just one epicycle for each planet, none for the Sun and Moon, and no eccentrics or equants. It would not have precisely agreed with all observations, because planets move not on circles but on nearly circular ellipses; their speed is only approximately constant; and the Sun is not at the center of each ellipse but at a point a little off-center, known as the focus. (See Technical Note 18.) Copernicus could have done even better by following Ptolemy and introducing an eccentric and equant for each planetary orbit, but now also including the orbit of the Earth; the discrepancy with observation would then have been almost too small for astronomers of the time to measure.

There is an episode in the development of quantum mechanics that shows the importance of not worrying too much about small conflicts with observation. In 1925 Erwin Schrödinger worked out a method for calculating the energies of the states of the simplest atom, that of hydrogen. His results were in good agreement with the gross pattern of these energies, but the fine details of his result, which took into account the departures of the mechanics of special relativity from Newtonian mechanics, did not agree with the fine details of the measured energies. Schrödinger sat on his results for a while, until he wisely realized that getting the gross pattern of the energy levels was a significant accomplishment, well worth publishing, and that the correct treatment of relativistic effects could wait. It was provided a few years later by Paul Dirac.

In addition to numerous epicycles, there was another complication adopted by Copernicus, one similar to the eccentric of Ptolemaic astronomy. The center of the Earth's orbit was taken to be, not the Sun, but a point at a relatively small distance from the Sun. These complications approximately accounted for various phenomena, such as the inequality of the seasons discovered by Euctemon, which are really consequences of the fact that the Sun is at the focus rather than the center of the Earth's elliptical orbit, and the Earth's speed in its orbit is not constant.

Another of the complications introduced by Copernicus was made necessary only by a misunderstanding. Copernicus seems to have thought that the revolution of the Earth around the Sun would give the axis of the Earth's rotation each year a 360° turn around the direction perpendicular to the plane of the Earth's orbit, somewhat as a finger at the end of the outstretched arm of a dancer executing a pirouette would undergo a 360° turn around the vertical direction for each rotation of the dancer. (He may have been influenced by the old idea that the planets ride on solid transparent spheres.) Of course, the direction of the Earth's axis does not in fact change appreciably in the course of a year, so Copernicus was forced to give the Earth a third motion, in addition to its revolution around the Sun and its rotation around its axis, which would almost cancel this swiveling of its axis. Copernicus assumed that the cancellation would not be perfect, so that the Earth's axis would swivel around over very many years, producing the slow precession of the equinoxes that had been discovered by Hipparchus. After Newton's work it became clear that the revolution of the Earth around the Sun in fact has no influence on the direction of the Earth's axis, aside from tiny effects due to the action of the gravity of the Sun and Moon on the Earth's equatorial bulge, and so (as Kepler argued) no cancellation of the sort arranged by Copernicus is actually necessary.

With all these complications, the theory of Copernicus was still simpler than that of Ptolemy, but not dramatically so. Though Copernicus could not have known it, his theory would have been closer to the truth if he had not bothered with epicycles, and had left the small inaccuracies of the theory to be dealt with in the future.

The *Commentariolus* did not give much in the way of technical details. These were supplied in his great work *De Revolutionibus Orbium Coelestium*,[5] commonly known as *De Revolutionibus*, finished in 1543 when Copernicus was on his deathbed. The book starts with a dedication to Alessandro Farnese, Pope Paul III. In it Copernicus raised again the old argument between the homocentric spheres of Aristotle and the eccentrics and epicycles of Ptolemy, pointing out that the former do not account for observations, while the latter "contradict the first principles of regularity of motion." In support of his daring in suggesting a moving Earth, Copernicus quoted a paragraph of Plutarch:

> Some think that the Earth remains at rest. But Philolaus the Pythagorean believes that, like the Sun and Moon, it revolves around the fire in an oblique circle. Heraclides of Pontus and Ecphantus the Pythagorean make the Earth move, not in a pro-

> gressive motion, but like a wheel in a rotation from west to east about its own center.

(In the standard edition of *De Revolutionibus* Copernicus makes no mention of Aristarchus, but his name had appeared originally, and had then been struck out.) Copernicus went on to explain that since others had considered a moving Earth, he too should be permitted to test the idea. He then described his conclusion:

> Having thus assumed the motions which I ascribe to the Earth later in the volume, by long and intense study I finally found that if the motions of the other planets are correlated with the orbiting of the Earth, and are computed for the revolution of each planet, not only do their phenomena follow therefrom but also the order and size of all the planets and spheres, and heaven itself is so linked together that in no portion of it can anything be shifted without disrupting the remaining parts and the universe as a whole.

As in the *Commentariolus*, Copernicus was appealing to the fact that his theory was more predictive than Ptolemy's; it dictated a unique order of planets and the sizes of their orbits required to account for observation, while Ptolemy's theory left these undetermined. Of course, Copernicus had no way of confirming that his orbital radii were correct without assuming the truth of his theory; this had to wait for Galileo's observations of planetary phases.

Most of *De Revolutionibus* is extremely technical, fleshing out the general ideas of the *Commentariolus*. One point worth special mention is that in Book 1 Copernicus states an a priori commitment to motion composed of circles. Thus Chapter 1 of Book I begins:

> First of all, we must note that the universe is spherical. The reason is either that, of all forms, the sphere is the most perfect, needing no joint and being a complete whole, which can

> neither be increased nor diminished [here Copernicus sounds like Plato]; or that it is the most capacious of figures, best suited to enclose and retain all things [that is, it has the greatest volume for a given surface area]; or even that all the separate parts of the universe, I mean the Sun, Moon, planets and stars are seen to be of this shape [how could he know anything about the shape of the stars?]; or that wholes strive to be circumscribed by this boundary, as is apparent in drops of water and other fluid bodies when they seek to be self-contained [this is an effect of surface tension, which is irrelevant on the scale of planets]. Hence no one will question that the attribution of this form belongs to the divine bodies.

He then went on to explain in Chapter 4 that in consequence the movement of the heavenly bodies is "uniform, eternal, and circular, or compounded of circular motions."

Later in Book 1, Copernicus pointed out one of the prettiest aspects of his heliocentric system: it explained why Mercury and Venus are never seen far in the sky from the Sun. For instance, the fact that Venus is never seen more than about 45° from the Sun is explained by the fact that its orbit around the Sun is about 70 percent the size of the orbit of the Earth. (See Technical Note 19.) As we saw in Chapter 11, in Ptolemy's theory this had required fine-tuning the motion of Mercury and Venus so that the centers of their epicycles are always on the line between the Earth and the Sun. The system of Copernicus also made unnecessary Ptolemy's fine-tuning of the motion of the outer planets, which kept the line between each planet and the center of its epicycle parallel to the line between the Earth and the Sun.

The Copernican system ran into opposition from religious leaders, beginning even before publication of *De Revolutionibus*. This conflict was exaggerated in a famous nineteenth-century polemic, *A History of the Warfare of Science with Theology in Christendom* by Cornell's first president, Andrew Dickson White,[6] which offers a number of unreliable quotations from Luther, Melanchthon, Calvin, and Wesley. But a conflict did exist.

There is a record of Martin Luther's conversations with his disciples at Wittenberg, known as *Tischreden*, or *Table Talk*.[7] The entry for June 4, 1539, reads in part:

> There was mention of a certain new astrologer who wanted to prove that the Earth moves and not the sky, the Sun, and the Moon. . . . [Luther remarked,] "So it goes now. Whoever wants to be clever must agree with nothing that others esteem. He must do something of his own. This is what that fool does who wishes to turn the whole of astronomy upside down. Even in these things that are thrown into disorder I believe in the Holy Scriptures, for Joshua commanded the Sun to stand still and not the Earth."[8]

A few years after the publication of *De Revolutionibus,* Luther's colleague Philipp Melanchthon (1497–1560) joined the attack on Copernicus, now citing Ecclesiastes 1:5—"The Sun also rises, and the Sun goes down, and hastens to his place where he rose."

Conflicts with the literal text of the Bible would naturally raise problems for Protestantism, which had replaced the authority of the pope with that of Scripture. Beyond this, there was a potential problem for all religions: man's home, the Earth, had been demoted to just one more planet among the other five.

Problems arose even with the printing of *De Revolutionibus*. Copernicus had sent his manuscript to a publisher in Nuremberg, and the publisher appointed as editor a Lutheran clergyman, Andreas Osiander, whose hobby was astronomy. Probably expressing his own views, Osiander added a preface that was thought to be by Copernicus until the substitution was unmasked in the following century by Kepler. In this preface Osiander had Copernicus disclaiming any intention to present the true nature of planetary orbits, as follows:[9]

> For it is the duty of an astronomer to compose the history of the [apparent] celestial motions through careful and expert study. Then he must conceive and devise the causes of these motions

> or hypotheses about them. Since he cannot in any way attain to the true cause, he will adopt whatever suppositions enable the motions to be computed correctly from the principles of geometry for the future as well as for the past.

Osiander's preface concludes:

> So far as hypotheses are concerned, let no one expect anything certain from astronomy, which cannot furnish it, lest he accept as the truth ideas conceived for another purpose, and depart from this study a greater fool than when he entered it.

This was in line with the views of Geminus around 70 BC (quoted here in Chapter 8), but it was quite contrary to the evident intention of Copernicus, in both the *Commentariolus* and *De Revolutionibus*, to describe the actual constitution of what is now called the solar system.

Whatever individual clergymen may have thought about a heliocentric theory, there was no general Protestant effort to suppress the works of Copernicus. Nor did Catholic opposition to Copernicus become organized until the 1600s. The famous execution of Giordano Bruno by the Roman Inquisition in 1600 was not for his defense of Copernicus, but for heresy, of which (by the standards of the time) he was surely guilty. But as we will see, the Catholic church did in the seventeenth century put in place a very serious suppression of Copernican ideas.

What was really important for the future of science was the reception of Copernicus among his fellow astronomers. The first to be convinced by Copernicus was his sole pupil, Rheticus, who in 1540 published an account of the Copernican theory, and who in 1543 helped to get *De Revolutionibus* into the hands of the Nuremberg publisher. (Rheticus was initially supposed to supply the preface to *De Revolutionibus*, but when he left to take a position in Leipzig the task unfortunately fell to Osiander.) Rheticus had earlier assisted Melanchthon in making the University of Wittenberg a center of mathematical and astronomical studies.

The theory of Copernicus gained prestige from its use in 1551 by Erasmus Reinhold, under the sponsorship of the duke of Prussia, to compile a new set of astronomical tables, the Prutenic Tables, which allow one to calculate the location of planets in the zodiac at any given date. These were a distinct improvement over the previously used Alfonsine Tables, which had been constructed in Castile in 1275 at the court of Alfonso X. The improvement was in fact due, not to the superiority of the theory of Copernicus, but rather to the accumulation of new observations in the centuries between 1275 and 1551, and perhaps also to the fact that the greater simplicity of heliocentric theories makes calculations easier. Of course, adherents of a stationary Earth could argue that *De Revolutionibus* provided only a convenient scheme for calculation, not a true picture of the world. Indeed, the Prutenic Tables were used by the Jesuit astronomer and mathematician Christoph Clavius in the 1582 calendar reform under Pope Gregory XIII that gave us our modern Gregorian calendar, but Clavius never gave up his belief in a stationary Earth.

One mathematician tried to reconcile this belief with the Copernican theory. In 1568, Melanchthon's son-in-law Caspar Peucer, professor of mathematics at Wittenberg, argued in *Hypotyposes orbium coelestium* that it should be possible by a mathematical transformation to rewrite the theory of Copernicus in a form in which the Earth rather than the Sun is stationary. This is precisely the result achieved later by one of Peucer's students, Tycho Brahe.

Tycho Brahe was the most proficient astronomical observer in history before the introduction of the telescope, and the author of the most plausible alternative to the theory of Copernicus. Born in 1546 in the province of Skåne, now in southern Sweden but until 1658 part of Denmark, Tycho was a son of a Danish nobleman. He was educated at the University of Copenhagen, where in 1560 he became excited by the successful prediction of a partial solar eclipse. He moved on to universities in Germany and Switzerland, at Leipzig, Wittenberg, Rostock, Basel, and Augsburg. During these years he studied the Prutenic Tables and was impressed by the fact that these tables predicted the date of the 1563 conjunction of Saturn and Jupiter to within a few days, while the older Alfonsine Tables were off by several months.

Back in Denmark, Tycho settled for a while in his uncle's house at Herrevad in Skåne. There in 1572 he observed in the constellation Cassiopeia what he called a "new star." (It is now recognized as the thermonuclear explosion, known as a type Ia supernova, of a preexisting star. The remnant of this explosion was discovered by radio astronomers in 1952 and found to be at a distance of about 9,000 light-years, too far for the star to have been seen without a telescope before the explosion.) Tycho observed the new star for months, using a sextant of his own construction, and found that it did not exhibit any diurnal parallax, the daily shift in position among the stars of the sort that would be expected to be caused by the rotation of the Earth (or the daily revolution around the Earth of everything else) if the new star were as close as the Moon, or closer. (See Technical Note 20.) He concluded, "This new star is not located in the upper regions of the air just under the lunar orb, nor in any place closer to Earth . . . but far above the sphere of the Moon in the very heavens."[10] This was a direct contradiction of the principle of Aristotle that the heavens beyond the orbit of the Moon can undergo no change, and it made Tycho famous.

In 1576 the Danish king Frederick II gave Tycho the lordship of the small island of Hven, in the strait between Skåne and the large Danish island of Zealand, along with a pension to support the building and maintenance of a residence and scientific establishment on Hven. There Tycho built Uraniborg, which included an observatory, library, chemical laboratory, and printing press. It was decorated with portraits of past astronomers—Hipparchus, Ptolemy, al-Battani, and Copernicus—and of a patron of the sciences: William IV, landgrave of Hesse-Cassel. On Hven Tycho trained assistants, and immediately began observations.

Already in 1577 Tycho observed a comet, and found that it had no observable diurnal parallax. Not only did this show, again contra Aristotle, that there was change in the heavens beyond the orbit of the Moon. Now Tycho could also conclude that the path of the comet would have taken it right through either Aristotle's supposed homocentric spheres or the spheres of the Ptolemaic theory. (This, of course, would be a problem only if the spheres were conceived as hard solids. This was the teaching of Aristotle, which we saw in Chapter 8 had been carried over to the Ptolemaic theory by the Hellenistic astronomers Adrastus and Theon. The idea of hard spheres was revived in early modern times,[11] not long before Tycho ruled it out.) Comets occur more frequently than supernovas, and Tycho was able to repeat these observations on other comets in the following years.

From 1583 on, Tycho worked on a new theory of the planets, based on the idea that the Earth is at rest, the Sun and Moon go around the Earth, and the five known planets go around the Sun. It was published in 1588 as Chapter 8 of Tycho's book on the comet of 1577. In this theory the Earth is not supposed to be moving or rotating, so in addition to having slower motions, the Sun, Moon, planets, and stars all revolve around the Earth from east to west once a day. Some astronomers adopted instead a "semi-Tychonic" theory, in which the planets revolve around the Sun, the Sun revolves around the Earth, but the Earth rotates and the stars are at rest. (The first advocate of a semi-Tychonic theory was Nicolas Reymers Bär, although he would not have called it a semi-Tychonic system, for he claimed Tycho had stolen the original Tychonic system from him.)[12]

As mentioned several times above, the Tychonic theory is identical to the version of Ptolemy's theory (never considered by Ptolemy) in which the deferents of the inner planets are taken to coincide with the orbit of the Sun around the Earth, and the epicycles of the outer planets have the same radius as the Sun's orbit around the Earth. As far as the *relative* separations and velocities of the heavenly bodies are concerned, it is also equivalent to the theory of Copernicus, differing only in the point of view: a stationary Sun for Copernicus, or a stationary and nonrotating Earth for Tycho. Regarding observations, Tycho's theory had the advantage that it automatically predicted no annual stellar parallax, without having to assume that the stars are very much farther from Earth than the Sun or planets (which, of course, we now know they are). It also made unnecessary Oresme's answer to the classic problem that had misled Ptolemy and Buridan: that objects thrown upward would seemingly be left behind by a rotating or moving Earth.

For the future of astronomy, the most important contribution of Tycho was not his theory, but the unprecedented accuracy of his observations. When I visited Hven in the 1970s, I found no sign of Tycho's buildings, but there, still in the ground, were the massive stone foundations to which Tycho had anchored his instruments. (A museum and formal gardens have been put in place since my visit.) With these instruments, Tycho was able to locate objects in the sky with an uncertainty of only 1⁄15°. Also at the site of Uraniborg stands a granite statue, carved in 1936 by Ivar Johnsson, showing Tycho in a posture appropriate for an astronomer, facing up into the sky.[13]

Tycho's patron Frederick II died in 1588. He was succeeded by Christian IV, whom Danes today regard as one of their greatest kings, but who unfortunately had little interest in supporting work on astronomy. Tycho's last observations from Hven were made in 1597; he then left on a journey that took him to Hamburg, Dresden, Wittenberg, and Prague. In Prague, he became the imperial mathematician to the Holy Roman Emperor Rudolph II and started work on a new set of astronomical tables, the Rudolphine Tables. After Tycho's death in 1601, this work was continued by Kepler.

Johannes Kepler was the first to understand the nature of the departures from uniform circular motion that had puzzled astronomers since the time of Plato. As a five-year-old he was inspired by the sight of the comet of 1577, the first comet that Tycho had studied from his new observatory on Hven. Kepler attended the University of Tübingen, which under the leadership of Melanchthon had become eminent in theology and mathematics. At Tübingen Kepler studied both of these subjects, but became more interested in mathematics. He learned about the theory of

Copernicus from the Tübingen mathematics professor Michael Mästlin and became convinced of its truth.

In 1594 Kepler was hired to teach mathematics at a Lutheran school in Graz, in southern Austria. It was there that he published his first original work, the *Mysterium Cosmographicum* (*Mystery of the Description of the Cosmos*). As we have seen, one advantage of the theory of Copernicus was that it allowed astronomical observations to be used to find unique results for the order of planets outward from the Sun and for the sizes of their orbits. As was still common at the time, Kepler in this work conceived these orbits to be the circles traced out by planets carried on transparent spheres, revolving in the Copernican theory around the Sun. These spheres were not strictly two-dimensional surfaces, but thin shells whose inner and outer radii were taken to be the minimum and maximum distance of the planet from the Sun. Kepler conjectured that the radii of these spheres are constrained by an a priori condition, that each sphere (other than the outermost sphere, of Saturn) just fits inside one of the five regular polyhedrons, and each sphere (other than the innermost sphere, of Mercury) just fits outside one of these regular polyhedrons. Specifically, in order outward from the Sun, Kepler placed (1) the sphere of Mercury, (2) then an octahedron, (3) the sphere of Venus, (4) an icosahedron, (5) the sphere of Earth, (6) a dodecahedron, (7) the sphere of Mars, (8) a tetrahedron, (9) the sphere of Jupiter, (10) a cube, and finally (11) the sphere of Saturn, all fitting together tightly.

This scheme dictated the relative sizes of the orbits of all the planets, with no freedom to adjust the results, except by choosing the order of the five regular polyhedrons that fit into the spaces between the planets. There are 30 different ways of choosing the order of the regular polyhedrons,* so it is not surprising that

* There are 120 ways of choosing the order of any five different things; any of the five can be first, any of the remaining four can be second, any of the remaining three can be third, and any of the remaining two can be fourth, leaving only one possibility for the fifth, so the number of ways of arranging five things in order is $5 \times 4 \times 3 \times 2 \times 1 = 120$. But as far as the ratio of circum-

Kepler could find one way of choosing their order so that the predicted sizes of planetary orbits would roughly fit the results of Copernicus.

In fact, Kepler's original scheme worked badly for Mercury, requiring Kepler to do some fudging, and only moderately well for the other planets.* But like many others at the time of the Renaissance, Kepler was deeply influenced by Platonic philosophy, and like Plato he was intrigued by the theorem that regular polyhedrons exist in only five possible shapes, leaving room for only six planets, including the Earth. He proudly proclaimed, "Now you have the reason for the number of planets!"

No one today would take seriously a scheme like Kepler's, even if it had worked better. This is not because we have gotten over the old Platonic fascination with short lists of mathematically possible objects, like regular polyhedrons. There are other such short lists that continue to intrigue physicists. For instance, it is known that there are just four kinds of "numbers" for which a version of arithmetic including division is possible: the real numbers, complex numbers (involving the square root of –1), and more exotic quantities known as quaternions and octonions. Some physicists have expended much effort trying to incorporate quaternions and octonions as well as real and complex numbers in the fundamental laws of physics. What makes Kepler's scheme

---

scribed and inscribed spheres is concerned, the five regular polyhedrons are not all different; this ratio is the same for the cube and the octahedron, and for the icosahedron and the dodecahedron. Hence two arrangements of the five regular polyhedrons that differ only by the interchange of a cube and an octahedron, or of an icosahedron and a dodecahedron, give the same model of the solar system. The number of different models is therefore $120/(2 \times 2) = 30$.

* For instance, if a cube is inscribed within the inner radius of the sphere of Saturn, and circumscribed about the outer radius of the sphere of Jupiter, then the ratio of the minimum distance of Saturn from the Sun and the maximum distance of Jupiter from the Sun, which according to Copernicus was 1.586, should equal the distance from the center of a cube to any of its vertices divided by the distance from the center of the same cube to the center of any of its faces, or $\sqrt{3} = 1.732$, which is 9 percent too large.

so foreign to us today is not his attempt to find some fundamental physical significance for the regular polyhedrons, but that he did this in the context of planetary orbits, which are just historical accidents. Whatever the fundamental laws of nature may be, we can be pretty sure now that they do not refer to the radii of planetary orbits.

This was not just stupidity on Kepler's part. In his time no one knew (and Kepler did not believe) that the stars were suns with their own systems of planets, rather than just lights on a sphere somewhere outside the sphere of Saturn. The solar system was generally thought to be pretty much the whole universe, and to have been created at the beginning of time. It was perfectly natural then to suppose that the detailed structure of the solar system is as fundamental as anything else in nature.

We may be in a similar position in today's theoretical physics. It is generally supposed that what we call the expanding universe, the enormous cloud of galaxies that we observe rushing apart uniformly in all directions, is the whole universe. We think that the constants of nature we measure, such as the masses of the various elementary particles, will eventually all be deduced from the yet unknown fundamental laws of nature. But it may be that what we call the expanding universe is just a small part of a much larger "multiverse," containing many expanding parts like the one we observe, and that the constants of nature take different values in different parts of the multiverse. In this case, these constants are environmental parameters that will never be deduced from fundamental principles any more than we can deduce the distances of the planets from the Sun from fundamental principles. The best we could hope for would be an anthropic estimate. Of the billions of planets in our own galaxy, only a tiny minority have the right temperature and chemical composition to be suitable for life, but it is obvious that when life does begin and evolves into astronomers, they will find themselves on a planet belonging to this minority. So it is not really surprising that the planet on which we live is not twice or half as far from the Sun as the Earth actually is. In the same way, it seems likely that only a tiny minority of the subuniverses in the multiverse would have physical constants that allow the evolution of life, but of course any scientists will find themselves in a subuniverse belonging to this minority. This had been offered as an explanation of the order of magnitude of the dark energy mentioned in Chapter 8, before dark energy was discovered.[14] All this, of course, is highly speculative, but it serves as a warning that in trying to understand the constants of nature we may face the same sort of disappointment Kepler faced in trying to explain the dimensions of the solar system.

Some distinguished physicists deplore the idea of a multiverse, because they cannot reconcile themselves to the possibility that there are constants of nature that can never be calculated. It is true that the multiverse idea may be all wrong, and so it would certainly be premature to give up the effort to calculate all the physical constants we know about. But it is no argument against the multiverse idea that it would make us unhappy not to be able to do these calculations. Whatever the final laws of nature may be, there is no reason to suppose that they are designed to make physicists happy.

At Graz Kepler began a correspondence with Tycho Brahe, who had read the *Mysterium Cosmographicum*. Tycho invited Kepler to visit him in Uraniborg, but Kepler thought it would be too far to go. Then in February 1600 Kepler accepted Tycho's invitation to visit him in Prague, the capital since 1583 of the Holy Roman Empire. There Kepler began to study Tycho's data, especially on the motions of Mars, and found a discrepancy of 0.13° between these data and the theory of Ptolemy.*

* The motion of Mars is the ideal test case for planetary theories. Unlike Mercury or Venus, Mars can be seen high in the night sky, where observations are easiest. In any given span of years, it makes many more revolutions in its orbit than Jupiter or Saturn. And its orbit departs from a circle more than that of any other major planet except Mercury (which is never seen far from the Sun and hence is difficult to observe), so departures from circular motion at constant speed are much more conspicuous for Mars than for other planets.

Kepler and Tycho did not get along well, and Kepler returned to Graz. At just that time Protestants were being expelled from Graz, and in August 1600 Kepler and his family were forced to leave. Back in Prague, Kepler began a collaboration with Tycho, working on the Rudolphine Tables, the new set of astronomical tables intended to replace Reinhold's Prutenic Tables. After Tycho died in 1601, Kepler's career problems were solved for a while by his appointment as Tycho's successor as court mathematician to the emperor Rudolph II.

The emperor was enthusiastic about astrology, so Kepler's duties as court mathematician included the casting of horoscopes. This was an activity in which he had been employed since his student days at Tübingen, despite his own skepticism about astrological prediction. Fortunately, he also had time to pursue real science. In 1604 he observed a new star in the constellation Ophiuchus, the last supernova seen in or near our galaxy until 1987. In the same year he published *Astronomiae Pars Optica* (*The Optical Part of Astronomy*), a work on optical theory and its applications to astronomy, including the effect of refraction in the atmosphere on observations of the planets.

Kepler continued work on the motions of planets, trying and failing to reconcile Tycho's precise data with Copernican theory by adding eccentrics, epicycles, and equants. Kepler had finished this work by 1605, but publication was held up by a squabble with the heirs of Tycho. Finally in 1609 Kepler published his results in *Astronomia Nova* (*New Astronomy Founded on Causes, or Celestial Physics Expounded in a Commentary on the Movements of Mars*).

Part III of *Astronomia Nova* made a major improvement in the Copernican theory by introducing an equant and eccentric for the Earth, so that there is a point on the other side of the center of the Earth's orbit from the Sun around which the line to the Earth rotates at a constant rate. This removed most of the discrepancies that had bedeviled planetary theories since the time of Ptolemy, but Tycho's data were good enough so that Kepler could see that there were still some conflicts between theory and observation.

At some point Kepler became convinced that the task was hopeless, and that he had to abandon the assumption, common to Plato, Aristotle, Ptolemy, Copernicus, and Tycho, that planets move on orbits composed of circles. Instead, he concluded that planetary orbits have an oval shape. Finally, in Chapter 58 (of 70 chapters) of *Astronomia Nova*, Kepler made this precise. In what later became known as Kepler's first law, he concluded that planets (including the Earth) move on ellipses, with the Sun at a focus, not at the center. Just as a circle is completely described (apart from its location) by a single number, its radius, any ellipse can be completely described (aside from its location and orientation) by two numbers, which can be taken as the lengths of its longer and shorter axes, or equivalently as the length of the longer axis and a number known as the "eccentricity," which tells us how different the major and minor axes are. (See Technical Note 18.) The two foci of an ellipse are points on the longer axis, evenly spaced around the center, with a separation from each other equal to the eccentricity times the length of the longer axis of the ellipse. For zero eccentricity, the two axes of the ellipse have equal length, the two foci merge to a single central point, and the ellipse degenerates into a circle.

In fact, the orbits of all the planets known to Kepler have small eccentricities, as shown in the following table of modern values (projected back to the year 1900):

Planet	**Eccentricity**
Mercury	0.205615
Venus	0.006820
Earth	0.016750
Mars	0.093312
Jupiter	0.048332
Saturn	0.055890

This is why simplified versions of the Copernican and Ptolemaic theories (with no epicycles in the Copernican theory and only one epicyle for each of the five planets in the Ptolemaic theory) would have worked pretty well.*

The replacement of circles with ellipses had another far-reaching implication. Circles can be generated by the rotation of spheres, but there is no solid body whose rotation can produce an ellipse. This, together with Tycho's conclusions from the comet of 1577, went far to discredit the old idea that planets are carried on revolving spheres, an idea that Kepler himself had assumed in the *Mysterium Cosmographicum*. Instead, Kepler and his successors now conceived of planets as traveling on freestanding orbits in empty space.

The calculations reported in *Astronomia Nova* also used what later became known as Kepler's second law, though this law was not clearly stated until 1621, in his *Epitome of Copernican Astronomy*. The second law tells how the speed of a planet changes as the planet moves around its orbit. It states that as the planet moves, the line between the Sun and the planet sweeps out equal areas in equal times. A planet has to move farther along its orbit to sweep out a given area when it is near the Sun than when it is far from the Sun, so Kepler's second law has the consequence that each planet must move faster the closer it comes to the Sun. Aside from tiny corrections proportional to the square of the eccentricity, Kepler's second law is the same as the statement that the line to the planet from the *other* focus (the one where the Sun isn't) turns at a constant rate—that is, it turns by the same angle

* The main effect of the ellipticity of planetary orbits is not so much the ellipticity itself as the fact that the Sun is at a focus rather than the center of the ellipse. To be precise, the distance between either focus and the center of an ellipse is proportional to the eccentricity, while the variation in the distance of points on the ellipse from either focus is proportional to the *square* of the eccentricity, which for a small eccentricity makes it much smaller. For instance, for an eccentricity of 0.1 (similar to that of the orbit of Mars) the smallest distance of the planet from the Sun is only ½ percent smaller than the largest distance. On the other hand, the distance of the Sun from the center of this orbit is 10 percent of the average radius of the orbit.

in every second. (See Technical Note 21.) Thus to a good approximation, Kepler's second law gives the same planetary velocities as the old idea of an equant, a point on the opposite side of the center of the circle from the Sun (or, for Ptolemy, from the Earth), and at the same distance from the center, around which the line to the planet turns at a constant rate. The equant was thus revealed as nothing but the empty focus of the ellipse. Only Tycho's superb data for Mars allowed Kepler to conclude that eccentrics and equants are not enough; circular orbits had to be replaced with ellipses.[15]

The second law also had profound applications, at least for Kepler. In *Mysterium Cosmographicum* Kepler had conceived of the planets as being moved by a "motive soul." But now, with the speed of each planet found to decrease as its distance from the Sun increases, Kepler instead concluded that the planets are impelled in their orbits by some sort of force radiating from the Sun:

> If you substitute the word "force" [*vis*] for the word "soul" [*anima*], you have the very principle on which the celestial physics in the *Commentary on Mars* [*Astronomia Nova*] is based. For I formerly believed completely that the cause moving the planets is a soul, having indeed been imbued with the teaching of J. C. Scaliger* on motive intelligences. But when I recognized that this motive cause grows weaker as the distance from the Sun increases, just as the light of the Sun is attenuated, I concluded that this force must be as it were corporeal.[16]

Of course, the planets continue in their motion not because of a force radiating from the Sun, but rather because there is nothing to drain their momentum. But they are held in their orbits rather than flying off into interstellar space by a force radiating from the Sun, the force of gravitation, so Kepler was not entirely wrong. The idea of force at a distance was becoming popular at this

* This is Julius Caesar Scaliger, a passionate defender of Aristotle and opponent of Copernicus.

time, partly owing to the work on magnetism by the president of the Royal College of Surgeons and court physician to Elizabeth I, William Gilbert, to whom Kepler referred. If by "soul" Kepler had meant anything like its usual meaning, then the transition from a "physics" based on souls to one based on forces was an essential step in ending the ancient mingling of religion with natural science.

Astronomia Nova was not written with the aim of avoiding controversy. By using the word "physics" in the full title, Kepler was throwing out a challenge to the old idea, popular among followers of Aristotle, that astronomy should concern itself only with the mathematical description of appearances, while for true understanding one must turn to physics—that is, to the physics of Aristotle. Kepler was staking out a claim that it is astronomers like himself who do true physics. In fact, much of Kepler's thinking was inspired by a mistaken physical idea, that the Sun drives the planets around their orbits, by a force similar to magnetism.

Kepler also challenged all opponents of Copernicanism. The introduction to *Astronomia Nova* contains the paragraph:

> *Advice for idiots*. But whoever is too stupid to understand astronomical science, or too weak to believe Copernicus without [it] affecting his faith, I would advise him that, having dismissed astronomical studies, and having damned whatever philosophical studies he pleases, he mind his own business and betake himself home to scratch in his own dirt patch.[17]

Kepler's first two laws had nothing to say about the comparison of the orbits of different planets. This gap was filled in 1619 in *Harmonices mundi*, by what became known as Kepler's third law:[18] "the ratio which exists between the periodic times of any two planets is precisely the ratio of the $3/2$-power of the mean distances." * That is, the square of the sidereal period of each planet

* A subsequent discussion shows that by the mean distance Kepler meant, not the distance averaged over time, but rather the average of the minimum

(the time it takes to complete a full circuit of its orbit) is proportional to the cube of the longer axis of the ellipse. Thus if T is the sidereal period in years, and a is half the length of the longer axis of the ellipse in astronomical units (AU), with 1 AU defined as half the longer axis of the Earth's orbit, then Kepler's third law says that T^2 / a^3 is the same for all planets. Since the Earth by definition has T equal to 1 year and a equal to 1 AU, in these units it has T^2 / a^3 equal to 1, so according to Kepler's third law each planet should also have $T^2 / a^3 = 1$. The accuracy with which modern values follow this rule is shown in the following table:

Planet	a **(AU)**	T **(years)**	T^2 / a^3
Mercury	0.38710	0.24085	1.0001
Venus	0.72333	0.61521	0.9999
Earth	1.00000	1.00000	1.0000
Mars	1.52369	1.88809	1.0079
Jupiter	5.2028	11.8622	1.001
Saturn	9.540	29.4577	1.001

(The departures from perfect equality of T^2 / a^3 for the different planets are due to tiny effects of the gravitational fields of the planets themselves acting on each other.)

Never entirely emancipated from Platonism, Kepler tried to make sense of the sizes of the orbits, resurrecting his earlier use of regular polyhedrons in *Mysterium Cosmographicum*. He also played with the Pythagorean idea that the different planetary periods form a sort of musical scale. Like other scientists of the time, Kepler belonged only in part to the new world of science

---

and maximum distances of the planet from the Sun. As shown in Technical Note 18, the minimum and maximum distances of a planet from the Sun are $(1 - e)a$ and $(1 + e)a$, where e is the eccentricity and a is half the longer axis of the ellipse (that is, the semimajor axis), so the mean distance is just a. It is further shown in Technical Note 18 that this is also the distance of the planet from the Sun, averaged over the distance traveled by the planet in its orbit.

that was just coming into being, and in part also to an older philosophical and poetic tradition.

The Rudolphine Tables were finally completed in 1627. Based on Kepler's first and second laws, they represented a real improvement in accuracy over the previous Prutenic Tables. The new tables predicted that there would be a transit of Mercury (that is, that Mercury would be seen to pass across the face of the Sun) in 1631. Kepler did not see it. Forced once again as a Protestant to leave Catholic Austria, Kepler died in 1630 in Regensburg.

The work of Copernicus and Kepler made a case for a heliocentric solar system based on mathematical simplicity and coherence, not on its better agreement with observation. As we have seen, the simplest versions of the Copernican and Ptolemaic theories make the same predictions for the apparent motions of the Sun and planets, in pretty good agreement with observation, while the improvements in the Copernican theory introduced by Kepler were the sort that could have been matched by Ptolemy if he had used an equant and eccentric for the Sun as well as for the planets, and if he had added a few more epicycles. The first *observational* evidence that decisively favored heliocentrism over the old Ptolemaic system was provided by Galileo Galilei.

With Galileo, we come to one of the greatest scientists of history, in a class with Newton, Darwin, and Einstein. He revolutionized observational astronomy with his introduction and use of the telescope, and his study of motion provided a paradigm for modern experimental physics. Further, to an extent that is unique, his scientific career was attended by high drama, of which we can here give only a condensed account.

Galileo was a patrician though not wealthy Tuscan, born in Pisa in 1564, the son of the musical theorist Vincenzo Galilei. After studies at a Florentine monastery, he enrolled as a medical student at the University of Pisa in 1581. Unsurprisingly for a medical student, at this point in his life he was a follower of Aristotle. Galileo's interests then shifted from medicine to mathematics, and for a while he gave mathematics lessons in Florence, the capital of Tuscany. In 1589 Galileo was called back to Pisa to take the chair of mathematics.

While at the University of Pisa Galileo started his study of falling bodies. Some of his work is described in a book, *De Motu* (*On Motion*), which he never published. Galileo concluded, contrary to Aristotle, that the speed of a heavy falling body does not depend appreciably on its weight. It's a nice story that he tested this by dropping various weights from Pisa's Leaning Tower, but there is no evidence for this. While in Pisa Galileo published nothing about his work on falling bodies.

In 1591 Galileo moved to Padua to take the chair of mathematics at its university, which was then the university of the republic of Venice and the most intellectually distinguished university in Europe. From 1597 on he was able to supplement his university salary with the manufacture and sale of mathematical instruments, used in business and war.

In 1597 Galileo received two copies of Kepler's *Mysterium Cosmographicum*. He wrote to Kepler, acknowledging that he, like Kepler, was a Copernican, though as yet he had not made his views public. Kepler replied that Galileo should come out for Copernicus, urging, "Stand forth, O Galileo!"[19]

Soon Galileo came into conflict with the Aristotelians who dominated the teaching of philosophy at Padua, as elsewhere in Italy. In 1604 he lectured on the "new star" observed that year by Kepler. Like Tycho and Kepler he drew the conclusion that change does occur in the heavens, above the orbit of the Moon. He was attacked for this by his sometime friend Cesare Cremonini, professor of philosophy at Padua. Galileo replied with an attack on Cremonini, written in a rustic Paduan dialect as a dialogue between two peasants. The Cremonini peasant argued that the ordinary rules of measurement do not apply in the heavens; and Galileo's peasant replied that philosophers know nothing about measurement: for this one must trust mathematicians, whether for measurement of the heavens or of polenta.

A revolution in astronomy began in 1609, when Galileo first

heard of a new Dutch device known as a spyglass. The magnifying property of glass spheres filled with water was known in antiquity, mentioned for instance by the Roman statesman and philosopher Seneca. Magnification had been studied by al-Haitam, and in 1267 Roger Bacon had written about magnifying glasses in *Opus Maius.* With improvements in the manufacture of glass, reading glasses had become common in the fourteenth century. But to magnify distant objects, it is necessary to combine a *pair* of lenses, one to focus the parallel rays of light from any point on the object so that they converge, and the second to gather these light rays, either with a concave lens while they are still converging or with a convex lens after they begin to diverge again, in either case sending them on parallel directions into the eye. (When relaxed the lens of the eye focuses parallel rays of light to a single point on the retina; the location of that point depends on the direction of the parallel rays.) Spyglasses with such an arrangement of lenses were being produced in the Netherlands by the beginning of the 1600s, and in 1608 several Dutch makers of spectacles applied for patents on their spyglasses. Their applications were rejected, on the ground that the device was already widely known. Spyglasses were soon available in France and Italy, but capable of magnification by only three or four times. (That is, if the lines of sight to two distant points are separated by a certain small angle, then with these spyglasses they seemed to be separated by three or four times that angle.)

Sometime in 1609 Galileo heard of the spyglass, and soon made an improved version, with the first lens convex on the side facing forward and planar on the back, and with long focal length,* while the second was concave on the side facing the first

* Focal length is a length that characterizes the optical properties of a lens. For a convex lens, it is the distance behind the lens at which rays that enter the lens in parallel directions converge. For a concave lens that bends converging rays into parallel directions, the focal length is the distance behind the lens at which the rays would have converged if not for the lens. The focal length depends on the radius of curvature of the lens and on the ratio of the speeds of light in air and glass. (See Technical Note 22.)

lens and planar on the back side, and with shorter focal length. With this arrangement, to send the light from a point source at very large distances on parallel rays into the eye, the distance between the lenses must be taken as the difference of the focal lengths, and the magnification achieved is the focal length of the first lens divided by the focal length of the second lens. (See Technical Note 23.) Galileo was soon able to achieve a magnification by eight or nine times. On August 23, 1609, he showed his spyglass to the doge and notables of Venice and demonstrated that with it ships could be seen at sea two hours before they became visible to the naked eye. The value of such a device to a maritime power like Venice was obvious. After Galileo donated his spyglass to the Venetian republic, his professorial salary was tripled, and his tenure was guaranteed. By November Galileo had improved the magnification of his spyglass to 20 times, and he began to use it for astronomy.

With his spyglass, later known as a telescope, Galileo made six astronomical discoveries of historic importance. The first four of these he described in *Siderius Nuncius* (*The Starry Messenger*),[20] published in Venice in March 1610.

1. On November 20, 1609, Galileo first turned his telescope on the crescent Moon. On the bright side, he could see that its surface is rough:

> By oft-repeated observations of [lunar markings] we have been led to the conclusion that we certainly see the surface of the Moon to be not smooth, even, and perfectly spherical, as the great crowd of philosophers have believed about this and other heavenly bodies, but on the contrary, to be uneven, rough, and crowded with depressions and bulges. And it is like the face of the Earth itself, which is marked here and there with chains of mountains and depths of valleys.

On the dark side, near the terminator, the boundary with the bright side, he could see spots of light, which he interpreted as mountaintops illuminated by the Sun when it was just about to come over the lunar horizon. From the distance of these bright spots from the terminator he could even estimate that some of these mountains were at least four miles high. (See Technical Note 24.) Galileo also interpreted the observed faint illumination of the dark side of the Moon. He rejected various suggestions of Erasmus Reinhold and of Tycho Brahe that the light comes from the Moon itself or from Venus or the stars, and correctly argued that "this marvelous brightness" is due to the reflection of sunlight from the Earth, just as the Earth at night is faintly illuminated by sunlight reflected from the Moon. So a heavenly body like the Moon was seen to be not so very different from the Earth.

2. The spyglass allowed Galileo to observe "an almost inconceivable crowd" of stars much dimmer than stars of the sixth magnitude, and hence too dim to have been seen with the naked eye. The six visible stars of the Pleiades were found to be accompanied with more than 40 other stars, and in the constellation Orion he could see over 500 stars never seen before. Turning his telescope on the Milky Way, he could see that it is composed of many stars, as had been guessed by Albertus Magnus.

3. Galileo reported seeing the planets through his telescope as "exactly circular globes that appear as little moons," but he could not discern any such image of the stars. Instead, he found that, although all stars seemed much brighter when viewed with his telescope, they did not seem appreciably larger. His explanation was confused. Galileo did not know that the apparent size of stars is caused by the bending of light rays in various directions by random fluctuations in the Earth's atmosphere, rather than by anything intrinsic to the neighborhood of the stars. It is these fluctuations that cause stars to appear to twinkle.* Galileo con-

* The angular size of planets is large enough so that the lines of sight from different points on a planetary disk are farther apart as they pass through

cluded that, since it was not possible to make out the images of stars with his telescope, they must be much farther from us than are the planets. As Galileo noted later, this helped to explain why, if the Earth revolves around the Sun, we do not observe an annual stellar parallax.

4. The most dramatic and important discovery reported in *Siderius Nuncius* was made on January 7, 1610. Training his telescope on Jupiter, Galileo saw that "three little stars were positioned near him, small but very bright." At first Galileo thought that these were just another three fixed stars, too dim to have been seen before, though he was surprised that they seemed to be lined up along the ecliptic, two to the east of Jupiter and one to the west. But on the next night all three of these "stars" were to the west of Jupiter, and on January 10 only two could be seen, both to the east. Finally on January 13, he saw that four of these "stars" were now visible, still more or less lined up along the ecliptic. Galileo concluded that Jupiter is accompanied in its orbit with four satellites, similar to Earth's Moon, and like our Moon revolving in roughly the same plane as planetary orbits, which are close to the ecliptic, the plane of the Earth's orbit around the Sun. (They are now known as the four largest moons of Jupiter: Ganymede, Io, Callisto, and Europa, named after the god Jupiter's male and female lovers.)*

This discovery gave important support to the Copernican theory. For one thing, the system of Jupiter and its moons provided a miniature example of what Copernicus had conceived to be the system of the Sun and its surrounding planets, with celestial bod-

---

the Earth's atmosphere than the size of typical atmospheric fluctuations; as a result, the effects of the fluctuations on the light from different lines of sight are uncorrelated, and therefore tend to cancel rather than add coherently. This is why we do not see planets twinkle.

* It would have pained Galileo to know that these are the names that have survived to the present. They were given to the Jovian satellites in 1614 by Simon Mayr, a German astronomer who argued with Galileo over who had discovered the satellites first.

ies evidently in motion about a body other than the Earth. Also, the example of Jupiter's moons put to rest the objection to Copernicus that, if the Earth is moving, why is the Moon not left behind? Everyone agreed that Jupiter is moving, and yet its moons were evidently not being left behind.

Though the results were too late to be included in *Siderius Nuncius*, Galileo by the end of 1611 had measured the periods of revolution of the four Jovian satellites that he had discovered, and in 1612 he published these results on the first page of a work on other matters.[21] Galileo's results are given along with modern values in days (d), hours (h), and minutes (m) in the table below:

Jovian satellite	Period (Galileo)	Period (modern)
Io	1d 18h 30m	1d 18h 29m
Europa	3d 13h 20m	3d 13h 18m
Ganymede	7d 4h 0m	7d 4h 0m
Callisto	16d 18h 0m	16d 18h 5m

The accuracy of Galileo's measurements testifies to his careful observations and precise timekeeping.*

Galileo dedicated *Siderius Nuncius* to his former pupil Cosimo II di Medici, now the grand duke of Tuscany, and he named the four companions of Jupiter the "Medicean stars." This was a calculated compliment. Galileo had a good salary at Padua, but he had been told that it would not again be increased. Also, for this salary he had to teach, taking time away from his research. He was able to strike an agreement with Cosimo, who named him court mathematician and philosopher, with a professorship at Pisa that carried no teaching duties. Galileo insisted on the title "court philosopher" because despite the exciting progress made

* Presumably Galileo was not using a clock, but rather observing the apparent motions of stars. Since the stars seem to go 360° around the Earth in a sidereal day of nearly 24 hours, a 1° change in the position of a star indicates a passage of time equal to 1/360 times 24 hours, or 4 minutes.

in astronomy by mathematicians such as Kepler, and despite the arguments of professors like Clavius, mathematicians continued to have a lower status than that enjoyed by philosophers. Also, Galileo wanted his work to be taken seriously as what philosophers called "physics," an explanation of the nature of the Sun and Moon and planets, not just a mathematical account of appearances.

In the summer of 1610 Galileo left Padua for Florence, a decision that turned out eventually to be disastrous. Padua was in the territory of the republic of Venice, which at the time was less under Vatican influence than any other state in Italy, having successfully resisted a papal interdict a few years before Galileo's departure. Moving to Florence made Galileo much more vulnerable to control by the church. A modern university dean might feel that this danger was a just punishment for Galileo's evasion of teaching duties. But for a while the punishment was deferred.

5. In September 1610 Galileo made the fifth of his great astronomical discoveries. He turned his telescope on Venus, and found that it has phases, like those of the Moon. He sent Kepler a coded message: "The Mother of Loves [Venus] emulates the shapes of Cynthia [the Moon]." The existence of phases would be expected in both the Ptolemaic and the Copernican theories, but the phases would be different. In the Ptolemaic theory, Venus is always more or less between the Earth and the Sun, so it can never be as much as half full. In the Copernican theory, on the other hand, Venus is fully illuminated when it is on the other side of its orbit from the Earth.

This was the first direct evidence that the Ptolemaic theory is wrong. Recall that the Ptolemaic theory gives the same appearance of solar and planetary motions seen from the Earth as the Copernican theory, whatever we choose for the size of each planet's deferent. But it does not give the same appearance as the Copernican theory of solar and planetary motions *as seen from the planets*. Of course, Galileo could not go to any planet to see how the motions of the Sun and other planets appear from there. But the phases of Venus did tell him the direction of the Sun as seen from Venus—the bright side is the side facing the Sun. Only one special case of Ptolemy's theory could give that correctly, the case in which the deferents of Venus and Mercury are identical with the orbit of the Sun, which as already remarked is just the theory of Tycho. That version had never been adopted by Ptolemy, or by any of his followers.

6. At some time after coming to Florence, Galileo found an ingenious way to study the face of the Sun, by using a telescope to project its image on a screen. With this he made his sixth discovery: dark spots were seen to move across the Sun. His results were published in 1613 in his *Sunspot Letters,* about which more later.

There are moments in history when a new technology opens up large possibilities for pure science. The improvement of vacuum pumps in the nineteenth century made possible experiments on electrical discharges in evacuated tubes that led to the discovery of the electron. The Ilford Corporation's development of photographic emulsions allowed the discovery of a host of new elementary particles in the decade following World War II. The development of microwave radar during that war allowed microwaves to be used as a probe of atoms, providing a crucial test of quantum electrodynamics in 1947. And we should not forget the gnomon. But none of these new technologies led to scientific results as impressive as those that flowed from the telescope in the hands of Galileo.

The reactions to Galileo's discoveries ranged from caution to enthusiasm. Galileo's old adversary at Padua, Cesare Cremonini, refused to look through the telescope, as did Giulio Libri, professor of philosophy at Pisa. On the other hand, Galileo was elected a member of the Lincean Academy, founded a few years earlier as Europe's first scientific academy. Kepler used a telescope sent to him by Galileo, and confirmed Galileo's discoveries. (Kepler worked out the theory of the telescope and soon invented his own version, with two convex lenses.)

At first, Galileo had no trouble with the church, perhaps because his support for Copernicus was still not explicit. Copernicus is mentioned only once in *Siderius Nuncias*, near the end, in connection with the question why, if the Earth is moving, it does not leave the Moon behind. At the time, it was not Galileo but Aristotelians like Cremonini who were in trouble with the Roman Inquisition, on much the same grounds that had led to the 1277 condemnation of various tenets of Aristotle. But Galileo managed to get into squabbles with both Aristotelian philosophers and Jesuits, which in the long run did him no good.

In July 1611, shortly after taking up his new position in Florence, Galileo entered into a debate with philosophers who, following what they supposed to be a doctrine of Aristotle, argued that solid ice had a greater density (weight per volume) than liquid water. The Jesuit cardinal Roberto Bellarmine, who had been on the panel of the Roman Inquisition that sentenced Bruno to death, took Galileo's side, arguing that since ice floats, it must be less dense than water. In 1612 Galileo made his conclusions about floating bodies public in his *Discourse on Bodies in Water.*[22]

In 1613 Galileo antagonized the Jesuits, including Christoph Scheiner, in an argument over a peripheral astronomical issue: Are sunspots associated with the Sun itself—perhaps as clouds immediately above its surface, as Galileo thought, which would provide an example (like lunar mountains) of the imperfections of heavenly bodies? Or are they little planets orbiting the Sun more closely than Mercury? If it could be established that they are clouds, then those who claimed that the Sun goes around the Earth could not also claim that the Earth's clouds would be left behind if the Earth went around the Sun. In his *Sunspot Letters* of 1613, Galileo argued that sunspots seemed to narrow as they approach the edge of the Sun's disk, showing that near the disk's edge they were being seen at a slant, and hence were being carried around with the Sun's surface as it rotates. There was also an argument over who had first discovered sunspots. This was only one episode in an increasing conflict with the Jesuits, in which unfairness was not all on one side.[23] Most important for the future, in *Sunspot Letters* Galileo at last came out explicitly for Copernicus.

Galileo's conflict with the Jesuits heated up in 1623 with the publication of *The Assayer.* This was an attack on the Jesuit mathematician Orazio Grassi for Grassi's perfectly correct conclusion, in agreement with Tycho, that the lack of diurnal parallax shows that comets are beyond the orbit of the Moon. Galileo instead offered a peculiar theory, that comets are reflections of the sun's light from linear disturbances of the atmosphere, and do not show diurnal parallax because the disturbances move with the Earth as it rotates. Perhaps the real enemy for Galileo was not Orazio Grassi but Tycho Brahe, who had presented a geocentric theory of the planets that observation could not then refute.

In these years it was still possible for the church to tolerate the Copernican system as a purely mathematical device for calculating apparent motions of planets, though not as a theory of the real nature of the planets and their motions. For instance, in 1615 Bellarmine wrote to the Neapolitan monk Paolo Antonio Foscarini with both a reassurance and a warning about Foscarini's advocacy of the Copernican system:

> It seems to me that Your Reverence and Signor Galileo would act prudently by contenting yourselves with speaking hypothetically and not absolutely, as I have always believed Copernicus to have spoken. [Was Bellarmine taken in by Osiander's preface? Galileo certainly was not.] To say that by assuming the Earth in motion and the Sun immobile saves all the appearances better than the eccentrics and epicycles ever did is to speak well indeed. [Bellarmine apparently did not realize that Copernicus like Ptolemy had employed epicycles, only not so many.] This holds no danger and it suffices for the mathemati-

> cian. But to want to affirm that the Sun really remains at rest at the world's center, that it turns only on itself without running from East to West, and that the Earth is situated in the third heaven and turns very swiftly around the Sun, that is a very dangerous thing. Not only may it irritate all the philosophers and scholastic theologians, it may also injure the faith and render Holy Scripture false.[24]

Sensing the trouble that was gathering over Copernicanism, Galileo in 1615 wrote a celebrated letter about the relation of science and religion to Christina of Lorraine, grand duchess of Tuscany, whose wedding to the late grand duke Ferdinando I Galileo had attended.[25] As Copernicus had in *De Revolutionibus*, Galileo mentioned the rejection of the spherical shape of the Earth by Lactantius as a horrible example of the use of Scripture to contradict the discoveries of science. He also argued against a literal interpretation of the text from the Book of Joshua that Luther had earlier invoked against Copernicus to show the motion of the Sun. Galileo reasoned that the Bible was hardly intended as a text on astronomy, since of the five planets it mentions only Venus, and that just a few times. The most famous line in the letter to Christina reads, "I would say here something that was heard from an ecclesiastic of the most eminent degree: 'That the intention of the Holy Ghost is to teach us how one goes to heaven, not how heaven goes.' " (A marginal note by Galileo indicated that the eminent ecclesiastic was the scholar Cardinal Caesar Baronius, head of the Vatican library.) Galileo also offered an interpretation of the statement in Joshua that the Sun had stood still: it was the *rotation* of the Sun, revealed to Galileo by the motion of sunspots, that had stopped, and this in turn stopped the orbital motion and rotation of the Earth and other planets, which as described in the Bible extended the day of battle. It is not clear whether Galileo actually believed this nonsense or was merely seeking political cover.

Against the advice of friends, Galileo in 1615 went to Rome to argue against the suppression of Copernicanism. Pope Paul V was anxious to avoid controversy and, on the advice of Bellarmine, decided to submit the Copernican theory to a panel of theologians. Their verdict was that the Copernican system is "foolish and absurd in Philosophy, and formally heretical inasmuch as it contradicts the express position of Holy Scripture in many places."[26]

In February 1616 Galileo was summoned to the Inquisition and received two confidential orders. A signed document ordered him not to hold or defend Copernicanism. An unsigned document went further, ordering him not to hold, defend, or teach Copernicanism in any way. In March 1616 the Inquisition issued a public formal order, not mentioning Galileo but banning Foscarini's book, and calling for the writings of Copernicus to be expurgated. *De Revolutionibus* was put on the Index of books forbidden to Catholics. Instead of returning to Ptolemy or Aristotle, some Catholic astronomers, such as the Jesuit Giovanni Battista Riccioli in his 1651 *Almagestum Novum,* argued for Tycho's system, which could not then be refuted by observation. *De Revolutionibus* remained on the Index until 1835, blighting the teaching of science in some Catholic countries, such as Spain.

Galileo hoped for better things after 1624, when Maffeo Barberini became Pope Urban VIII. Barberini was a Florentine and an admirer of Galileo. He welcomed Galileo to Rome and granted him half a dozen audiences. In these conversations Galileo explained his theory of the tides, on which he had been working since before 1616.

Galileo's theory depended crucially on the motion of the Earth. In effect, the idea was that the waters of the oceans slosh back and forth as the Earth rotates while it goes around the Sun, during which movement the net speed of a spot on the Earth's surface along the direction of the Earth's motion in its orbit is continually increasing and decreasing. This sets up a periodic ocean wave with a one-day period, and as with any other oscillation, there are overtones, with periods of half a day, a third of a day, and so on. So far, this leaves out any influence of the Moon, but it had been known since antiquity that the higher "spring" tides occur at full and new moon, while the lower "neap" tides are at the times of half-moon. Galileo tried to explain the influence of the Moon by supposing that for some reason the Earth's orbital speed is increased at new moon, when the Moon is between the Earth and the Sun, and decreased at full moon, when the Moon is on the other side of the Earth from the Sun.

This was not Galileo at his best. It's not so much that his theory was wrong. Without a theory of gravitation there was no way that Galileo could have correctly understood the tides. But Galileo should have known that a speculative theory of tides that had no significant empirical support could not be counted as a verification of the Earth's motion.

The pope said that he would permit publication of this theory of tides if Galileo would treat the motion of the Earth as a mathematical hypothesis, not as something likely to be true. Urban explained that he did not approve of the Inquisition's public order of 1616, but he was not ready to rescind it. In these conversations Galileo did not mention to the pope the Inquisition's private orders to him.

In 1632 Galileo was ready to publish his theory of the tides, which had grown into a comprehensive defense of Copernicanism. As yet, the church had made no public criticism of Galileo, so when he applied to the local bishop for permission to publish a new book it was granted. This was his *Dialogo* (*Dialogue Concerning the Two Chief Systems of the World—Ptolemaic and Copernican*).

The title of Galileo's book is peculiar. There were at the time not two but *four* chief systems of the world: not just the Ptolemaic and Copernican, but also the Aristotelian, based on homocentric spheres revolving around the Earth, and the Tychonic, with the Sun and Moon going around a stationary Earth but all other planets going around the Sun. Why did Galileo not consider the Aristotelian and Tychonic systems?

About the Aristotelian system, one can say that it did not agree with observation, but it had been known to disagree with observation for two thousand years without losing all its adherents. Just look back at the argument made by Fracastoro at the beginning of the sixteenth century, quoted in Chapter 10. Galileo a century later evidently thought such arguments not worth answering, but it is not clear how that came about.

On the other hand, the Tychonic system worked too well for it to be justly dismissed. Galileo certainly knew about Tycho's system. Galileo may have thought his own theory of the tides showed that the Earth does move, but this theory was not supported by any quantitative successes. Or perhaps Galileo just did not want to expose Copernicus to competition with the formidable Tycho.

The *Dialogo* took the form of a conversation among three characters: Salviati, a stand-in for Galileo named for Galileo's friend the Florentine nobleman Filippo Salviati; Simplicio, an Aristotelian, perhaps named for Simplicius (and perhaps intended to represent a simpleton); and Sagredo, named for Galileo's Venetian friend the mathematician Giovanni Francesco Sagredo, to judge wisely between them. The first three days of the conversation showed Salviati demolishing Simplicio, with the tides brought in only on the fourth day. This certainly violated the Inquisition's unsigned order to Galileo, and arguably the less stringent signed order (not to hold or defend Copernicanism) as well. To make matters worse, the *Dialogo* was in Italian rather than Latin, so that it could be read by any literate Italian, not just by scholars.

At this point, Pope Urban was shown the unsigned 1616 order of the Inquisition to Galileo, perhaps by enemies that Galileo had made in the earlier arguments over sunspots and comets. Urban's anger may have been amplified by a suspicion that he was the model for Simplicio. It didn't help that some of the pope's words when he was Cardinal Barberini showed up in the mouth of Simplicio. The Inquisition ordered sales of the *Dialogo* to be banned, but it was too late—the book was already sold out.

Galileo was put on trial in April 1633. The case against him hinged on his violation of the Inquisition's orders of 1616. Galileo was shown the instruments of torture and tried a plea bargain, admitting that personal vanity had led him to go too far. But he was nevertheless declared under "vehement suspicion of heresy," condemned to eternal imprisonment, and forced to abjure his view that the Earth moves around the Sun. (An apocryphal story has it that as Galileo left the court, he muttered under his breath, "Eppur si muove," that is, "But it does move.")

Fortunately Galileo was not treated as roughly as he might have been. He was allowed to begin his imprisonment as a guest of the archbishop of Siena, and then to continue it in his own villa at Arcetri, near Florence, and near the convent residence of his daughters, Sister Maria Celeste and Sister Arcangela.[27] As we will see in Chapter 12, Galileo was able during these years to return to his work on the problem of motion, begun a half century earlier at Pisa.

Galileo died in 1642 while still under house arrest in Arcetri. It was not until 1835 that books like Galileo's that advocated the Copernican system were removed from the Index of books banned by the Catholic church, though long before that Copernican astronomy had become widely accepted in most Catholic as well as Protestant countries. Galileo was rehabilitated by the church in the twentieth century.[28] In 1979 Pope John Paul II referred to Galileo's *Letter to Christina* as having "formulated important norms of an epistemological character, which are indispensable to reconcile Holy Scripture and science."[29] A commission was convened to look into the case of Galileo, and reported that the church in Galileo's time had been mistaken. The pope responded, "The error of the theologians of the time, when they maintained the centrality of the Earth, was to think that our understanding of the physical world's structure was, in some way, imposed by the literal sense of the Sacred Scripture."[30]

My own view is that this is quite inadequate. The church of course cannot avoid the knowledge, now shared by everyone, that it had been wrong about the motion of the Earth. But suppose the church had been correct and Galileo mistaken about astronomy. The church would still have been wrong to sentence Galileo to imprisonment and to deny his right to publish, just as it had been wrong to burn Giordano Bruno, heretic as he was.[31] Fortunately, although I don't know if this has been explicitly acknowledged by the church, it would not today dream of such actions. With the exception of those Islamic countries that punish blasphemy or apostasy, the world has generally learned the lesson that governments and religious authorities have no business imposing criminal penalties on religious opinions, whether true or false.

From the calculations and observations of Copernicus, Tycho Brahe, Kepler, and Galileo there had emerged a correct description of the solar system, encoded in Kepler's three laws. An *explanation* of *why* the planets obey these laws had to wait a generation, until the advent of Newton.

12

Experiments Begun

No one can manipulate heavenly bodies, so the great achievements in astronomy described in Chapter 11 were necessarily based on passive observation. Fortunately the motions of planets in the solar system are simple enough so that after many centuries of observation with increasingly sophisticated instruments these motions could at last be correctly described. For the solution of other problems it was necessary to go beyond observation and measurement and perform experiments, in which general theories are tested or suggested by the artificial manipulation of physical phenomena.

In a sense people have always experimented, using trial and error in order to discover ways to get things done, from smelting ores to baking cakes. In speaking here of the beginnings of experiment, I am concerned only with experiments carried out to discover or test general theories about nature.

It is not possible to be precise about the beginning of experimentation in this sense.[1] Archimedes may have tested his theory of hydrostatics experimentally, but his treatise *On Floating Bodies* followed the purely deductive style of mathematics, and gave no hint of the use of experiment. Hero and Ptolemy did experiments to test their theories of reflection and refraction, but their example was not followed until centuries later.

One new thing about experimentation in the seventeenth century was eagerness to make public use of its results in judging the validity of physical theories. This appears early in the century in work on hydrostatics, as is shown in Galileo's *Discourse on Bodies in Water* of 1612. More important was the quantitative study of the motion of falling bodies, an essential prerequisite to the work of Newton. It was work on this problem, and also on the nature of air pressure, that marked the real beginning of modern experimental physics.

Like much else, the experimental study of motion begins with Galileo. His conclusions about motion appeared in *Dialogues Concerning Two New Sciences*, finished in 1635, when he was under house arrest at Arcetri. Publication was forbidden by the church's Congregation of the Index, but copies were smuggled out of Italy. In 1638 the book was published in the Protestant university town of Leiden by the firm of Louis Elzevir. The cast of *Two New Sciences* again consists of Salviati, Simplicio, and Sagredo, playing the same roles as before.

Among much else, the "First Day" of *Two New Sciences* contains an argument that heavy and light bodies fall at the same rate, contradicting Aristotle's doctrine that heavy bodies fall faster than light ones. Of course, because of air resistance, light bodies do fall a little more slowly than heavy ones. In dealing with this, Galileo demonstrates his understanding of the need for scientists to live with approximations, running counter to the Greek emphasis on precise statements based on rigorous mathematics. As Salviati explains to Simplicio:[2]

> Aristotle says, "A hundred pound iron ball falling from the height of a hundred braccia hits the ground before one of just one pound has descended a single braccio." I say that they arrive at the same time. You find, on making the experiment, that the larger anticipates the smaller by two inches; that is, when the larger one strikes the ground, the other is two inches behind it. And now you want to hide, behind those two inches, the ninety-nine braccia of Aristotle, and speaking only of my tiny error, remain silent about his enormous one.

Galileo also shows that air has positive weight; estimates its density; discusses motion through resisting media; explains musical harmony; and reports on the fact that a pendulum will take the same time for each swing, whatever the amplitude of the swings.* This is the principle that decades later was to lead to the invention of pendulum clocks and to the accurate measurement of the rate of acceleration of falling bodies.

The "Second Day" of *Two New Sciences* deals with the strengths of bodies of various shapes. It is on the "Third Day" that Galileo returns to the problem of motion, and makes his most interesting contribution. He begins the Third Day by reviewing some trivial properties of uniform motion, and then goes on to define uniform acceleration along the same lines as the fourteenth-century Merton College definition: the speed increases by equal amounts in each equal interval of time. Galileo also gives a proof of the mean speed theorem, along the same lines as Oresme's proof, but he makes no reference to Oresme or to the Merton dons. Unlike his medieval predecessors, Galileo goes beyond this mathematical theorem and argues that freely falling bodies undergo uniform acceleration, but he declines to investigate the cause of this acceleration.

As already mentioned in Chapter 10, there was at the time a widely held alternative to the theory that bodies fall with uniform acceleration. According to this other view, the speed that freely falling bodies acquire in any interval of time is proportional to the *distance* fallen in that interval, not to the time.† Galileo gives

* This is actually true only for swings of the pendulum through small angles, though Galileo did not note this qualification. Indeed he speaks of swings of 50° or 60° (degrees of arc) taking the same time as much smaller swings, and this suggests that he did not actually do all the experiments on the pendulum that he reported.

† Taken literally, this would mean that a body dropped from rest would never fall, since with zero initial velocity at the end of the first infinitesimal instant it would not have moved, and hence with speed proportional to distance would still have zero velocity. Perhaps the doctrine that the speed is proportional to the distance fallen was intended to apply only after a brief initial period of acceleration.

various arguments against this view,* but the verdict regarding these different theories of the acceleration of falling bodies had to come from experiment.

With the distance fallen from rest equal (according to the mean speed theorem) to half the velocity attained times the elapsed time, and with that velocity itself proportional to the time elapsed, the distance traveled in free fall should be proportional to the *square* of the time. (See Technical Note 25.) This is what Galileo sets out to verify.

Freely falling bodies move too rapidly for Galileo to have been able to check this conclusion by following how far a falling body falls in any given time, so he had the idea of slowing the fall by studying balls rolling down an inclined plane. For this to be relevant, he had to show how the motion of a ball rolling down an inclined plane is related to a body in free fall. He did this by noting that the speed a ball reaches after rolling down an inclined plane depends only on the *vertical* distance through which the ball has rolled, not on the angle with which the plane is tilted.† A freely falling ball can be regarded as one that rolls down a vertical plane, and so if the speed of a ball rolling down an inclined plane is proportional to the time elapsed, then the same ought to be true for a freely falling ball. For a plane inclined at a small angle the speed is of course much less than the speed of a body falling freely (that is the point of using an inclined plane) but the two speeds are proportional, and so the distance traveled along

* One of Galileo's arguments is fallacious, because it applies to the *average* speed during an interval of time, not to the speed acquired by the end of that interval.

† This is shown in Technical Note 25. As explained there, though Galileo did not know it, the speed of the ball rolling down the plane is not equal to the speed of a body that would have fallen freely the same vertical difference, because some of the energy released by the vertical descent goes into the rotation of the ball. But the speeds are proportional, so Galileo's qualitative conclusion that the speed of a falling body is proportional to the time elapsed is not changed when we take into account the ball's rotation.

the plane is proportional to the distance that a freely falling body would have traveled in the same time.

In *Two New Sciences* Galileo reports that the distance rolled *is* proportional to the square of the time. Galileo had done these experiments at Padua in 1603 with a plane at a less than 2° angle to the horizontal, ruled with lines marking intervals of about 1 millimeter.[3] He judged the time by the equality of the intervals between sounds made as the ball reached marks along its path, whose distances from the starting point are in the ratios $1^2 = 1 : 2^2 = 4 : 3^2 = 9$, and so on. In the experiments reported in *Two New Sciences* he instead measured relative intervals of time with a water clock. A modern reconstruction of this experiment shows that Galileo could very well have achieved the accuracy he claimed.[4]

Galileo had already considered the acceleration of falling bodies in the work discussed in Chapter 11, the *Dialogue Concerning the Two Chief World Systems*. On the Second Day of this previous *Dialogue*, Salviati in effect claims that the distance fallen is proportional to the square of the time, but gives only a muddled explanation. He also mentions that a cannonball dropped from a height of 100 *braccia* will reach the ground in 5 seconds. It is pretty clear that Galileo did not actually measure this time,[5] but is here presenting only an illustrative example. If one *braccio* is taken as 21.5 inches, then using the modern value of the acceleration due to gravity, the time for a heavy body to drop 100 *braccia* is 3.3 seconds, not 5 seconds. But Galileo apparently never attempted a serious measurement of the acceleration due to gravity.

The "Fourth Day" of *Dialogues Concerning Two New Sciences* takes up the trajectory of projectiles. Galileo's ideas were largely based on an experiment he did in 1608[6] (discussed in detail in Technical Note 26). A ball is allowed to roll down an inclined plane from various initial heights, then rolls along the horizontal tabletop on which the inclined plane sits, and finally shoots off into the air from the table edge. By measuring the distance traveled when the ball reaches the floor, and by observation of the ball's path in the air, Galileo concluded that the trajectory is a parabola. Galileo does not describe this experiment in *Two New Sciences*, but instead gives the theoretical argument for a parabola. The crucial point, which turned out to be essential in Newton's mechanics, is that each component of a projectile's motion is separately subject to the corresponding component of the force acting on the projectile. Once a projectile rolls off a table edge or is shot out of a cannon, there is nothing but air resistance to change its horizontal motion, so the horizontal distance traveled is very nearly proportional to the time elapsed. On the other hand, during the same time, like any freely falling body, the projectile is accelerated downward, so that the vertical distance fallen is proportional to the square of the time elapsed. It follows that the vertical distance fallen is proportional to the square of the horizontal distance traveled. What sort of curve has this property? Galileo shows that the path of the projectile is a parabola, using Apollonius' definition of a parabola as the intersection of a cone with a plane parallel to the cone's surface. (See Technical Note 26.)

The experiments described in *Two New Sciences* made a historic break with the past. Instead of limiting himself to the study of free fall, which Aristotle had regarded as natural motion, Galileo turned to artificial motions, of balls constrained to roll down an inclined plane or projectiles thrown forward. In this sense, Galileo's inclined plane is a distant ancestor of today's particle accelerators, with which we artificially create particles found nowhere in nature.

Galileo's work on motion was carried forward by Christiaan Huygens, perhaps the most impressive figure in the brilliant generation between Galileo and Newton. Huygens was born in 1629 into a family of high civil servants who had worked in the administration of the Dutch republic under the House of Orange. From 1645 to 1647 he studied both law and mathematics at the University of Leiden, but he then turned full-time to mathematics and eventually to natural science. Like Descartes, Pascal, and Boyle, Huygens was a polymath, working on a wide range of problems in mathematics, astronomy, statics, hydrostatics, dynamics, and optics.

Huygens' most important work in astronomy was his telescopic study of the planet Saturn. In 1655 he discovered its largest moon, Titan, revealing thereby that not only the Earth and Jupiter have satellites. He also explained that Saturn's peculiar noncircular appearance, noticed by Galileo, is due to rings surrounding the planet.

In 1656–1657 Huygens invented the pendulum clock. It was based on Galileo's observation that the time a pendulum takes for each swing is independent of the swing's amplitude. Huygens recognized that this is true only in the limit of very small swings, and found ingenious ways to preserve the amplitude-independence of the times even for swings of appreciable amplitudes. While previous crude mechanical clocks would gain or lose about 5 minutes a day, Huygens' pendulum clocks generally gained or lost no more than 10 seconds a day, and one of them lost only about ½ second per day.[7]

From the period of a pendulum clock of a given length, Huygens the next year was able to infer the value of the acceleration of freely falling bodies near the Earth's surface. In the *Horologium oscillatorium*—published later, in 1673—Huygens was able to show that "the time of one small oscillation is related to the time of perpendicular fall from half the height of the pendulum as the circumference of a circle is related to its diameter."[8] That is, the time for a pendulum to swing through a small angle from one side to the other equals π times the time for a body to fall a distance equal to half the length of the pendulum. (Not an easy result to obtain as Huygens did, without calculus.) Using this principle, and measuring the periods of pendulums of various lengths, Huygens was able to calculate the acceleration due to gravity, something that Galileo could not measure accurately with the means he had at hand. As Huygens expressed it, a freely falling body falls 15$\frac{1}{12}$ "Paris feet" in the first second. The ratio of the Paris foot to the modern English foot is variously estimated as between 1.06 and 1.08; if we take 1 Paris foot to equal 1.07

English feet, then Huygens' result was that a freely falling body falls 16.1 feet in the first second, which implies an acceleration of 32.2 feet/second per second, in excellent agreement with the standard modern value of 32.17 feet/second per second. (As a good experimentalist, Huygens checked that the acceleration of falling bodies actually does agree within experimental error with the acceleration he inferred from his observations of pendulums.) As we will see, this measurement, later repeated by Newton, was essential in relating the force of gravity on Earth to the force that keeps the Moon in its orbit.

The acceleration due to gravity could have been inferred from earlier measurements by Riccioli of the time for weights to fall various distances.[9] To measure time accurately, Riccioli used a pendulum that had been carefully calibrated by counting its strokes in a solar or sidereal day. To his surprise, his measurements confirmed Galileo's conclusion that the distance fallen is proportional to the square of the time. From these measurements, published in 1651, it could have been calculated (though Riccioli did not do so) that the acceleration due to gravity is 30 Roman feet/second per second. It is fortunate that Riccioli recorded the height of the Asinelli tower in Bologna, from which many of the weights were dropped, as 312 Roman feet. The tower still stands, and its height is known to be 323 modern English feet, so Riccioli's Roman foot must have been 323/312 = 1.035 English feet, and 30 Roman feet/second per second therefore corresponds to 31 English feet/second per second, in fair agreement with the modern value. Indeed, if Riccioli had known Huygens' relation between the period of a pendulum and the time required for a body to fall half its length, he could have used his calibration of pendulums to calculate the acceleration due to gravity, without having to drop anything off towers in Bologna.

In 1664 Huygens was elected to the new Académie Royale des Sciences, with an accompanying stipend, and he moved to Paris for the next two decades. His great work on optics, the *Treatise on Light*, was written in Paris in 1678 and set out the wave theory of light. It was not published until 1690, perhaps because Huygens had hoped to translate it from French to Latin but had never found the time before his death in 1695. We will come back to Huygens' wave theory in Chapter 14.

In a 1669 article in the *Journal des Sçavans*, Huygens gave the correct statement of the rules governing collisions of hard bodies (which Descartes had gotten wrong): it is the conservation of what are now called momentum and kinetic energy.[10] Huygens claimed that he had confirmed these results experimentally, presumably by studying the impact of colliding pendulum bobs, for which initial and final velocities could be precisely calculated. And as we shall see in Chapter 14, Huygens in the *Horologium oscillatorium* calculated the acceleration associated with motion on a curved path, a result of great importance to Newton's work.

The example of Huygens shows how far science had come from the imitation of mathematics—from the reliance on deduction and the aim of certainty characteristic of mathematics. In the preface to the *Treatise on Light* Huygens explains:

> There will be seen [in this book] demonstrations of those kinds which do not produce as great a certitude as those of Geometry, and which even differ much therefrom, since whereas the Geometers prove their Propositions by fixed and incontestable Principles, here the Principles are verified by the conclusions to be drawn from them; the nature of these things not allowing of this being done otherwise.[11]

It is about as good a description of the methods of modern physical science as one can find.

In the work of Galileo and Huygens on motion, experiment was used to refute the physics of Aristotle. The same can be said of the contemporaneous study of air pressure. The impossibility of a vacuum was one of the doctrines of Aristotle that came into question in the seventeenth century. It was eventually understood that phenomena such as suction, which seemed to arise from nature's abhorrence of a vacuum, actually represent effects of the pressure of the air. Three figures played a key role in this discovery, in Italy, France, and England.

Well diggers in Florence had known for some time that suction pumps cannot lift water to a height more than about 18 *braccia*, or 32 feet. (The actual value at sea level is closer to 33.5 feet.) Galileo and others had thought that this showed a limitation on nature's abhorrence of a vacuum. A different interpretation was offered by Evangelista Torricelli, a Florentine who worked on geometry, projectile motion, fluid mechanics, optics, and an early version of calculus. Torricelli argued that this limitation on suction pumps arises because the weight of the air pressing down on the water in the well could support only a column of water no more than 18 *braccia* high. This weight is diffused through the air, so any surface whether horizontal or not is subjected by the air to a force proportional to its area; the force per area, or *pressure*, exerted by air at rest is equal to the weight of a vertical column of air, going up to the top of the atmosphere, divided by the cross-sectional area of the column. This pressure acts on the surface of water in a well, and adds to the pressure of the water, so that when air pressure at the top of a vertical pipe immersed in the water is reduced by a pump, water rises in the pipe, but by only an amount limited by the finite pressure of the air.

In the 1640s Torricelli set out on a series of experiments to prove this idea. He reasoned that since the weight of a volume of mercury is 13.6 times the weight of the same volume of water, the maximum height of a column of mercury in a vertical glass tube closed on top that can be supported by the air—whether by the air pressing down on the surface of a pool of mercury in which the tube is standing, or on the open bottom of the tube when exposed to the air—should be 18 *braccia* divided by 13.6, or using more accurate modern values, 33.5 feet/13.6 = 30 inches = 760 millimeters. In 1643 he observed that if a vertical glass tube longer than this and closed at the top end is filled with mercury, then some mercury will flow out until the height of the mercury in the tube is about 30 inches. This leaves empty space on top, now known as a "Torricellian vacuum." Such a tube can then serve as a barometer, to measure changes in ambient air pressure; the higher the air pressure, the higher the column of mercury that it can support.

The French polymath Blaise Pascal is best known for his work of Christian theology, the *Pensées*, and for his defense of the Jansenist sect against the Jesuit order, but he also contributed to geometry and to the theory of probability, and explored the pneumatic phenomena studied by Torricelli. Pascal reasoned that if the column of mercury in a glass tube open at the bottom is held up by the pressure of the air, then the height of the column should decrease when the tube is carried to high altitude on a mountain, where there is less air overhead and hence lower air pressure. After this prediction was verified in a series of expeditions from 1648 to 1651, Pascal concluded, "All the effects ascribed to [the abhorrence of a vacuum] are due to the weight and pressure of the air, which is the only real cause."[12]

Pascal and Torricelli have been honored by having modern units of pressure named after them. One pascal is the pressure that produces a force of 1 newton (the force that gives a mass of 1 kilogram an acceleration of 1 meter per second in a second) when exerted on an area of 1 square meter. One torr is the pressure that will support a column of 1 millimeter of mercury. Standard atmospheric pressure is 760 torr, which equals a little more than 100,000 pascals.

The work of Torricelli and Pascal was carried further in England by Robert Boyle. Boyle was a son of the earl of Cork, and hence an absentee member of the "ascendancy," the Protestant upper class that dominated Ireland in his time. He was educated at Eton College, took a grand tour of the Continent, and fought on the side of Parliament in the civil wars that raged in England in the 1640s. Unusually for a member of his class, he became fascinated by science. He was introduced to the new ideas revolutionizing astronomy in 1642, when he read Galileo's *Dialogue Concerning the Two Chief World Systems*. Boyle insisted on naturalistic explanations of natural phenomena, declaring, "None is more willing [than myself] to acknowledge and venerate Divine Omnipotence, [but] our controversy is not about what God can do, but about what can be done by natural agents, not elevated above the sphere of nature."[13] But, like many before Darwin and some even after, he argued that the wonderful capabilities of animals and men showed that they must have been designed by a benevolent creator.

Boyle's work on air pressure was described in 1660 in *New Experiments Physico-Mechanical Touching the Spring of the Air*. In his experiments, he used an improved air pump, invented by his assistant Robert Hooke, about whom more in Chapter 14. By pumping air out of vessels, Boyle was able to establish that air is needed for the propagation of sound, for fire, and for life. He found that the level of mercury in a barometer drops when air is pumped out of its surroundings, adding a powerful argument in favor of Torricelli's conclusion that air pressure is responsible for phenomena previously attributed to nature's abhorrence of a vacuum. By using a column of mercury to vary both the pressure and the volume of air in a glass tube, not letting air in or out and keeping the temperature constant, Boyle was able to study the relation between pressure and volume. In 1662, in a second edition of *New Experiments*, he reported that the pressure varies with the volume in such a way as to keep the pressure times the volume fixed, a rule now known as Boyle's law.

Not even Galileo's experiments with inclined planes illustrate so well the new aggressive style of experimental physics as these experiments on air pressure. No longer were natural philosophers relying on nature to reveal its principles to casual observers. Instead Mother Nature was being treated as a devious adversary, whose secrets had to be wrested from her by the ingenious construction of artificial circumstances.

13

Method Reconsidered

By the end of the sixteenth century the Aristotelian model for scientific investigation had been severely challenged. It was natural then to seek a new approach to the method for gathering reliable knowledge about nature. The two figures who became best known for attempts to formulate a new method for science are Francis Bacon and René Descartes. They are, in my opinion, the two individuals whose importance in the scientific revolution is most overrated.

Francis Bacon was born in 1561, the son of Nicholas Bacon, Lord Keeper of the Privy Seal of England. After an education at Trinity College, Cambridge, he was called to the bar, and followed a career in law, diplomacy, and politics. He rose to become Baron Verulam and lord chancellor of England in 1618, and later Viscount St. Albans, but in 1621 he was found guilty of corruption and declared by Parliament to be unfit for public office.

Bacon's reputation in the history of science is largely based on his book *Novum Organum* (*New Instrument, or True Directions Concerning the Interpretation of Nature*), published in 1620. In this book Bacon, neither a scientist nor a mathematician, expressed an extreme empiricist view of science, rejecting not only Aristotle but also Ptolemy and Copernicus. Discoveries were to emerge directly from careful, unprejudiced observation of nature, not by deduction from first principles. He also disparaged any research that did not serve an immediate practical purpose. In *The New Atlantis*, he imagined a cooperative research institute, "Solomon's House," whose members would devote themselves to collecting useful facts about nature. In this way, man would supposedly regain the dominance over nature that was lost after the expulsion from Eden. Bacon died in 1626. There is a story that, true to his empirical principles, he succumbed to pneumonia after an experimental study of the freezing of meat.

Bacon and Plato stand at opposite extremes. Of course, both extremes were wrong. Progress depends on a blend of observation or experiment, which may suggest general principles, and of deductions from these principles that can be tested against new observations or experiments. The search for knowledge of practical value can serve as a corrective to uncontrolled speculation, but explaining the world has value in itself, whether or not it leads directly to anything useful. Scientists in the seventeenth and eighteenth centuries would invoke Bacon as a counterweight to Plato and Aristotle, somewhat as an American politician might invoke Jefferson without ever having been influenced by anything Jefferson said or did. It is not clear to me that anyone's scientific work was actually changed for the better by Bacon's writing. Galileo did not need Bacon to tell him to do experiments, and neither I think did Boyle or Newton. A century before Galileo, another Florentine, Leonardo da Vinci, was doing experiments on falling bodies, flowing liquids, and much else.[1] We know about this work only from a pair of treatises on painting and on fluid motion that were compiled after his death, and from notebooks that have been discovered from time to time since then, but if Leonardo's experiments had no influence on the advance of science, at least they show that experiment was in the air long before Bacon.

René Descartes was an altogether more noteworthy figure than Bacon. Born in 1596 into the juridical nobility of France, the *noblesse de robe*, he was educated at the Jesuit college of La Flèche, studied law at the University of Poitiers, and served in the army of Maurice of Nassau in the Dutch war of independence.

In 1619 Descartes decided to devote himself to philosophy and mathematics, work that began in earnest after 1628, when he settled permanently in Holland.

Descartes put his views about mechanics into *Le Monde*, written in the early 1630s but not published until 1664, after his death. In 1637 he published a philosophical work, *Discours de la méthode pour bien conduire sa raison, et chercher la vérité dans les sciences* (*Discourse on the Method of Rightly Conducting One's Reason and of Seeking Truth in the Sciences*). His ideas were further developed in his longest work, the *Principles of Philosophy*, published in Latin in 1644 and then in a French translation in 1647. In these works Descartes expresses skepticism about knowledge derived from authority or the senses. For Descartes the only certain fact is that he exists, deduced from the observation that he is thinking about it. He goes on to conclude that the world exists, because he perceives it without exerting an effort of will. He rejects Aristotelian teleology—things are as they are, not because of any purpose they might serve. He gives several arguments (all unconvincing) for the existence of God, but rejects the authority of organized religion. He also rejects occult forces at a distance—things interact with each other through direct pulling and pushing.

Descartes was a leader in bringing mathematics into physics, but like Plato he was too much impressed by the certainty of mathematical reasoning. In Part I of the *Principles of Philosophy*, titled "On the Principles of Human Knowledge," Descartes described how fundamental scientific principles could be deduced with certainty by pure thought. We can trust in the "natural enlightenment or the faculty of knowledge given to us by God" because "it would be completely contradictory for Him to deceive us."[2] It is odd that Descartes thought that a God who allowed earthquakes and plagues would not allow a philosopher to be deceived.

Descartes did accept that the application of fundamental physical principles to specific systems might involve uncertainty and call for experimentation, if one did not know all the details of what the system contains. In his discussion of astronomy in Part III of *Principles of Philosophy*, he considers various hypotheses about the nature of the planetary system, and cites Galileo's observations of the phases of Venus as reason for preferring the hypotheses of Copernicus and Tycho to that of Ptolemy.

This brief summary barely touches on Descartes' views. His philosophy was and is much admired, especially in France and among specialists in philosophy. I find this puzzling. For someone who claimed to have found the true method for seeking reliable knowledge, it is remarkable how wrong Descartes was about so many aspects of nature. He was wrong in saying that the Earth is prolate (that is, that the distance through the Earth is greater from pole to pole than through the equatorial plane). He, like Aristotle, was wrong in saying that a vacuum is impossible. He was wrong in saying that light is transmitted instantaneously.* He was wrong in saying that space is filled with material vortices that carry the planets around in their paths. He was wrong in saying that the pineal gland is the seat of a soul responsible for human consciousness. He was wrong about what quantity is conserved in collisions. He was wrong in saying that the speed of a freely falling body is proportional to the distance fallen. Finally, on the basis of observation of several lovable pet cats, I am convinced that Descartes was also wrong in saying that animals are machines without true consciousness. Voltaire had similar reservations about Descartes:[3]

* Descartes compared light to a rigid stick, which when pushed at one end instantaneously moves at the other end. He was wrong about sticks too, though for reasons he could not then have known. When a stick is pushed at one end, nothing happens at the other end until a wave of compression (essentially a sound wave) has traveled from one end of the stick to the other. The speed of this wave increases with the rigidity of the stick, but Einstein's special theory of relativity does not allow anything to be perfectly rigid; no wave can have a speed exceeding that of light. Descartes' use of this sort of comparison is discussed by Peter Galison, "Descartes Comparisons: From the Invisible to the Visible," *Isis* 75, 311 (1984).

> He erred on the nature of the soul, on the proofs of the existence of God, on the subject of matter, on the laws of motion, on the nature of light. He admitted innate ideas, he invented new elements, he created a world, he made man according to his own fashion—in fact, it is rightly said that man according to Descartes is Descartes' man, far removed from man as he actually is.

Descartes' scientific misjudgments would not matter in assessing the work of someone who wrote about ethical or political philosophy, or even metaphysics; but because Descartes wrote about "the method of rightly conducting one's reason and of seeking truth in the sciences," his repeated failure to get things right must cast a shadow on his philosophical judgment. Deduction simply cannot carry the weight that Descartes placed on it.

Even the greatest scientists make mistakes. We have seen how Galileo was wrong about the tides and comets, and we will see how Newton was wrong about diffraction. For all his mistakes, Descartes, unlike Bacon, did make significant contributions to science. These were published as a supplement to the *Discourse on Method*, under three headings: geometry, optics, and meteorology.[4] These, rather than his philosophical writings, in my view represent his positive contributions to science.

Descartes' greatest contribution was the invention of a new mathematical method, now known as analytic geometry, in which curves or surfaces are represented by equations that are satisfied by the coordinates of points on the curve or surface. "Coordinates" in general can be any numbers that give the location of a point, such as longitude, latitude, and altitude, but the particular kind known as "Cartesian coordinates" are the distances of the point from a center along a set of fixed perpendicular directions. For instance, in analytic geometry a circle of radius R is a curve on which the coordinates x and y are distances measured from the center of the circle along any two perpendicular directions, and satisfy the equation $x^2 + y^2 = R^2$. (Technical Note 18 gives a similar description of an ellipse.) This very important use of letters of the alphabet to represent unknown distances or other numbers originated in the sixteenth century with the French mathematician, courtier, and cryptanalyst François Viète, but Viète still wrote out equations in words. The modern formalism of algebra and its application to analytic geometry are due to Descartes.

Using analytic geometry, we can find the coordinates of the point where two curves intersect, or the equation for the curve where two surfaces intersect, by solving the pair of equations that define the curves or the surfaces. Most physicists today solve geometric problems in this way, using analytic geometry, rather than the classic methods of Euclid.

In physics Descartes' significant contributions were in the study of light. First, in his *Optics*, Descartes presented the relation between the angles of incidence and refraction when light passes from medium A to medium B (for example, from air to water): if the angle between the incident ray and the perpendicular to the surface separating the media is i, and the angle between the refracted ray and this perpendicular is r, then the sine* of i divided by the sine of r is an angle-independent constant n:

$$\text{sine of } i \,/\, \text{sine of } r = n$$

In the common case where medium A is the air (or, strictly speaking, empty space), n is the constant known as the "index of refraction" of medium B. For instance, if A is air and B is water then n is the index of refraction of water, about 1.33. In any case like this, where n is larger than 1, the angle of refraction r is smaller than the angle of incidence i, and the ray of light entering the denser medium is bent toward the direction perpendicular to the surface.

* Recall that the sine of an angle is the side opposite that angle in a right triangle, divided by the hypotenuse of the triangle. It increases as the angle increases from zero to 90°, in proportion to the angle for small angles, and then more slowly.

Unknown to Descartes, this relation had already been obtained empirically in 1621 by the Dane Willebrord Snell and even earlier by the Englishman Thomas Harriot; and a figure in a manuscript by the tenth-century Arab physicist Ibn Sahl suggests that it was also known to him. But Descartes was the first to publish it. Today the relation is usually known as Snell's law, except in France, where it is more commonly attributed to Descartes.

Descartes' derivation of the law of refraction is difficult to follow, in part because neither in his account of the derivation nor in the statement of the result did Descartes make use of the trigonometric concept of the sine of an angle. Instead, he wrote in purely geometric terms, though as we have seen the sine had been introduced from India almost seven centuries earlier by al-Battani, whose work was well known in medieval Europe. Descartes' derivation is based on an analogy with what Descartes imagined would happen when a tennis ball is hit through a thin fabric; the ball will lose some speed, but the fabric can have no effect on the component of the ball's velocity *along* the fabric. This assumption leads (as shown in Technical Note 27) to the result cited above: the ratio of the sines of the angles that the tennis ball makes with the perpendicular to the screen before and after it hits the screen is an angle-independent constant n. Though it is hard to see this result in Descartes' discussion, he must have understood this result, because with a suitable value for n he gets more or less the right numerical answers in his theory of the rainbow, discussed below.

There are two things clearly wrong with Descartes' derivation. Obviously, light is not a tennis ball, and the surface separating air and water or glass is not a thin fabric, so this is an analogy of dubious relevance, especially for Descartes, who thought that light, unlike tennis balls, always travels at infinite speed.[5] In addition, Descartes' analogy also leads to a wrong value for n. For tennis balls (as shown in Technical Note 27) his assumption implies that n equals the ratio of the speed of the ball v_B in medium B after it passes through the screen to its speed v_A in medium A before it hits the screen. Of course, the ball would be slowed by passing through the screen, so v_B would be less than v_A and their ratio n would be less than 1. If this applied to light, it would mean that the angle between the refracted ray and the perpendicular to the surface would be *greater* than the angle between the incident ray and this perpendicular. Descartes knew this, and even supplied a diagram showing the path of the tennis ball being bent away from the perpendicular. Descartes also knew that this is wrong for light, for as had been observed at least since the time of Ptolemy, a ray of light entering water from the air is bent *toward* the perpendicular to the water's surface, so that the sine of i is greater than the sine of r, and hence n is greater than 1. In a thoroughly muddled discussion that I cannot understand, Descartes somehow argues that light travels more easily in water than in air, so that for light n is greater than 1. For Descartes' purposes his failure to explain the value of n didn't really matter, because he could and did take the value of n from experiment (perhaps from the data in Ptolemy's *Optics*), which of course gives n greater than 1.

A more convincing derivation of the law of refraction was given by the mathematician Pierre de Fermat (1601–1665), along the lines of the derivation by Hero of Alexandria of the equal-angles rule governing reflection, but now making the assumption that light rays take the path of least *time*, rather than of least distance. This assumption (as shown in Technical Note 28) leads to the correct formula, that n is the ratio of the speed of light in medium A to its speed in medium B, and is therefore greater than 1 when A is air and B is glass or water. Descartes could never have derived this formula for n, because for him light traveled instantaneously. (As we will see in Chapter 14, yet another derivation of the correct result was given by Christiaan Huygens, a derivation based on Huygens' theory of light as a traveling disturbance, which did not rely on Fermat's a priori assumption that the light ray travels the path of least time.)

Descartes made a brilliant application of the law of refraction: in his *Meteorology* he used his relation between angles of incidence and refraction to explain the rainbow. This was Descartes at his best as a scientist. Aristotle had argued that the colors of the rainbow are produced when light is reflected by small particles of water suspended in the air.[6] Also, as we have seen in Chapters 9 and 10, in the Middle Ages both al-Farisi and Dietrich of Freiburg had recognized that rainbows are due to the refraction of rays of light when they enter and leave drops of rain suspended in the air. But no one before Descartes had presented a detailed quantitative description of how this works.

Descartes first performed an experiment, using a thin-walled spherical glass globe filled with water as a model of a raindrop. He observed that when rays of sunlight were allowed to enter the globe along various directions, the light that emerged at an angle of about 42° to the incident direction was "completely red, and incomparably more brilliant than the rest." He concluded that a rainbow (or at least its red edge) traces the arc in the sky for which the angle between the line of sight to the rainbow and the direction from the rainbow to the sun is about 42°. Descartes assumed that the light rays are bent by refraction when entering a drop, are reflected from the back surface of the drop, and then are bent again by refraction when emerging from the drop back into the air. But what explains this property of raindrops, of preferentially sending light back at an angle of about 42° to the incident direction?

To answer this, Descartes considered rays of light that enter a spherical drop of water along 10 different parallel lines. He labeled these rays by what is today called their impact parameter b, the closest distance to the center of the drop that the ray would reach if it went straight through the drop without being refracted. The first ray was chosen so that if not refracted it would pass the center of the drop at a distance equal to 10 percent of the drop's radius R (that is, with $b = 0.1\ R$), while the tenth ray was chosen to graze the drop's surface (so that $b = R$), and the intermediate rays were taken to be equally spaced between these two. Descartes worked out the path of each ray as it was refracted entering the drop, reflected by the back surface of the drop, and then refracted again as it left the drop, using the equal-angles law of reflection of Euclid and Hero, and his own law of refraction, and taking the index of refraction n of water to be $4/3$. The following table gives values found by Descartes for the angle φ (phi) between the emerging ray and its incident direction for each ray, along with the results of my own calculation using the same index of refraction:

b/*R*	φ (Descartes)	φ (recalculated)
0.1	5° 40'	5° 44'
0.2	11° 19'	11° 20'
0.3	17° 56'	17° 6'
0.4	22° 30'	22° 41'
0.5	27° 52'	28° 6'
0.6	32° 56'	33° 14'
0.7	37° 26'	37° 49'
0.8	40° 44'	41° 13'
0.9	40° 57'	41° 30'
1.0	13° 40'	14° 22'

The inaccuracy of some of Descartes' results can be set down to the limited mathematical aids available in his time. I don't know if he had access to a table of sines, and he certainly had nothing like a modern pocket calculator. Still, Descartes would have shown better judgment if he had quoted results only to the nearest 10 minutes of arc, rather than to the nearest minute.

As Descartes noticed, there is a relatively wide range of values of the impact parameter *b* for which the angle φ is close to 40°. He then repeated the calculation for 18 more closely spaced rays with values of *b* between 80 percent and 100 percent of the drop's radius, where φ is around 40°. He found that the angle φ for 14 of these 18 rays was between 40° and a maximum of 41° 30'. So these theoretical calculations explained his experimental observation mentioned earlier, of a preferred angle of roughly 42°.

Technical Note 29 gives a modern version of Descartes' calculation. Instead of working out the numerical value of the angle φ between the incoming and outgoing rays for each ray in an ensemble of rays, as Descartes did, we derive a simple formula that gives φ for any ray, with any impact parameter b, and for any value of the ratio n of the speed of light in air to the speed of light in water. This formula is then used to find the value of φ where the emerging rays are concentrated.* For n equal to $4/3$ the favored value of φ, where the emerging light is somewhat concentrated, turns out to be 42.0°, just as found by Descartes. Descartes even calculated the corresponding angle for the secondary rainbow, produced by light that is reflected twice within a raindrop before it emerges.

Descartes saw a connection between the separation of colors that is characteristic of the rainbow and the colors exhibited by refraction of light in a prism, but he was unable to deal with either quantitatively, because he did not know that the white light of the sun is composed of light of all colors, or that the index of refraction of light depends slightly on its color. In fact, while Descartes had taken the index for water to be $4/3 = 1.333\ldots$, it is actually closer to 1.330 for typical wavelengths of red light and closer to 1.343 for blue light. One finds (using the general formula derived in Technical Note 29) that the maximum value for the angle φ between the incident and emerging rays is 42.8° for red light and 40.7° for blue light. This is why Descartes saw bright red light when he looked at his globe of water at an angle of 42° to the direction of the Sun's rays. That value of the angle φ is above the maximum value 40.7° of the angle that can emerge

* This is done by finding the value of b/R where an infinitesimal change in b produces no change in φ, so that at that value of φ the graph of φ versus b/R is flat. This is the value of b/R where φ reaches its maximum value. (Any smooth curve like the graph of φ against b/R that rises to a maximum and then falls again must be flat at the maximum. A point where the curve is not flat cannot be the maximum, since if the curve at some point rises to the right or left there will be points to the right or left where the curve is higher.) Values of φ in the range where the curve of φ versus b/R is nearly flat vary only slowly as we vary b/R, so there are relatively many rays with values of φ in this range.

from the globe of water for blue light, so no light from the blue end of the spectrum could reach Descartes; but it is just below the maximum value 42.8° of φ for red light, so (as explained in the footnote, page 211) this would make the red light particularly bright.

The work of Descartes on optics was very much in the mode of modern physics. Descartes made a wild guess that light crossing the boundary between two media behaves like a tennis ball penetrating a thin screen, and used it to derive a relation between the angles of incidence and refraction that (with a suitable choice of the index of refraction n) agreed with observation. Next, using a globe filled with water as a model of a raindrop, he made observations that suggested a possible origin of rainbows, and he then showed mathematically that these observations followed from his theory of refraction. He didn't understand the colors of the rainbow, so he sidestepped the issue, and published what he did understand. This is just about what a physicist would do today, but aside from its application of mathematics to physics, what does it have to do with Descartes' *Discourse on Method*? I can't see any sign that he was following his own prescriptions for "Rightly Conducting One's Reason and of Seeking Truth in the Sciences."

I should add that in his *Principles of Philosophy* Descartes offered a significant qualitative improvement to Buridan's notion of impetus.[7] He argued that "all movement is, of itself, along straight lines," so that (contrary to both Aristotle and Galileo) a force is required to keep planetary bodies in their curved orbits. But Descartes made no attempt at a calculation of this force. As we will see in Chapter 14, it remained for Huygens to calculate the force required to keep a body moving at a given speed on a circle of given radius, and for Newton to explain this force, as the force of gravitation.

In 1649 Descartes traveled to Stockholm to serve as a teacher of the reigning Queen Christina. Perhaps as a result of the cold Swedish weather, and having to get up to meet Christina at an unwontedly early hour, Descartes in the next year, like Bacon, died of pneumonia. Fourteen years later his works joined those of Copernicus and Galileo on the Index of books forbidden to Roman Catholics.

The writings of Descartes on scientific method have attracted much attention among philosophers, but I don't think they have had much positive influence on the practice of scientific research (or even, as argued above, on Descartes' own most successful scientific work). His writings did have one negative effect: they delayed the reception of Newtonian physics in France. The program set out in the *Discourse on Method,* of deriving scientific principles by pure reason, never worked, and never could have worked. Huygens when young considered himself a follower of Descartes, but he came to understand that scientific principles were only hypotheses, to be tested by comparing their consequences with observation.[8]

On the other hand, Descartes' work on optics shows that he too understood that this sort of scientific hypothesis is sometimes necessary. Laurens Laudan has found evidence for the same understanding in Descartes' discussion of chemistry in the *Principles of Philosophy.*[9] This raises the question whether any scientists actually learned from Descartes the practice of making hypotheses to be tested experimentally, as Laudan thought was true of Boyle. My own view is that this hypothetical practice was widely understood before Descartes. How else would one describe what Galileo did, in using the hypothesis that falling bodies are uniformly accelerated to derive the consequence that projectiles follow parabolic paths, and then testing it experimentally?

According to the biography of Descartes by Richard Watson,[10] "Without the Cartesian method of analyzing material things into their primary elements, we would never have developed the atom bomb. The seventeenth-century rise of Modern Science, the eighteenth-century Enlightenment, the nineteenth-century Industrial Revolution, your twentieth-century personal computer, and the twentieth-century deciphering of the brain—all Cartesian." Descartes did make a great contribution to mathematics, but it is absurd to suppose that it is Descartes' writing on scientific method that has brought about any of these happy advances.

Descartes and Bacon are only two of the philosophers who over the centuries have tried to prescribe rules for scientific research. It never works. We learn how to do science, not by making rules about how to do science, but from the experience of doing science, driven by desire for the pleasure we get when our methods succeed in explaining something.

14

The Newtonian Synthesis

With Newton we come to the climax of the scientific revolution. But what an odd bird to be cast in such a historic role! Newton never traveled outside a narrow strip of England, linking London, Cambridge, and his birthplace in Lincolnshire, not even to see the sea, whose tides so much interested him. Until middle age he was never close to any woman, not even to his mother.* He was deeply concerned with matters having little to do with science, such as the chronology of the Book of Daniel. A catalog of Newton manuscripts put on sale at Sotheby's in 1936 shows 650,000 words on alchemy, and 1.3 million words on religion. With those who might be competitors Newton could be devious and nasty. Yet he tied up strands of physics, astronomy, and mathematics whose relations had perplexed philosophers since Plato.

Writers about Newton sometimes stress that he was not a modern scientist. The best-known statement along these lines

* In his fifties, Newton hired his half sister's beautiful daughter, Catherine Barton, as his housekeeper, but though they were close friends they do not seem to have been romantically attached. Voltaire, who was in England at the time of Newton's death, reported that Newton's doctor and "the surgeon in whose arms he died" confirmed to Voltaire that Newton never had intimacies with a woman (see Voltaire, *Philosophical Letters*, Bobbs-Merrill Educational Publishing, Indianapolis, Ind., 1961, p. 63). Voltaire did not say how the doctor and surgeon could have known this.

is that of John Maynard Keynes (who had bought some of the Newton papers in the 1936 auction at Sotheby's): "Newton was not the first of the age of reason. He was the last of the magicians, the last of the Babylonians and Sumerians, the last great mind which looked out on the visible and intellectual world with the same eyes as those who began to build our intellectual inheritance rather less than 10,000 years ago." * But Newton was not a talented holdover from a magical past. Neither a magician nor an entirely modern scientist, he crossed the frontier between the natural philosophy of the past and what became modern science. Newton's achievements, if not his outlook or personal behavior, provided the paradigm that all subsequent science has followed, as it became modern.

Isaac Newton was born on Christmas Day 1642 at a family farm, Woolsthorpe Manor, in Lincolnshire. His father, an illiterate yeoman farmer, had died shortly before Newton's birth. His mother was higher in social rank, a member of the gentry, with a brother who had graduated from the University of Cambridge and become a clergyman. When Newton was three his mother remarried and left Woolsthorpe, leaving him behind with his grandmother. When he was 10 years old Newton went to the one-room King's School at Grantham, eight miles from Woolsthorpe, and lived there in the house of an apothecary. At Grantham he learned Latin and theology, arithmetic and geometry, and a little Greek and Hebrew.

At the age of 17 Newton was called home to take up his duties as a farmer, but for these he was found to be not well suited. Two years later he was sent up to Trinity College, Cambridge, as a sizar, meaning that he would pay for his tuition and room and board by waiting on fellows of the college and on those students who had been able to pay their fees. Like Galileo at Pisa, he began his education with Aristotle, but he soon turned away to

* This is from a speech, "Newton, the Man," that Keynes was to give at a meeting at the Royal Society in 1946. Keynes died three months before the meeting, and the speech was given by his brother.

his own concerns. In his second year he started a series of notes, *Questiones quandam philosophicae*, in a notebook that had previously been used for notes on Aristotle, and which fortunately is still extant.

In December 1663 the University of Cambridge received a donation from a member of Parliament, Henry Lucas, establishing a professorship in mathematics, the Lucasian chair, with a stipend of £100 a year. Beginning in 1664 the chair was occupied by Isaac Barrow, the first professor of mathematics at Cambridge, 12 years older than Newton. Around then Newton began his study of mathematics, partly with Barrow and partly alone, and received his bachelor of arts degree. In 1665 the plague struck Cambridge, the university largely shut down, and Newton went home to Woolsthorpe. In those years, from 1664 on, Newton began his scientific research, to be described below.

Back in Cambridge, in 1667 Newton was elected a fellow of Trinity College; the fellowship brought him £2 a year and free access to the college library. He worked closely with Barrow, helping to prepare written versions of Barrow's lectures. Then in 1669 Barrow resigned the Lucasian chair in order to devote himself entirely to theology. At Barrow's suggestion, the chair went to Newton. With financial help from his mother, Newton began to spread himself, buying new clothes and furnishings and doing a bit of gambling.[1]

A little earlier, immediately after the restoration of the Stuart monarchy in 1660, a society had been formed by a few Londoners including Boyle, Hooke, and the astronomer and architect Christopher Wren, who would meet to discuss natural philosophy and observe experiments. At the beginning it had just one foreign member, Christiaan Huygens. The society received a royal charter in 1662 as the Royal Society of London, and has remained Britain's national academy of science. In 1672 Newton was elected to membership in the Royal Society, which he later served as president.

In 1675 Newton faced a crisis. Eight years after beginning his fellowship, he had reached the point at which fellows of a Cambridge college were supposed to take holy orders in the Church of England. This would require swearing to belief in the doctrine of the Trinity, but that was impossible for Newton, who rejected the decision of the Council of Nicaea that the Father and the Son are of one substance. Fortunately, the deed that had established the Lucasian chair included a stipulation that its holder should not be active in the church, and on that basis King Charles II was induced to issue a decree that the holder of the Lucasian chair would thenceforth never be required to take holy orders. So Newton was able to continue at Cambridge.

Let's now take up the great work that Newton began at Cambridge in 1664. This research centered on optics, mathematics, and what later came to be called dynamics. His work in any one of these three areas would qualify him as one of the great scientists of history.

Newton's chief experimental achievements were concerned with optics.* His undergraduate notes, the *Questiones quandam philosophicae*, show him already concerned with the nature of light. Newton concluded, contrary to Descartes, that light is not a pressure on the eyes, for if it were then the sky would seem brighter to us when we are running. At Woolsthorpe in 1665 he developed his greatest contribution to optics, his theory of color. It had been known since antiquity that colors appear when light passes through a curved piece of glass, but it had generally been thought that these colors were somehow produced by the glass. Newton conjectured instead that white light consists of all the colors, and that the angle of refraction in glass or water depends slightly on the color, red light being bent somewhat less than blue light, so that the colors are separated when light passes through

* Newton devoted a comparable effort to experiments in alchemy. This could just as well be called chemistry, as between the two there was then no meaningful distinction. As remarked in connection with Jabir ibn Hayyan in Chapter 9, until the late eighteenth century there was no established chemical theory that would rule out the aims of alchemy, like the transmutation of base metals into gold. Although Newton's work on alchemy thus did not represent an abandonment of science, it led to nothing important.

a prism or a raindrop.* This would explain what Descartes had not understood, the appearance of colors in the rainbow. To test this idea, Newton carried out two decisive experiments. First, after using a prism to create separate rays of blue and red light, he directed these rays separately into other prisms, and found no further dispersion into different colors. Next, with a clever arrangement of prisms, he managed to recombine all the different colors produced by refraction of white light, and found that when these colors are combined they produce white light.

The dependence of the angle of refraction on color has the unfortunate consequence that the glass lenses in telescopes like those of Galileo, Kepler, and Huygens focus the different colors in white light differently, blurring the images of distant objects. To avoid this chromatic aberration Newton in 1669 invented a telescope in which light is initially focused by a curved mirror rather than by a glass lens. (The light rays are then deflected by a plane mirror out of the telescope to a glass eyepiece, so not all chromatic aberration was eliminated.) With a reflecting telescope only six inches long, he was able to achieve a magnification by 40 times. All major astronomical light-gathering telescopes are now reflecting telescopes, descendants of Newton's invention. On my first visit to the present quarters of the Royal Society in Carlton House Terrace, as a treat I was taken down to the basement to look at Newton's little telescope, the second one he made.

In 1671 Henry Oldenburg, the secretary and guiding spirit of the Royal Society, invited Newton to publish a description of his telescope. Newton submitted a letter describing it and his work on color to *Philosophical Transactions of the Royal Society*

* A flat piece of glass does not separate the colors, because although each color is bent by a slightly different angle on entering the glass, they are all bent back to their original direction on leaving it. Because the sides of a prism are not parallel, light rays of different color that are refracted differently on entering the glass reach the prism's surface on leaving the prism at angles that are not equal to the angles of refraction on entering, so when these rays are bent back on leaving the prism the different colors are still separated by small angles.

early in 1672. This began a controversy over the originality and significance of Newton's work, especially with Hooke, who had been curator of experiments at the Royal Society since 1662, and holder of a lectureship endowed by Sir John Cutler since 1664. No feeble opponent, Hooke had made significant contributions to astronomy, microscopy, watchmaking, mechanics, and city planning. He claimed that he had performed the same experiments on light as Newton, and that they proved nothing—colors were simply added to white light by the prism.

Newton lectured on his theory of light in London in 1675. He conjectured that light, like matter, is composed of many small particles—contrary to the view, proposed at about the same time by Hooke and Huygens, that light is a wave. This was one place where Newton's scientific judgment failed him. There are many observations, some even in Newton's time, that show the wave nature of light. It is true that in modern quantum mechanics light is described as an ensemble of massless particles, called photons, but in the light encountered in ordinary experience the number of photons is enormous, and in consequence light does behave as a wave.

In his 1678 *Treatise on Light*, Huygens described light as a wave of disturbance in a medium, the ether, which consists of a vast number of tiny material particles in close proximity. Just as in an ocean wave in deep water it is not the water that moves along the surface of the ocean but the disturbance of the water, so likewise in Huygens' theory it is the wave of disturbance in the particles of the ether that moves in a ray of light, not the particles themselves. Each disturbed particle acts as a new source of disturbance, which contributes to the total amplitude of the wave. Of course, since the work of James Clerk Maxwell in the nineteenth century we have known that (even apart from quantum effects) Huygens was only half right—light *is* a wave, but a wave of disturbances in electric and magnetic fields, not a wave of disturbance of material particles.

Using this wave theory of light, Huygens was able to derive the result that light in a homogeneous medium (or empty space) behaves as if it travels in straight lines, as it is only along these lines that the waves produced by all the disturbed particles add up constructively. He gave a new derivation of the equal-angles rule for reflection, and of Snell's law for refraction, without Fermat's a priori assumption that light rays take the path of least time. (See Technical Note 30.) In Huygens' theory of refraction, a ray of light is bent in passing at an oblique angle through the boundary between two media with different light speeds in much the way the direction of march of a line of soldiers will change when the leading edge of the line enters a swampy terrain, in which their marching speed is reduced.

To digress a bit, it was essential to Huygens' wave theory that light travels at a finite speed, contrary to what had been thought by Descartes. Huygens argued that effects of this finite speed are hard to observe simply because light travels so fast. If for instance it took light an hour to travel the distance of the Moon from the Earth, then at the time of an eclipse of the Moon the Moon would be seen not directly opposite the Sun, but lagging behind by about 33°. From the fact that no lag is seen, Huygens concluded that the speed of light must be at least 100,000 times as fast as the speed of sound. This is correct; the actual ratio is about 1 million.

Huygens went on to describe recent observations of the moons of Jupiter by the Danish astronomer Ole Rømer. These observations showed that the period of Io's revolution appears shorter when Earth and Jupiter are approaching each other and longer when they are moving apart. (Attention focused on Io, because it has the shortest orbital period of any of Jupiter's Galilean moons—only 1.77 days.) Huygens interpreted this as what later became known as a "Doppler effect": when Jupiter and the Earth are moving closer together or farther apart, their separation at each successive completion of a whole period of revolution of Io is respectively decreasing or increasing, and so if light travels at a finite speed, the time interval between observations of complete periods of Io should be respectively less or greater than if Jupiter and the Earth were at rest. Specifically, the fractional shift in the apparent period of Io should be the ratio of the relative speed of Jupiter and the Earth along the direction separating them to the speed of light, with the relative speed taken as positive or negative if Jupiter and the Earth are moving farther apart or closer together, respectively. (See Technical Note 31.) Measuring the apparent changes in the period of Io and knowing the relative speed of Earth and Jupiter, one could calculate the speed of light. Because the Earth moves much faster than Jupiter, it is chiefly the Earth's velocity that dominates the relative speed. The scale of the solar system was then not well known, so neither was the numerical value of the relative speed of separation of the Earth and Jupiter, but using Rømer's data Huygens was able to calculate that it takes 11 minutes for light to travel a distance equal to the radius of the Earth's orbit, a result that did not depend on knowing the size of the orbit. To put it another way, since the astronomical unit (AU) of distance is defined as the mean radius of the Earth's orbit, the speed of light was found by Huygens to be 1 AU per 11 minutes. The modern value is 1 AU per 8.32 minutes.

There already was experimental evidence of the wave nature of light that would have been available to Newton and Huygens: the discovery of diffraction by the Bolognese Jesuit Francesco Maria Grimaldi (a student of Riccioli), published posthumously in 1665. Grimaldi had found that the shadow of a narrow opaque rod in sunlight is not perfectly sharp, but is bordered by fringes. The fringes are due to the fact that the wavelength of light is not negligible compared with the thickness of the rod, but Newton argued that they were instead the result of some sort of refraction at the surface of the rod. The issue of light as corpuscle or wave was settled for most physicists when, in the early nineteenth century, Thomas Young discovered interference, the pattern of reinforcement or cancellation of light waves that arrive at given spots along different paths. As already mentioned, in the twentieth century it was discovered that the two views are not incompatible. Einstein in 1905 realized that although light for most purposes behaves as a wave, the energy in light comes in small packets, later called photons, each with a tiny energy and momentum proportional to the frequency of the light.

Newton finally presented his work on light in his book *Opticks*, written (in English) in the early 1690s. It was published in 1704, after he had already become famous.

Newton was not only a great physicist but also a creative mathematician. He began in 1664 to read works on mathematics, including Euclid's *Elements* and Descartes' *Geometrie*. He soon started to work out the solutions to a variety of problems, many involving infinities. For instance, he considered infinite series, such as $x - x^2/2 + x^3/3 - x^4/4 + \ldots$, and showed that this adds up to the logarithm* of $1 + x$.

In 1665 Newton began to think about infinitesimals. He took up a problem: suppose we know the distance $D(t)$ traveled in any time t; how do we find the velocity at any time? He reasoned that in nonuniform motion, the velocity at any instant is the ratio of the distance traveled to the time elapsed in an infinitesimal interval of time at that instant. Introducing the symbol o for an infinitesimal interval of time, he defined the velocity at time t as the ratio to o of the distance traveled between time t and time $t + o$, that is, the velocity is $[D(t + o) - D(t)]/o$. For instance, if $D(t) = t^3$, then $D(t + o) = t^3 + 3t^2o + 3to^2 + o^3$. For o infinitesimal, we can neglect the terms proportional to o^2 and o^3, and take $D(t + o) = t^3 + 3t^2 o$, so that $D(t + o) - D(t) = 3t^2 o$ and the velocity is just $3t^2$. Newton called this the "fluxion" of $D(t)$, but it became known as the "derivative," the fundamental tool of modern differential calculus.†

* This is the "natural logarithm" of $1 + x$, the power to which the constant $e = 2.71828\ldots$ must be raised to give the result $1 + x$. The reason for this peculiar definition is that the natural logarithm has some properties that are much simpler than those of the "common logarithm," in which 10 takes the place of e. For instance, Newton's formula shows that the natural logarithm of 2 is given by the series $1 - \frac{1}{2} + \frac{1}{3} - \frac{1}{4} + \ldots$, while the formula for the common logarithm of 2 is more complicated.

† The neglect of the terms $3to^2$ and o^3 in this calculation may make it seem that the calculation is only approximate, but that is misleading. In the nine-

Newton then took up the problem of finding the areas bounded by curves. His answer was the fundamental theorem of calculus; one must find the quantity whose fluxion is the function described by the curve. For instance, as we have seen, $3x^2$ is the fluxion of x^3, so the area under the parabola $y = 3x^2$ between $x = 0$ and any other x is x^3. Newton called this the "inverse method of fluxions," but it became known as the process of "integration."

Newton had invented the differential and integral calculus, but for a long while this work did not become widely known. Late in 1671 he decided to publish it along with an account of his work on optics, but apparently no London bookseller was willing to undertake the publication without a heavy subsidy.[2]

In 1669 Barrow gave a manuscript of Newton's *De analysi per aequationes numero terminorum infinitas* to the mathematician John Collins. A copy made by Collins was seen on a visit to London in 1676 by the philosopher and mathematician Gottfried Wilhelm Leibniz, a former student of Huygens and a few years younger than Newton, who had independently discovered the essentials of the calculus in the previous year. In 1676 Newton revealed some of his own results in letters that were meant to be seen by Leibniz. Leibniz published his work on calculus in articles in 1684 and 1685, without acknowledging Newton's work. In these publications Leibniz introduced the word "calculus," and presented its modern notation, including the integration sign $\int$.

To establish his claim to calculus, Newton described his own methods in two papers included in the 1704 edition of *Opticks*. In January 1705 an anonymous review of *Opticks* hinted that these methods were taken from Leibniz. As Newton guessed, this review had been written by Leibniz. Then in 1709 *Philosophical*

---

teenth century mathematicians learned to dispense with the rather vague idea of an infinitesimal o, and to speak instead of precisely defined *limits*: the velocity is the number to which $[D(t + o) - D(t)]/o$ can be made as close as we like by taking o sufficiently small. As we will see, Newton later moved away from infinitesimals and toward the modern idea of limits.

Transactions of the Royal Society published an article by John Keill defending Newton's priority of discovery, and Leibniz replied in 1711 with an angry complaint to the Royal Society. In 1712 the Royal Society convened an anonymous committee to look into the controversy. Two centuries later the membership of this committee was made public, and it turned out to have consisted almost entirely of Newton's supporters. In 1715 the committee reported that Newton deserved credit for the calculus. This report had been drafted for the committee by Newton. Its conclusions were supported by an anonymous review of the report, also written by Newton.

The judgment of contemporary scholars[3] is that Leibniz and Newton had discovered the calculus independently. Newton accomplished this a decade earlier than Leibniz, but Leibniz deserves great credit for publishing his work. In contrast, after his original effort in 1671 to find a publisher for his treatise on calculus, Newton allowed this work to remain hidden until he was forced into the open by the controversy with Leibniz. The decision to go public is generally a critical element in the process of scientific discovery.[4] It represents a judgment by the author that the work is correct and ready to be used by other scientists. For this reason, the credit for a scientific discovery today usually goes to the first to publish. But though Leibniz was the first to publish on calculus, as we shall see it was Newton rather than Leibniz who applied calculus to problems in science. Though, like Descartes, Leibniz was a great mathematician whose philosophical work is much admired, he made no important contributions to natural science.

It was Newton's theories of motion and gravitation that had the greatest historical impact. It was an old idea that the force of gravity that causes objects to fall to the Earth decreases with distance from the Earth's surface. This much was suggested in the ninth century by a well-traveled Irish monk, Duns Scotus (Johannes Scotus Erigena, or John the Scot), but with no suggestion of any connection of this force with the motion of the planets. The suggestion that the force that holds the planets in their orbits decreases with the inverse square of the distance from the Sun may have been first made in 1645 by a French priest, Ismael Bullialdus, who was later quoted by Newton and elected to the Royal Society. But it was Newton who made this convincing, and related the force to gravity.

Writing about 50 years later, Newton described how he began to study gravitation. Even though his statement needs a good deal of explanation, I feel I have to quote it here, because it describes in Newton's own words what seems to have been a turning point in the history of civilization. According to Newton, it was in 1666 that:

> I began to think of gravity extending to the orb of the Moon & (having found out how to estimate the force with which [a] globe revolving within a sphere presses the surface of the sphere) from Kepler's rule of the periodical times of the Planets being in sesquialterate proportion of their distances from the center of their Orbs, I deduced that the forces which keep the Planets in their Orbs must [be] reciprocally as the squares of their distances from the centers about which they revolve & thereby compared the Moon in her Orb with the force of gravity at the surface of the Earth & found them answer pretty nearly. All this [including his work on infinite series and calculus] was in the two plague years of 1665–1666. For in those days I was in the prime of my age for invention and minded Mathematicks and Philosophy more than at any time since.[5]

As I said, this takes some explaining.

First, Newton's parenthesis "having found out how to estimate the force with which [a] globe revolving within a sphere presses the surface of the sphere" refers to the calculation of centrifugal force, a calculation that had already been done (probably unknown to Newton) around 1659 by Huygens. For Huygens and Newton (as for us), acceleration had a broader definition than just a number giving the change of speed per time elapsed; it is a *directed* quantity, giving the change per time elapsed in the direction as well as in the magnitude of the velocity. There is an acceleration in circular motion even at constant speed—it is the "centripetal acceleration," consisting of a continual turning toward the center of the circle. Huygens and Newton concluded that a body moving at a constant speed v around a circle of radius r is accelerating toward the center of the circle, with acceleration v^2/r, so the force needed to keep it moving on the circle rather than flying off in a straight line into space is proportional to v^2/r. (See Technical Note 32.) The resistance to this centripetal acceleration is experienced as what Huygens called centrifugal force, as when a weight at the end of a cord is swung around in a circle. For the weight, the centrifugal force is resisted by tension in the cord. But planets are not attached by cords to the Sun. What is it that resists the centrifugal force produced by a planet's nearly circular motion around the Sun? As we will see, the answer to this question led to Newton's discovery of the inverse square law of gravitation.

Next, by "Kepler's rule of the periodical times of the Planets being in sesquialterate proportion of their distances from the center of their Orbs" Newton meant what we now call Kepler's third law: that the square of the periods of the planets in their orbits is proportional to the cubes of the mean radii of their orbits, or in other words, the periods are proportional to the $3/2$ power (the "sesquialterate proportion") of the mean radii.* The period of a body moving with speed v around a circle of radius r is the circumference $2\pi r$ divided by the speed v, so for circular orbits Kepler's third law tells us that r^2/v^2 is proportional to r^3, and therefore their inverses are proportional: v^2/r^2 is proportional to $1/r^3$. It follows that the force keeping the planets in orbit, which is proportional to v^2/r, must be proportional to $1/r^2$. This is the inverse square law of gravity.

* Kepler's three laws of planetary motion were not well established before Newton, though the first law—that each planetary orbit is an ellipse with the Sun at one focus—was widely accepted. It was Newton's derivation of these laws in the *Principia* that led to the general acceptance of all three laws.

This in itself might be regarded as just a way of restating Kepler's third law. Nothing in Newton's consideration of the planets makes any connection between the force holding the planets in their orbits and the commonly experienced phenomena associated with gravity on the Earth's surface. This connection was provided by Newton's consideration of the Moon. Newton's statement that he "compared the Moon in her Orb with the force of gravity at the surface of the Earth & found them answer pretty nearly" indicates that he had calculated the centripetal acceleration of the Moon, and found that it was less than the acceleration of falling bodies on the surface of the Earth by just the ratio one would expect if these accelerations were inversely proportional to the square of the distance from the center of the Earth.

To be specific, Newton took the radius of the Moon's orbit (well known from observations of the Moon's diurnal parallax) to be 60 Earth radii; it is actually about 60.2 Earth radii. He used a crude estimate of the Earth's radius,* which gave a crude value for the radius of the Moon's orbit, and knowing that the sidereal period of the Moon's revolution around the Earth is 27.3 days, he could estimate the Moon's velocity and from that its centripetal acceleration. This acceleration turned out to be less than the acceleration of falling bodies on the surface of the Earth by a factor roughly (very roughly) equal to $1/(60)^2$, as expected if the force holding the Moon in its orbit is the same that attracts bodies on the Earth's surface, but reduced in accordance with the inverse square law. (See Technical Note 33.) This is what Newton meant when he said that he found that the forces "answer pretty nearly."

This was the climactic step in the unification of the celestial and terrestrial in science. Copernicus had placed the Earth among the planets, Tycho had shown that there is change in the heavens, and Galileo had seen that the Moon's surface is rough,

* The first reasonably precise measurement of the circumference of the Earth was made around 1669 by Jean-Félix Picard (1620–1682), and was used by Newton in 1684 to improve this calculation.

like the Earth's, but none of this related the motion of planets to forces that could be observed on Earth. Descartes had tried to understand the motions of the solar system as the result of vortices in the ether, not unlike vortices in a pool of water on Earth, but his theory had no success. Now Newton had shown that the force that keeps the Moon in its orbit around the Earth and the planets in their orbits around the Sun is the same as the force of gravity that causes an apple to fall to the ground in Lincolnshire, all governed by the same quantitative laws. After this the distinction between the celestial and terrestrial, which had constrained physical speculation from Aristotle on, had to be forever abandoned. But this was still far short of a principle of universal gravitation, which would assert that every body in the universe, not just the Earth and Sun, attracts every other body with a force that decreases as the inverse square of the distance between them.

There were still four large holes in Newton's arguments:

1. In comparing the centripetal acceleration of the Moon with the acceleration of falling bodies on the surface of the Earth, Newton had assumed that the force producing these accelerations decreases with the inverse square of the distance, but the distance from what? This makes little difference for the motion of the Moon, which is so far from the Earth that the Earth can be taken as almost a point particle as far as the Moon's motion is concerned. But for an apple falling to the ground in Lincolnshire, the Earth extends from the bottom of the tree, a few feet away, to a point at the antipodes, 8,000 miles away. Newton had assumed that the distance relevant to the fall of any object near the Earth's surface is its distance to the center of the Earth, but this was not obvious.

2. Newton's explanation of Kepler's third law ignored the obvious differences between the planets. Somehow it does not matter that Jupiter is much bigger than Mercury; the difference in their centripetal accelerations is just a matter of their distances from the Sun. Even more dramatically, Newton's comparison of

the centripetal acceleration of the Moon and the acceleration of falling bodies on the surface of the Earth ignored the conspicuous difference between the Moon and a falling body like an apple. Why do these differences not matter?

3. In the work he dated to 1665–1666, Newton interpreted Kepler's third law as the statement that the products of the centripetal accelerations of the various planets with the squares of their distances from the Sun are the same for all planets. But the common value of this product is not at all equal to the product of the centripetal acceleration of the Moon with the square of its distance from the Earth; it is much greater. What accounts for this difference?

4. Finally, in this work Newton had taken the orbits of the planets around the Sun and of the Moon around the Earth to be circular at constant speed, even though as Kepler had shown they are not precisely circular but instead elliptical, the Sun and Earth are not at the centers of the ellipses, and the Moon's and planets' speeds are only approximately constant.

Newton struggled with these problems in the years following 1666. Meanwhile, others were coming to the same conclusions that Newton had already reached. In 1679 Newton's old adversary Hooke published his Cutlerian lectures, which contained some suggestive though nonmathematical ideas about motion and gravitation:

> First, that all Coelestial Bodies whatsoever, have an attraction or gravitating power towards their own Centers, whereby they attract not only their own parts, and keep them from flying from them, as we may observe the Earth to do, but that they do also attract all the other Coelestial Bodies that are within the sphere of their activity—The second supposition is this, That all bodies whatsoever that are put into a direct and simple motion, will so continue to move forward in a straight line,

> till they are by some other effectual powers deflected and bent into a Motion, describing a Circle, Ellipsis, or some other more compounded Curve Line. The third supposition is, That these attractive powers are so much the more powerful in operating, by how much the nearer the body wrought upon is to their own Centers.[6]

Hooke wrote to Newton about his speculations, including the inverse square law. Newton brushed him off, replying that he had not heard of Hooke's work, and that the "method of indivisibles"[7] (that is, calculus) was needed to understand planetary motions.

Then in August 1684 Newton received a fateful visit in Cambridge from the astronomer Edmund Halley. Like Newton and Hooke and also Wren, Halley had seen the connection between the inverse square law of gravitation and Kepler's third law for *circular* orbits. Halley asked Newton what would be the actual shape of the orbit of a body moving under the influence of a force that decreases with the inverse square of the distance. Newton answered that the orbit would be an ellipse, and promised to send a proof. Later that year Newton submitted a 10-page document, *On the Motion of Bodies in Orbit*, which showed how to treat the general motion of bodies under the influence of a force directed toward a central body.

Three years later the Royal Society published Newton's *Philosophiae Naturalis Principia Mathematica* (*Mathematical Principles of Natural Philosophy*), doubtless the greatest book in the history of physical science.

A modern physicist leafing through the *Principia* may be surprised to see how little it resembles any of today's treatises on physics. There are many geometrical diagrams, but few equations. It seems almost as if Newton had forgotten his own development of calculus. But not quite. In many of his diagrams one sees features that are supposed to become infinitesimal or infinitely numerous. For instance, in showing that Kepler's equal-area rule follows for any force directed toward a fixed center,

Newton imagines that the planet receives infinitely many impulses toward the center, each separated from the next by an infinitesimal interval of time. This is just the sort of calculation that is made not only respectable but quick and easy by the general formulas of calculus, but nowhere in the *Principia* do these general formulas make their appearance. Newton's mathematics in the *Principia* is not very different from what Archimedes had used in calculating the areas of circles, or what Kepler had used in calculating the volumes of wine casks.

The style of the *Principia* reminds the reader of Euclid's *Elements*. It begins with definitions:[8]

Definition I

Quantity of matter is a measure of matter that arises from its density and volume jointly.

What appears in English translation as "quantity of matter" was called *massa* in Newton's Latin, and is today called "mass." Newton here defines it as the product of density and volume. Even though Newton does not define density, his definition of mass is still useful because his readers could take it for granted that bodies made of the same substances, such as iron at a given temperature, will have the same density. As Archimedes had shown, measurements of specific gravity give values for density relative to that of water. Newton notes that we measure the mass of a body from its weight, but does not confuse mass and weight.

Definition II

Quantity of motion is a measure of motion that arises from the velocity and the quantity of matter jointly.

What Newton calls "quantity of motion" is today termed "momentum." It is defined here by Newton as the product of the velocity and the mass.

Definition III

> Inherent force of matter [*vis insita*] is the power of resisting by which every body, so far as [it] is able, perseveres in its state either of resting or of moving uniformly straight forward.

Newton goes on to explain that this force arises from the body's mass, and that it "does not differ in any way from the inertia of the mass." We sometimes now distinguish mass, in its role as the quantity that resists changes in motion, as "inertial mass."

Definition IV

> Impressed force is the action exerted on a body to change its state either of resting or of uniformly moving straight forward.

This defines the general concept of force, but does not yet give meaning to any numerical value we might assign to a given force. Definitions V through VIII go on to define centripetal acceleration and its properties.

After the definitions comes a scholium, or annotation, in which Newton declines to define space and time, but does offer a description:

> I. Absolute, true, and mathematical time, in and of itself, and of its own nature, without relation to anything external, flows uniformly. . . .

> II. Absolute space, of its own nature without relation to anything external, always remains homogeneous and immovable.

Both Leibniz and Bishop George Berkeley criticized this view of time and space, arguing that only relative positions in space and time have any meaning. Newton had recognized in this scholium that we normally deal only with relative positions and velocities, but now he had a new handle on absolute space: in Newton's mechanics, acceleration (unlike position or velocity) has an absolute significance. How could it be otherwise? It is a matter of common experience that acceleration has effects; there is no need to ask, "Acceleration relative to what?" From the forces pressing us back in our seats, we know that we are being accelerated when we are in a car that suddenly speeds up, whether or not we look out the car's window. As we will see, in the twentieth century the views of space and time of Leibniz and Newton were reconciled in the general theory of relativity.

Then at last come Newton's famous three laws of motion:

> **Law I**
>
> Every body perseveres in its state of being at rest or of moving uniformly straight forward, except insofar as it is compelled to change its state by forces impressed.

This was already known to Gassendi and Huygens. It is not clear why Newton bothered to include it as a separate law, since the first law is a trivial (though important) consequence of the second.

> **Law II**
>
> A change in motion is proportional to the motive force impressed and takes place along the straight line in which that force is impressed.

By "change of motion" here Newton means the change in the momentum, which he called the "quantity of motion" in Definition II. It is actually the *rate* of change of momentum that is proportional to the force. We conventionally define the units in which force is measured so that the rate of change of momentum is actually equal to the force. Since momentum is mass times velocity, its rate of change is mass times acceleration. Newton's second law is thus the statement that mass times acceleration equals the force producing the acceleration. But the famous equation $F = ma$ does not appear in the *Principia*; the second law was re-expressed in this way by Continental mathematicians in the eighteenth century.

Law III

> To any action there is always an opposite and equal reaction; in other words, the actions of two bodies upon each other are always equal, and always opposite in direction.

In true geometric style, Newton then goes on to present a series of corollaries deduced from these laws. Notable among them was Corollary III, which gives the law of the conservation of momentum. (See Technical Note 34.)

After completing his definitions, laws, and corollaries, Newton begins in Book I to deduce their consequences. He proves that central forces (forces directed toward a single central point) and only central forces give a body a motion that sweeps out equal areas in equal times; that central forces proportional to the inverse square of the distance and only such central forces produce motion on a conic section, that is, a circle, an ellipse, a parabola, or a hyperbola; and that for motion on an ellipse such a force gives periods proportional to the $3/2$ power of the major axis of the ellipse (which, as mentioned in Chapter 11, is the distance of the planet from the Sun averaged over the length of its path). So a central force that goes as the inverse square of the distance can account for all of Kepler's laws. Newton also fills in the gap in his comparison of lunar centripetal acceleration and the acceleration of falling bodies, proving in Section XII of Book I that a spherical body, composed of particles that each produce a force that goes as the inverse square of the distance to that particle, produces a total force that goes as the inverse square of the distance to the center of the sphere.

There is a remarkable scholium at the end of Section I of

Book I, in which Newton remarks that he is no longer relying on the notion of infinitesimals. He explains that "fluxions" such as velocities are not the ratios of infinitesimals, as he had earlier described them; instead, "Those ultimate ratios with which quantities vanish are not actually ratios of ultimate quantities, but limits which the ratios of quantities decreasing without limit are continually approaching, and which they can approach so closely that their difference is less than any given quantity." This is essentially the modern idea of a limit, on which calculus is now based. What is not modern about the *Principia* is Newton's idea that limits have to be studied using the methods of geometry.

Book II presents a long treatment of the motion of bodies through fluids; the primary goal of this discussion was to derive the laws governing the forces of resistance on such bodies.[9] In this book he demolishes Descartes' theory of vortices. He then goes on to calculate the speed of sound waves. His result in Proposition 49 (that the speed is the square root of the ratio of the pressure and the density) is correct only in order of magnitude, because no one then knew how to take account of the changes in temperature during expansion and compression. But (together with his calculation of the speed of ocean waves) this was an amazing achievement: the first time that anyone had used the principles of physics to give a more or less realistic calculation of the speed of any sort of wave.

At last Newton comes to the evidence from astronomy in Book III, *The System of the World*. At the time of the first edition of the *Principia*, there was general agreement with what is now called Kepler's first law, that the planets move on elliptical orbits; but there was still considerable doubt about the second and third laws: that the line from the Sun to each planet sweeps out equal areas in equal times, and that the squares of the periods of the various planetary motions go as the cubes of the major axes of these orbits. Newton seems to have fastened on Kepler's laws not because they were well established, but because they fitted so well with his theory. In Book III he notes that Jupiter's and Saturn's moons obey Kepler's second and third laws, that the observed phases of the five planets other than Earth show that they revolve around the Sun, that all six planets obey Kepler's laws, and that the Moon satisfies Kepler's second law.* His own careful observations of the comet of 1680 showed that it too moved on a conic section: an ellipse or hyperbola, in either case very close to a parabola. From all this (and his earlier comparison of the centripetal acceleration of the Moon and the acceleration of falling bodies on the Earth's surface), he comes to the conclusion that it is a central force obeying an inverse square law by which the moons of Jupiter and Saturn and the Earth are attracted to their planets, and all the planets and comets are attracted to the Sun. From the fact that the accelerations produced by gravitation are independent of the nature of the body being accelerated, whether it is a planet or a moon or an apple, depending only on the nature of the body producing the force and the distance between them, together with the fact that the acceleration produced by any force is inversely proportional to the mass of the body on which it acts, he concludes that the force of gravity on any body must be proportional to the mass of that body, so that all dependence on the body's mass cancels when we calculate the acceleration. This makes a clear distinction between gravitation and magnetism, which acts very differently on bodies of different composition, even if they have the same mass.

Newton then, in Proposition 7, uses his third law of motion to find out how the force of gravity depends on the nature of the body producing the force. Consider two bodies, 1 and 2, with masses m_1 and m_2. Newton had shown that the gravitational force exerted by body 1 on body 2 is proportional to m_2, and that the force that body 2 exerts on body 1 is proportional to m_1. But according to the third law, these forces are equal in magnitude, and so they must each be proportional to *both* m_1 and m_2. New-

* Newton was unable to solve the three-body problem of the Earth, Sun, and Moon with enough accuracy to calculate the peculiarities in the motion of the Moon that had worried Ptolemy, Ibn al-Shatir, and Copernicus. This was finally accomplished in 1752 by Alexis-Claude Clairaut, who used Newton's theories of motion and gravitation.

ton was able to confirm the third law in collisions but not in gravitational interactions. As George Smith has emphasized, it was many years before it became possible to confirm the proportionality of gravitational force to the inertial mass of the attracting as well as the attracted body. Nevertheless, Newton concluded, "Gravity exists in all bodies universally and is proportional to the quantity of matter in each." This is why the products of the centripetal accelerations of the various planets with the squares of their distances from the Sun are much greater than the product of the centripetal acceleration of the Moon with the square of its distances from the Earth: it is just that the Sun, which produces the gravitational force on the planets, is much more massive than the Earth.

These results of Newton are commonly summarized in a formula for the gravitational force *F* between two bodies, of masses m_1 and m_2, separated by a distance *r*:

$$F = G \times m_1 \times m_2 / r^2$$

where *G* is a universal constant, known today as Newton's constant. Neither this formula nor the constant *G* appears in the *Principia*, and even if Newton had introduced this constant he could not have found a value for it, because he did not know the mass of the Sun or the Earth. In calculating the motion of the Moon or the planets, *G* appears only as a factor multiplying the mass of the Earth or the Sun, respectively.

Even without knowing the value of *G*, Newton could use his theory of gravitation to calculate the *ratios* of the masses of various bodies in the solar system. (See Technical Note 35.) For instance, knowing the ratios of the distances of Jupiter and Saturn from their moons and from the Sun, and knowing the ratios of the orbital periods of Jupiter and Saturn and their moons, he could calculate the ratios of the centripetal accelerations of the moons of Jupiter and Saturn toward their planets and the centripetal acceleration of these planets toward the Sun, and from this he could calculate the ratios of the masses of Jupiter, Saturn, and the Sun. Since the Earth also has a Moon, the same technique could in principle be used to calculate the ratio of the masses of the Earth and the Sun. Unfortunately, although the distance of the Moon from the Earth was well known from the Moon's diurnal parallax, the Sun's diurnal parallax was too small to measure, and so the ratio of the distances from the Earth to the Sun and to the Moon was not known. (As we saw in Chapter 7, the data used by Aristarchus and the distances he inferred from those data were hopelessly inaccurate.) Newton went ahead anyway, and calculated the ratio of masses, using a value for the distance of the Earth from the Sun that was no better than a lower limit on this distance, and actually about half the true value. Here are Newton's results for ratios of masses, given as a corollary to Theorem VIII in Book III of the *Principia*, together with modern values:[10]

Ratio	Newton's value	Modern value
m(Sun)/*m*(Jupiter)	1,067	1,048
m(Sun)/*m*(Saturn)	3,021	3,497
m(Sun)/*m*(Earth)	169,282	332,950

As can be seen from this table, Newton's results were pretty good for Jupiter, not bad for Saturn, but way off for the Earth, because the distance of the Earth from the Sun was not known. Newton was quite aware of the problems posed by observational uncertainties, but like most scientists until the twentieth century, he was cavalier about giving the resulting range of uncertainty in his calculated results. Also, as we have seen with Aristarchus and al-Biruni, he quoted results of calculations to a much greater precision than was warranted by the accuracy of the data on which the calculations were based.

Incidentally, the first serious estimate of the size of the solar system was carried out in 1672 by Jean Richer and Giovanni Domenico Cassini. They measured the distance to Mars by observing the difference in the direction to Mars as seen from Paris and

Cayenne; since the ratios of the distances of the planets from the Sun were already known from the Copernican theory, this also gave the distance of the Earth from the Sun. In modern units, their result for this distance was 140 million kilometers, reasonably close to the modern value of 149.5985 million kilometers for the mean distance. A more accurate measurement was made later by comparing observations at different locations on Earth of the transits of Venus across the face of the Sun in 1761 and 1769, which gave an Earth–Sun distance of 153 million kilometers.[11]

In 1797–1798 Henry Cavendish was at last able to measure the gravitational force between laboratory masses, from which a value of G could be inferred. But Cavendish did not refer to his measurement this way. Instead, using the well-known acceleration of 32 feet/second per second due to the Earth's gravitational field at its surface, and the known volume of the Earth, Cavendish calculated that the average density of the Earth was 5.48 times the density of water.

This was in accord with a long-standing practice in physics: to report results as ratios or proportions, rather than as definite magnitudes. For instance, as we have seen, Galileo showed that the distance a body falls on the surface of the Earth is proportional to the square of the time, but he never said that the constant multiplying the square of the time that gives the distance fallen was half of 32 feet/second per second. This was due at least in part to the lack of any universally recognized unit of length. Galileo could have given the acceleration due to gravity as so many *braccia*/second per second, but what would this mean to Englishmen, or even to Italians outside Tuscany? The international standardization of units of length and mass[12] began in 1742, when the Royal Society sent two rulers marked with standard English inches to the French Académie des Sciences; the French marked these with their own measures of length, and sent one back to London. But it was not until the gradual international adoption of the metric system, starting in 1799, that scientists had a universally understood system of units. Today we cite a value for

G of 66.724 trillionths of a meter/second^2 per kilogram: that is, a small body of mass 1 kilogram at a distance of 1 meter produces a gravitational acceleration of 66.724 trillionths of a meter/second per second.

After laying out Newton's theories of motion and gravitation, the *Principia* goes on to work out some of their consequences. These go far beyond Kepler's three laws. For instance, in Proposition 14 Newton explains the precession of planetary orbits measured (for the Earth) by al-Zarqali, though Newton does not attempt a quantitative calculation.

In Proposition 19 Newton notes that the planets must all be oblate, because their rotation produces centrifugal forces that are largest at the equator and vanish at the poles. For instance, the Earth's rotation produces a centripetal acceleration at its equator equal to 0.11 feet/second per second, as compared with the acceleration 32 feet/second per second of falling bodies, so the centrifugal force produced by the Earth's rotation is much less than its gravitational attraction, but not entirely negligible, and the Earth is therefore nearly spherical, but slightly oblate. Observations in the 1740s finally showed that the same pendulum will swing more slowly near the equator than at higher latitudes, just as would be expected if at the equator the pendulum is farther from the center of the Earth, because the Earth is oblate.

In Proposition 39 Newton shows that the effect of gravity on the oblate Earth causes a precession of its axis of rotation, the "precession of the equinoxes" first noted by Hipparchus. (Newton had an extracurricular interest in this precession; he used its values along with ancient observations of the stars in an attempt to date supposed historical events, such as the expedition of Jason and the Argonauts.)[13] In the first edition of the *Principia* Newton calculates in effect that the annual precession due to the Sun is 6.82° (degrees of arc), and that the effect of the Moon is larger by a factor 6⅓, giving a total of 50.0" (seconds of arc) per year, in perfect agreement with the precession of 50" per year then measured, and close to the modern value of 50.375" per year.

Very impressive, but Newton later realized that his result for the precession due to the Sun and hence for the total precession was 1.6 times too small. In the second edition he corrected his result for the effect of the Sun, and also corrected the ratio of the effects of the Moon and Sun, so that the total was again close to 50" per year, still in good agreement with what was observed.[14] Newton had the correct qualitative explanation of the precession of the equinoxes, and his calculation gave the right order of magnitude for the effect, but to get an answer in precise agreement with observation he had to make many artful adjustments.

This is just one example of Newton fudging his calculations to get answers in close agreement with observation. Along with this example, R. S. Westfall[15] has given others, including Newton's calculation of the speed of sound, and his comparison of the centripetal acceleration of the Moon with the acceleration of falling bodies on the Earth's surface mentioned earlier. Perhaps Newton felt that his real or imagined adversaries would never be convinced by anything but nearly perfect agreement with observation.

In Proposition 24, Newton presents his theory of the tides. Gram for gram, the Moon attracts the ocean beneath it more strongly than it attracts the solid Earth, whose center is farther away, and it attracts the solid Earth more strongly than it attracts the ocean on the side of the Earth away from the Moon. Thus there is a tidal bulge in the ocean both below the Moon, where the Moon's gravity pulls water away from the Earth, and on the opposite side of the Earth, where the Moon's gravity pulls the Earth away from the water. This explained why in some locations high tides are separated by roughly 12 rather than 24 hours. But the effect is too complicated for this theory of tides to have been verified in Newton's time. Newton knew that the Sun as well as the Moon plays a role in raising the tides. The highest and lowest tides, known as spring tides, occur when the Moon is new or full, so that the Sun, Moon, and Earth are on the same line, intensifying the effects of gravitation. But the worst complications come from the fact that any gravitational effects on the oceans are greatly influenced by the shape of the continents and the topography of the ocean bottom, which Newton could not possibly take into account.

This is a common theme in the history of physics. Newton's theory of gravitation made successful predictions for simple phenomena like planetary motion, but it could not give a quantitative account of more complicated phenomena, like the tides. We are in a similar position today with regard to the theory of the strong forces that hold quarks together inside the protons and neutrons inside the atomic nucleus, a theory known as quantum chromodynamics. This theory has been successful in accounting for certain processes at high energy, such as the production of various strongly interacting particles in the annihilation of energetic electrons and their antiparticles, and its successes convince us that the theory is correct. We cannot use the theory to calculate precise values for other things that we would like to explain, like the masses of the proton and neutron, because the calculation is too complicated. Here, as for Newton's theory of the tides, the proper attitude is patience. Physical theories are validated when they give us the ability to calculate enough things that are sufficiently simple to allow reliable calculations, even if we can't calculate everything that we might want to calculate.

Book III of *Principia* presents calculations of things already measured, and new predictions of things not yet measured, but even in the final third edition of *Principia* Newton could point to no predictions that had been verified in the 40 years since the first edition. Still, taken all together, the evidence for Newton's theories of motion and gravitation was overwhelming. Newton did not need to follow Aristotle and explain why gravity exists, and he did not try. In his "General Scholium" Newton concluded:

> Thus far I have explained the phenomena of the heavens and of our sea by the force of gravity, but I have not yet assigned a cause to gravity. Indeed, this force arises from some cause that penetrates as far as the centers of the Sun and planets without any diminution of its power to act, and that acts not in propor-

> tion to the quantity of the surfaces of the particles on which it acts (as mechanical causes are wont to do), but in proportion to the quantity of *solid* matter, and whose action is extended everywhere to immense distances, always decreasing as the inverse squares of the distances. . . . I have not as yet been able to deduce from phenomena the reasons for these properties of gravity, and I do not "feign" hypotheses.

Newton's book appeared with an appropriate ode by Halley. Here is its final stanza:

> *Then ye who now on heavenly nectar fare,*
> *Come celebrate with me in song the name*
> *Of Newton, to the Muses dear; for he*
> *Unlocked the hidden treasuries of Truth:*
> *So richly through his mind had Phoebus cast*
> *The radius of his own divinity,*
> *Nearer the gods no mortal may approach.*

The *Principia* established the laws of motion and the principle of universal gravitation, but that understates its importance. Newton had given to the future a model of what a physical theory can be: a set of simple mathematical principles that precisely govern a vast range of different phenomena. Though Newton knew very well that gravitation was not the only physical force, as far as it went his theory was universal—every particle in the universe attracts every other particle with a force proportional to the product of their masses and inversely proportional to the square of their separation. The *Principia* not only deduced Kepler's rules of planetary motion as an exact solution of a simplified problem, the motion of point masses in response to the gravitation of a single massive sphere; it went on to explain (even if only qualitatively in some cases) a wide variety of other phenomena: the precession of equinoxes, the precession of perihelia, the paths of comets, the motions of moons, the rise and fall of the tides, and the fall of apples.[16] By comparison, all past successes of physical theory were parochial.

After the publication of the *Principia* in 1686–1687, Newton became famous. He was elected a member of parliament for the University of Cambridge in 1689 and again in 1701. In 1694 he became warden of the Mint, where he presided over a reform of the English coinage while still retaining his Lucasian professorship. When Czar Peter the Great came to England in 1698, he made a point of visiting the Mint, and hoped to talk with Newton, but I can't find any account of their actually meeting. In 1699 Newton was appointed master of the Mint, a much better-paid position. He gave up his professorship, and became rich. In 1703, after the death of his old enemy Hooke, Newton became president of the Royal Society. He was knighted in 1705. When in 1727 Newton died of a kidney stone, he was given a state funeral in Westminster Abbey, even though he had refused to take the sacraments of the Church of England. Voltaire reported that Newton was "buried like a king who had benefited his subjects."[17]

Newton's theory did not meet universal acceptance.[18] Despite Newton's own commitment to Unitarian Christianity, some in England, like the theologian John Hutchinson and Bishop Berkeley, were appalled by the impersonal naturalism of Newton's theory. This was unfair to the devout Newton. He even argued that only divine intervention could explain why the mutual gravitational attraction of the planets does not destabilize the solar system,* and why some bodies like the Sun and stars shine by their own light, while others like the planets and their satellites are

* In Book III of the *Opticks*, Newton expressed the view that the solar system is unstable, and requires occasional readjustment. The question of the stability of the solar system remained controversial for centuries. In the late 1980s Jacques Laskar showed that the solar system is chaotic; it is impossible to predict the motions of Mercury, Venus, Earth, and Mars for more than about 5 million years into the future. Some initial conditions lead to planets colliding or being ejected from the solar system after a few billion years, while others that are nearly indistinguishable do not. For a review, see J. Laskar, "Is the Solar System Stable?," www.arxiv.org/1209.5996 (2012).

themselves dark. Today of course we understand the light of the Sun and stars in a naturalistic way—they shine because they are heated by nuclear reactions in their cores.

Though unfair to Newton, Hutchinson and Berkeley were not entirely wrong about Newtonianism. Following the example of Newton's work, if not of his personal opinions, by the late eighteenth century physical science had become thoroughly divorced from religion.

Another obstacle to the acceptance of Newton's work was the old false opposition between mathematics and physics that we have seen in a comment of Geminus of Rhodes quoted in Chapter 8. Newton did not speak the Aristotelian language of substances and qualities, and he did not try to explain the cause of gravitation. The priest Nicolas de Malebranche (1638–1715) in reviewing the *Principia* said that it was the work of a geometer, not of a physicist. Malebranche clearly was thinking of physics in the mode of Aristotle. What he did not realize is that Newton's example had revised the definition of physics.

The most formidable criticism of Newton's theory of gravitation came from Christiaan Huygens.[19] He greatly admired the *Principia*, and did not doubt that the motion of planets is governed by a force decreasing as the inverse square of the distance, but Huygens had reservations about whether it is true that every particle of matter attracts every other particle with such a force, proportional to the product of their masses. In this, Huygens seems to have been misled by inaccurate measurements of the rates of pendulums at various latitudes, which seemed to show that the slowing of pendulums near the equator could be entirely explained as an effect of the centrifugal force due to the Earth's rotation. If true, this would imply that the Earth is not oblate, as it would be if the particles of the Earth attract each other in the way prescribed by Newton.

Starting already in Newton's lifetime, his theory of gravitation was opposed in France and Germany by followers of Descartes and by Newton's old adversary Leibniz. They argued that an attraction operating over millions of miles of empty space would be an occult element in natural philosophy, and they further insisted that the action of gravity should be given a rational explanation, not merely assumed.

In this, natural philosophers on the Continent were hanging on to an old ideal for science, going back to the Hellenic age, that scientific theories should ultimately be founded solely on reason. We have learned to give this up. Even though our very successful theory of electrons and light can be deduced from the modern standard model of elementary particles, which may (we hope) in turn eventually be deduced from a deeper theory, however far we go we will never come to a foundation based on pure reason. Like me, most physicists today are resigned to the fact that we will always have to wonder why our deepest theories are not something different.

The opposition to Newtonianism found expression in a famous exchange of letters during 1715 and 1716 between Leibniz and Newton's disciple, the Reverend Samuel Clarke, who had translated Newton's *Opticks* into Latin. Much of their argument focused on the nature of God: Did He intervene in the running of the world, as Newton thought, or had He set it up to run by itself from the beginning?[20] The controversy seems to me to have been supremely futile, for even if its subject were real, it is something about which neither Clarke nor Leibniz could have had any knowledge whatever.

In the end the opposition to Newton's theories didn't matter, for Newtonian physics went from success to success. Halley fitted the observations of the comets observed in 1531, 1607, and 1682 to a single nearly parabolic elliptical orbit, showing that these were all recurring appearances of the same comet. Using Newton's theory to take into account gravitational perturbations due to the masses of Jupiter and Saturn, the French mathematician Alexis-Claude Clairaut and his collaborators predicted in November 1758 that this comet would return to perihelion in mid-April 1759. The comet was observed on Christmas Day 1758, 15 years after Halley's death, and reached perihelion on March 13, 1759. Newton's theory was promoted in the mid-eighteenth century by the French translations of the *Principia* by Clairaut and by Émilie du Châtelet, and through the influence of du Châtelet's lover Voltaire. It was another Frenchman, Jean d'Alembert, who in 1749 published the first correct and accurate calculation of the precession of the equinoxes, based on Newton's ideas. Eventually Newtonianism triumphed everywhere.

This was not because Newton's theory satisfied a preexisting metaphysical criterion for a scientific theory. It didn't. It did not answer the questions about purpose that were central in Aristotle's physics. But it provided universal principles that allowed the successful calculation of a great deal that had previously seemed mysterious. In this way, it provided an irresistible model for what a physical theory should be, and could be.

This is an example of a kind of Darwinian selection operating in the history of science. We get intense pleasure when something has been successfully explained, as when Newton explained Kepler's laws of planetary motion along with much else. The scientific theories and methods that survive are those that provide such pleasure, whether or not they fit any preexisting model of how science ought to be done.

The rejection of Newton's theories by the followers of Descartes and Leibniz suggests a moral for the practice of science: it is never safe simply to reject a theory that has as many impressive successes in accounting for observation as Newton's had. Successful theories may work for reasons not understood by their creators, and they always turn out to be approximations to more successful theories, but they are never simply mistakes.

This moral was not always heeded in the twentieth century. The 1920s saw the advent of quantum mechanics, a radically new framework for physical theory. Instead of calculating the trajectories of a planet or a particle, one calculates the evolution of waves of probability, whose intensity at any position and time tells us the probability of finding the planet or particle then and there. The abandonment of determinism so appalled some of the founders of quantum mechanics, including Max Planck, Erwin Schrödinger, Louis de Broglie, and Albert Einstein, that they did no further work on quantum mechanical theories, except to point out the unacceptable consequences of these theories. Some of the criticisms of quantum mechanics by Schrödinger and Einstein were troubling, and continue to worry us today, but by the end of the 1920s quantum mechanics had already been so successful in accounting for the properties of atoms, molecules, and photons that it had to be taken seriously. The rejection of quantum mechanical theories by these physicists meant that they were unable to participate in the great progress in the physics of solids, atomic nuclei, and elementary particles of the 1930s and 1940s.

Like quantum mechanics, Newton's theory of the solar system had provided what later came to be called a Standard Model. I introduced this term in 1971[21] to describe the theory of the structure and evolution of the expanding universe as it had developed up to that time, explaining:

> Of course, the standard model may be partly or wholly wrong. However, its importance lies not in its certain truth, but in the common meeting ground that it provides for an enormous variety of cosmological data. By discussing this data in the context of a standard cosmological mode, we can begin to appreciate their cosmological relevance, whatever model ultimately proves correct.

A little later, I and other physicists started using the term Standard Model also to refer to our emerging theory of elementary particles and their various interactions. Of course, Newton's successors did not use this term to refer to the Newtonian theory of the solar system, but they well might have. The Newtonian theory certainly provided a common meeting ground for astronomers trying to explain observations that went beyond Kepler's laws.

The methods for applying Newton's theory to problems involving more than two bodies were developed by many authors in the late eighteenth and early nineteenth centuries. There was one innovation of great future importance that was explored especially by Pierre-Simon Laplace in the early nineteenth century. Instead of adding up the gravitational forces exerted by all the bodies in an ensemble like the solar system, one calculates a "field," a condition of space that at every point gives the magnitude and direction of the acceleration produced by all the masses in the ensemble. To calculate the field, one solves certain differential equations that it obeys. (These equations set conditions on the way that the field varies when the point at which it is measured is moved in any of three perpendicular directions.) This approach makes it nearly trivial to prove Newton's theorem that the gravitational forces exerted outside a spherical mass go as the inverse square of the distance from the sphere's center. More important, as we will see in Chapter 15, the field concept was to play a crucial role in the understanding of electricity, magnetism, and light.

These mathematical tools were used most dramatically in 1846 to predict the existence and location of the planet Neptune from irregularities in the orbit of the planet Uranus, independently by John Couch Adams and Jean-Joseph Leverrier. Neptune was discovered soon afterward, in the expected place.

Small discrepancies between theory and observation remained, in the motion of the Moon and of Halley's and Encke's comets, and in a precession of the perihelia of the orbit of Mercury that was observed to be 43" (seconds of arc) per century greater than could be accounted for by gravitational forces produced by the other planets. The discrepancies in the motion of the Moon and comets were eventually traced to nongravitational forces, but the excess precession of Mercury was not explained until the advent in 1915 of the general theory of relativity of Albert Einstein.

In Newton's theory the gravitational force at a given point and a given time depends on the positions of all masses at the same time, so a sudden change of any of these positions (such as a flare on the surface of the Sun) produces an instantaneous change in gravitational forces everywhere. This was in conflict with the principle of Einstein's 1905 special theory of relativity, that no influence can travel faster than light. This pointed to a clear need to seek a modified theory of gravitation. In Einstein's general theory a sudden change in the position of a mass will produce a change in the gravitational field in the immediate neighborhood of the mass, which then propagates at the speed of light to greater distances.

General relativity rejects Newton's notion of absolute space and time. Its underlying equations are the same in all reference frames, whatever their acceleration or rotation. Thus far, Leibniz would have been pleased, but in fact general relativity justifies Newtonian mechanics. Its mathematical formulation is based on a property that it shares with Newton's theory: that all bodies at a given point undergo the same acceleration due to gravity. This means that one can eliminate the effects of gravitation at any point by using a frame of reference, known as an inertial frame, that shares this acceleration. For instance, one does not feel the effects of the Earth's gravity in a freely falling elevator. It is in these inertial frames of reference that Newton's laws apply, at least for bodies whose speeds do not approach that of light.

The success of Newton's treatment of the motion of planets and comets shows that the inertial frames in the neighborhood of the solar system are those in which the Sun rather than the Earth is at rest (or moving with constant velocity). According to general relativity, this is because that is the frame of reference in which the matter of distant galaxies is not revolving around the solar system. In this sense, Newton's theory provided a solid basis for preferring the Copernican theory to that of Tycho. But in general relativity we can use any frame of reference we like, not just inertial frames. If we were to adopt a frame of reference like Tycho's in which the Earth is at rest, then the distant galaxies would seem to be executing circular turns once a year, and in general relativity this enormous motion would create forces akin to gravitation, which would act on the Sun and planets and give them the motions of the Tychonic theory. Newton seems to have had a hint of this. In an unpublished "Proposition 43" that did not make it into the *Principia*, Newton acknowledged that Tycho's theory could be true if some other force besides ordinary gravitation acted on the Sun and planets.[22]

When Einstein's theory was confirmed in 1919 by the observation of a predicted bending of rays of light by the gravitational field of the Sun, the *Times* of London declared that Newton had been shown to be wrong. This was a mistake. Newton's theory can be regarded as an approximation to Einstein's, one that becomes increasingly valid for objects moving at velocities much less than that of light. Not only does Einstein's theory not disprove Newton's; relativity explains why Newton's theory works, when it does work. General relativity itself is doubtless an approximation to a more satisfactory theory.

In general relativity a gravitational field can be fully described by specifying at every point in space and time the inertial frames in which the effects of gravitation are absent. This is mathematically similar to the fact that we can make a map of a small region about any point on a curved surface in which the surface appears flat, like the map of a city on the surface of the Earth; the curvature of the whole surface can be described by compiling an atlas of overlapping local maps. Indeed, this mathematical similarity allows us to describe any gravitational field as a curvature of space and time.

The conceptual basis of general relativity is thus different from that of Newton. The notion of gravitational force is largely replaced in general relativity with the concept of curved spacetime. This was hard for some people to swallow. In 1730 Alexander Pope had written a memorable epitaph for Newton:

Nature and nature's laws lay hid in night;
God said, "Let Newton be!" And all was light.

In the twentieth century the British satirical poet J. C. Squire[23] added two more lines:

It did not last: the Devil howling "Ho,
Let Einstein be," restored the status quo.

Do not believe it. The general theory of relativity is very much in the style of Newton's theories of motion and gravitation: it is based on general principles that can be expressed as mathematical equations, from which consequences can be mathematically deduced for a broad range of phenomena, which when compared with observation allow the theory to be verified. The difference between Einstein's and Newton's theories is far less than the difference between Newton's theories and anything that had gone before.

A question remains: why did the scientific revolution of the sixteenth and seventeenth centuries happen when and where it did? There is no lack of possible explanations. Many changes occurred in fifteenth-century Europe that helped to lay the foundation for the scientific revolution. National governments were consolidated in France under Charles VII and Louis XI and in England under Henry VII. The fall of Constantinople in 1453 sent Greek scholars fleeing westward to Italy and beyond. The Renaissance intensified interest in the natural world and set higher standards for the accuracy of ancient texts and their translation. The invention of printing with movable type made scholarly communication far quicker and cheaper. The discovery and exploration of America reinforced the lesson that there is much that the ancients did not know. In addition, according to the "Merton thesis," the Protestant Reformation of the early sixteenth century set the stage for the great scientific breakthroughs of seventeenth-century England. The sociologist Robert Merton supposed that Protestantism created social attitudes favorable to science and promoted a combination of rationalism and empiricism and a belief in an understandable order in nature—attitudes and beliefs that he found in the actual behavior of Protestant scientists.[24]

It is not easy to judge how important were these various external influences on the scientific revolution. But although I cannot tell why it was Isaac Newton in late-seventeenth-century England who discovered the classical laws of motion and gravitation, I think I know why these laws took the form they did. It is, very simply, because to a very good approximation the world really does obey Newton's laws.

Having surveyed the history of physical science from Thales to Newton, I would like now to offer some tentative thoughts on what drove us to the modern conception of science, represented by the achievements of Newton and his successors. Nothing like modern science was conceived as a goal in the ancient world or the medieval world. Indeed, even if our predecessors could have imagined science as it is today, they might not have liked it very much. Modern science is impersonal, without room for supernatural intervention or (outside the behavioral sciences) for human values; it has no sense of purpose; and it offers no hope for certainty. So how did we get here?

Faced with a puzzling world, people in every culture have sought explanations. Even where they abandoned mythology, most attempts at explanation did not lead to anything satisfying. Thales tried to understand matter by guessing that it is all water, but what could he do with this idea? What new information did it give him? No one at Miletus or anywhere else could build anything on the notion that everything is water.

But every once in a while someone finds a way of explaining some phenomenon that fits so well and clarifies so much that it gives the finder intense satisfaction, especially when the new understanding is quantitative, and observation bears it out in detail. Imagine how Ptolemy must have felt when he realized that, by adding an equant to the epicycles and eccentrics of Apollonius and Hipparchus, he had found a theory of planetary motions that allowed him to predict with fair accuracy where any planet would be found in the sky at any time. We can get a sense of his joy from the lines of his that I quoted earlier: "When I search out the massed wheeling circles of the stars, my feet no longer touch the Earth, but, side by side with Zeus himself, I take my fill of ambrosia, the food of the gods."

The joy was flawed—it always is. You didn't have to be a follower of Aristotle to be repelled by the peculiar looping motion of planets moving on epicycles in Ptolemy's theory. There was also the nasty fine-tuning: it had to take precisely one year for the centers of the epicycles of Mercury and Venus to move around the Earth, and for Mars, Jupiter, and Saturn to move around their epicycles. For over a thousand years philosophers argued about the proper role of astronomers like Ptolemy—really to understand the heavens, or merely to fit the data.

What pleasure Copernicus must then have felt when he was able to explain that the fine-tuning and the looping orbits of Ptolemy's scheme arose simply because we view the solar system from a moving Earth. Still flawed, the Copernican theory did not quite fit the data without ugly complications. How much then the mathematically gifted Kepler must have enjoyed replacing the Copernican mess with motion on ellipses, obeying his three laws.

So the world acts on us like a teaching machine, reinforcing our good ideas with moments of satisfaction. After centuries we learn what kinds of understanding are possible, and how to find them. We learn not to worry about purpose, because such worries never lead to the sort of delight we seek. We learn to abandon the search for certainty, because the explanations that make us happy never are certain. We learn to do experiments, not worrying about the artificiality of our arrangements. We develop an aesthetic sense that gives us clues to what theories will work, and that adds to our pleasure when they do work. Our understandings accumulate. It is all unplanned and unpredictable, but it leads to reliable knowledge, and gives us joy along the way.

15

Epilogue: The Grand Reduction

Newton's great achievement left plenty yet to be explained. The nature of matter, the properties of forces other than gravitation that act on matter, and the remarkable capabilities of life were all still mysterious. Enormous progress was made in the years after Newton,[1] far too much to cover in one book, let alone a single chapter. This epilogue aims at making just one point, that as science progressed after Newton a remarkable picture began to take shape: it turned out that the world is governed by natural laws far simpler and more unified than had been imagined in Newton's time.

Newton himself in Book III of his *Opticks* sketched the outline of a theory of matter that would at least encompass optics and chemistry:

> Now the smallest particles of matter may cohere by the strongest attractions, and compose bigger particles of weaker virtue; and many of these may cohere and compose bigger particles whose virtue is still weaker, and so on for diverse successions, until the progression ends in the biggest particles on which the operations in chemistry, and the colors of natural bodies depend, and which by cohering compose bodies of a sensible magnitude.[2]

He also focused attention on the forces acting on these particles:

> For we must learn from the phenomena of nature what bodies attract one another, and what are the laws and properties of the attraction, before we inquire the cause by which the attraction is perform'd. The attractions of gravity, magnetism, and electricity, reach to very sensible distances, and so have been observed by vulgar eyes, and there may be others which reach to so small distances as to escape observation.[3]

As this shows, Newton was well aware that there are other forces in nature besides gravitation. Static electricity was an old story. Plato had mentioned in the *Timaeus* that when a piece of amber (in Greek, *electron*) is rubbed it can pick up light bits of matter. Magnetism was known from the properties of naturally magnetic lodestones, used by the Chinese for geomancy and studied in detail by Queen Elizabeth's physician, William Gilbert. Newton here also hints at the existence of forces not yet known because of their short range, a premonition of the weak and strong nuclear forces discovered in the twentieth century.

In the early nineteenth century the invention of the electric battery by Alessandro Volta made it possible to carry out detailed quantitative experiments in electricity and magnetism, and it soon became known that these are not entirely separate phenomena. First, in 1820 Hans Christian Ørsted in Copenhagen found that a magnet and a wire carrying an electric current exert forces on each other. After hearing of this result, André-Marie Ampère in Paris discovered that wires carrying electric currents also exert forces on one another. Ampère conjectured that these various phenomena are all much the same: the forces exerted by and on pieces of magnetized iron are due to electric currents circulating within the iron.

Just as happened with gravitation, the notion of currents and magnets exerting forces on each other was replaced with the idea of a field, in this case a magnetic field. Each magnet and current-carrying wire contributes to the total magnetic field at any point in its vicinity, and this magnetic field exerts a force on any magnet or electric current at that point. Michael Faraday attributed the magnetic forces produced by an electric current to lines of magnetic field encircling the wire. He also described the electric forces produced by a piece of rubbed amber as due to an electric field, pictured as lines emanating radially from the electric charges on the amber. Most important, Faraday in the 1830s showed a connection between electric and magnetic fields: a changing magnetic field, like that produced by the electric current in a rotating coil of wire, produces an electric field, which can drive electric currents in another wire. It is this phenomenon that is used to generate electricity in modern power plants.

The final unification of electricity and magnetism was achieved a few decades later, by James Clerk Maxwell. Maxwell thought of electric and magnetic fields as tensions in a pervasive medium, the ether, and expressed what was known about electricity and magnetism in equations relating the fields and their rates of change to each other. The new thing added by Maxwell was that, just as a changing magnetic field generates an electric field, so also a changing electric field generates a magnetic field. As often happens in physics, the conceptual basis for Maxwell's equations in terms of an ether has been abandoned, but the equations survive, even on T-shirts worn by physics students.*

Maxwell's theory had a spectacular consequence. Since oscillating electric fields produce oscillating magnetic fields, and oscillating magnetic fields produce oscillating electric fields, it is possible to have a self-sustaining oscillation of both electric and magnetic fields in the ether, or as we would say today, in empty space. Maxwell found around 1862 that this electromagnetic oscillation would propagate at a speed that, according to his equa-

* Maxwell himself did not write equations governing electric and magnetic fields in the form known today as "Maxwell's equations." His equations instead involved other fields known as potentials, whose rates of change with time and position are the electric and magnetic fields. The more familiar modern form of Maxwell's equations was given around 1881 by Oliver Heaviside.

tions, had just about the same numerical value as the measured speed of light. It was natural for Maxwell to jump to the conclusion that light is nothing but a mutually self-sustaining oscillation of electric and magnetic fields. Visible light has a frequency far too high for it to be produced by currents in ordinary electric circuits, but in the 1880s Heinrich Hertz was able to generate waves in accordance with Maxwell's equations: radio waves that differed from visible light only in having much lower frequency. Electricity and magnetism had thus been unified not only with each other, but also with optics.

As with electricity and magnetism, progress in understanding the nature of matter began with quantitative measurement, here measurement of the weights of substances participating in chemical reactions. The key figure in this chemical revolution was a wealthy Frenchman, Antoine Lavoisier. In the late eighteenth century he identified hydrogen and oxygen as elements and showed that water is a compound of hydrogen and oxygen, that air is a mixture of elements, and that fire is due to the combination of other elements with oxygen. Also on the basis of such measurements, it was found a little later by John Dalton that the weights with which elements combine in chemical reactions can be understood on the hypothesis that pure chemical compounds like water or salt consist of large numbers of particles (later called molecules) that themselves consist of definite numbers of atoms of pure elements. The water molecule, for instance, consists of two hydrogen atoms and one oxygen atom. In the following decades chemists identified many elements: some familiar, like carbon, sulfur, and the common metals; and others newly isolated, such as chlorine, calcium, and sodium. Earth, air, fire, and water did not make the list. The correct chemical formulas for molecules like water and salt were worked out, in the first half of the nineteenth century, allowing the calculation of the ratios of the masses of the atoms of the different elements from measurements of the weights of substances participating in chemical reactions.

The atomic theory of matter scored a great success when Maxwell and Ludwig Boltzmann showed how heat could be understood as energy distributed among vast numbers of atoms or molecules. This step toward unification was resisted by some physicists, including Pierre Duhem, who doubted the existence of atoms and held that the theory of heat, thermodynamics, was at least as fundamental as Newton's mechanics and Maxwell's electrodynamics. But soon after the beginning of the twentieth century several new experiments convinced almost everyone that atoms are real. One series of experiments, by J. J. Thomson, Robert Millikan, and others, showed that electric charges are gained and lost only as multiples of a fundamental charge: the charge of the electron, a particle that had been discovered by Thomson in 1897. The random "Brownian" motion of small particles on the surface of liquids was interpreted by Albert Einstein in 1905 as due to collisions of these particles with individual molecules of the liquid, an interpretation confirmed by experiments of Jean Perrin. Responding to the experiments of Thomson and Perrin, the chemist Wilhelm Ostwald, who earlier had been skeptical about atoms, expressed his change of mind in 1908, in a statement that implicitly looked all the way back to Democritus and Leucippus: "I am now convinced that we have recently become possessed of experimental evidence of the discrete or grained nature of matter, which the atomic hypothesis sought in vain for hundreds and thousands of years." [4]

But what are atoms? A great step toward the answer was taken in 1911, when experiments in the Manchester laboratory of Ernest Rutherford showed that the mass of gold atoms is concentrated in a small heavy positively charged nucleus, around which revolve lighter negatively charged electrons. The electrons are responsible for the phenomena of ordinary chemistry, while changes in the nucleus release the large energies encountered in radioactivity.

This raised a new question: what keeps the orbiting atomic electrons from losing energy through the emission of radiation, and spiraling down into the nucleus? Not only would this rule out the existence of stable atoms; the frequencies of the radiation emitted in these little atomic catastrophes would form a continuum, in contradiction with the observation that atoms can emit and absorb radiation only at certain discrete frequencies, seen as bright or dark lines in the spectra of gases. What determines these special frequencies?

The answers were worked out in the first three decades of the twentieth century with the development of quantum mechanics, the most radical innovation in physical theory since the work of Newton. As its name suggests, quantum mechanics requires a quantization (that is, a discreteness) of the energies of various physical systems. Niels Bohr in 1913 proposed that an atom can exist only in states of certain definite energies, and gave rules for calculating these energies in the simplest atoms. Following earlier work of Max Planck, Einstein had already in 1905 suggested that the energy in light comes in quanta, particles later called photons, each photon with an energy proportional to the frequency of the light. As Bohr explained, when an atom loses energy by emitting a single photon, the energy of that photon must equal the difference in the energies of the initial and final atomic states, a requirement that fixes its frequency. There is always an atomic state of lowest energy, which cannot emit radiation and is therefore stable.

These early steps were followed in the 1920s with the development of general rules of quantum mechanics, rules that can be applied to any physical system. This was chiefly the work of Louis de Broglie, Werner Heisenberg, Wolfgang Pauli, Pascual Jordan, Erwin Schrödinger, Paul Dirac, and Max Born. The energies of allowed atomic states are calculated by solving an equation, the Schrödinger equation, of a general mathematical type that was already familiar from the study of sound and light waves. A string on a musical instrument can produce just those tones for which a whole number of half wavelengths fit on the string; analogously, Schrödinger found that the allowed energy levels of an atom are those for which the wave governed by the Schrödinger equation just fits around the atom without discontinuities. But as first recognized by Born, these waves are not waves of pressure or of electromagnetic fields, but waves of probability—a particle is most likely to be near where the wave function is largest.

Quantum mechanics not only solved the problem of the stability of atoms and the nature of spectral lines; it also brought chemistry into the framework of physics. With the electrical forces among electrons and atomic nuclei already known, the Schrödinger equation could be applied to molecules as well as to atoms, and allowed the calculation of the energies of their various states. In this way it became possible in principle to decide which molecules are stable and which chemical reactions are energetically allowed. In 1929 Dirac announced triumphantly that "the underlying physical laws necessary for the mathematical theory of a larger part of physics and the whole of chemistry are thus completely known."[5]

This did not mean that chemists would hand over their problems to physicists, and retire. As Dirac well understood, for all but the smallest molecules the Schrödinger equation is too complicated to be solved, so the special tools and insights of chemistry remain indispensable. But from the 1920s on, it would be understood that any general principle of chemistry, such as the rule that metals form stable compounds with halogen elements like chlorine, is what it is because of the quantum mechanics of nuclei and electrons acted on by electromagnetic forces.

Despite its great explanatory power, this foundation was itself far from being satisfactorily unified. There were particles: electrons and the protons and neutrons that make up atomic nuclei. And there were fields: the electromagnetic field, and whatever then-unknown short-range fields are presumably responsible for the strong forces that hold atomic nuclei together and for the weak forces that turn neutrons into protons or protons into neutrons in radioactivity. This distinction between particles and fields began to be swept away in the 1930s, with the advent of quantum field theory. Just as there is an electromagnetic field, whose energy and momentum are bundled in particles known as photons, so there is also an electron field, whose energy and momentum are bundled in electrons, and likewise for other types of elementary particles.

This was far from obvious. We can directly feel the effects of gravitational and electromagnetic fields because the quanta of these fields have zero mass, and they are particles of a type (known as bosons) that in large numbers can occupy the same state. These properties allow large numbers of photons to build up to form states that we observe as electric and magnetic fields that seem to obey the rules of classical (that is, non-quantum) physics. Electrons, in contrast, have mass and are particles of a type (known as fermions) no two of which can occupy the same state, so that electron fields are never apparent in macroscopic observations.

In the late 1940s quantum electrodynamics, the quantum field theory of photons, electrons, and antielectrons, scored stunning successes, with the calculation of quantities like the strength of the electron's magnetic field that agreed with experiment to many decimal places.* Following this achievement, it was natural to try to develop a quantum field theory that would encompass not only photons, electrons, and antielectrons but also the other particles being discovered in cosmic rays and accelerators and the weak and strong forces that act on them.

We now have such a quantum field theory, known as the Standard Model. The Standard Model is an expanded version of quantum electrodynamics. Along with the electron field there is a neutrino field, whose quanta are fermions like electrons but with zero electric charge and nearly zero mass. There is a pair of quark fields, whose quanta are the constituents of the protons and neutrons that make up atomic nuclei. For reasons that no one understands, this menu is repeated twice, with much heavier quarks and much heavier electron-like particles and their neutrino partners. The electromagnetic field appears in a unified "electroweak" picture along with other fields responsible for the

* Here and in what follows I will not cite individual physicists. So many are involved that it would take too much space, and many are still alive, so that I would risk giving offense by citing some physicists and not others.

weak nuclear interactions, which allow protons and neutrons to convert into one another in radioactive decays. The quanta of these fields are heavy bosons: the electrically charged W^+ and W^-, and the electrically neutral Z^0. There are also eight mathematically similar "gluon" fields responsible for the strong nuclear interactions, which hold quarks together inside protons and neutrons. In 2012 the last missing piece of the Standard Model was discovered: a heavy electrically neutral boson that had been predicted by the electroweak part of the Standard Model.

The Standard Model is not the end of the story. It leaves out gravitation; it does not account for the "dark matter" that astronomers tell us makes up five-sixths of the mass of the universe; and it involves far too many unexplained numerical quantities, like the ratios of the masses of the various quarks and electron-like particles. But even so, the Standard Model provides a remarkably unified view of all types of matter and force (except for gravitation) that we encounter in our laboratories, in a set of equations that can fit on a single sheet of paper. We can be certain that the Standard Model will appear as at least an approximate feature of any better future theory.

The Standard Model would have seemed unsatisfying to many natural philosophers from Thales to Newton. It is impersonal; there is no hint in it of human concerns like love or justice. No one who studies the Standard Model will be helped to be a better person, as Plato expected would follow from the study of astronomy. Also, contrary to what Aristotle expected of a physical theory, there is no element of purpose in the Standard Model. Of course, we live in a universe governed by the Standard Model and can imagine that electrons and the two light quarks are what they are to make us possible, but then what do we make of their heavier counterparts, which are irrelevant to our lives?

The Standard Model is expressed in equations governing the various fields, but it cannot be deduced from mathematics alone. Nor does it follow straightforwardly from observation of nature. Indeed, quarks and gluons are attracted to each other by forces that increase with distance, so these particles can never be observed in isolation. Nor can the Standard Model be deduced from philosophical preconceptions. Rather, the Standard Model is a product of guesswork, guided by aesthetic judgment, and validated by the success of many of its predictions. Though the Standard Model has many unexplained aspects, we expect that at least some of these features will be explained by whatever deeper theory succeeds it.

The old intimacy between physics and astronomy has continued. We now understand nuclear reactions well enough not only to calculate how the Sun and stars shine and evolve, but also to understand how the lightest elements were produced in the first few minutes of the present expansion of the universe. And as in the past, astronomy now presents physics with a formidable challenge: the expansion of the universe is speeding up, presumably owing to dark energy that is contained not in particle masses and motions, but in space itself.

There is one aspect of experience that at first sight seems to defy understanding on the basis of any unpurposeful physical theory like the Standard Model. We cannot avoid teleology in talking of living things. We describe hearts and lungs and roots and flowers in terms of the purpose they serve, a tendency that was only increased with the great expansion after Newton of information about plants and animals due to naturalists like Carl Linnaeus and Georges Cuvier. Not only theologians but also scientists including Robert Boyle and Isaac Newton have seen the marvelous capabilities of plants and animals as evidence for a benevolent Creator. Even if we can avoid a supernatural explanation of the capabilities of plants and animals, it long seemed inevitable that an understanding of life would rest on teleological principles very different from those of physical theories like Newton's.

The unification of biology with the rest of science first began to be possible in the mid-nineteenth century, with the independent proposals by Charles Darwin and Alfred Russel Wallace of the theory of evolution through natural selection. Evolution was already a familiar idea, suggested by the fossil record. Many of those who accepted the reality of evolution explained it as a result of a fundamental principle of biology, an inherent tendency of living things to improve, a principle that would have ruled out any unification of biology with physical science. Darwin and Wallace instead proposed that evolution acts through the appearance of inheritable variations, with favorable variations no more likely than unfavorable ones, but with the variations that improve the chances of survival and reproduction being the ones that are likely to spread.*

It took a long time for natural selection to be accepted as the mechanism for evolution. No one in Darwin's time knew the mechanism for inheritance, or for the appearance of inheritable variations, so there was room for biologists to hope for a more purposeful theory. It was particularly distasteful to imagine that humans are the result of millions of years of natural selection acting on random inheritable variations. Eventually the discovery of the rules of genetics and of the occurrence of mutations led in the twentieth century to a "neo-Darwinian synthesis" that put the theory of evolution through natural selection on a firmer basis. Finally this theory was grounded on chemistry, and thereby on physics, through the realization that genetic information is carried by the double helix molecules of DNA.

So biology joined chemistry in a unified view of nature based on physics. But it is important to acknowledge the limitations of this unification. No one is going to replace the language and methods of biology with a description of living things in terms of individual molecules, let alone quarks and electrons. For one thing, even more than the large molecules of organic chemistry, living things are too complicated for such a description. More important, even if we could follow the motion of every atom in a plant or animal, in that immense mass of data we would lose

*I am here lumping sexual selection together with natural selection, and punctuated equilibrium along with steady evolution; and I am not distinguishing between mutations and genetic drift as a source of inheritable variations. These distinctions are very important to biologists, but they do not affect the point that concerns me here: there is no independent law of biology that makes inheritable variations more likely to be improvements.

the things that interest us—a lion hunting antelope or a flower attracting bees.

For biology, like geology but unlike chemistry, there is another problem. Living things are what they are not only because of the principles of physics, but also because of a vast number of historical accidents, including the accident that a comet or meteor hit the Earth 65 million years ago with enough impact to kill off the dinosaurs, and going back to the fact that the Earth formed at a certain distance from the Sun and with a certain initial chemical composition. We can understand some of these accidents statistically, but not individually. Kepler was wrong; no one will ever be able to calculate the distance of the Earth from the Sun solely from the principles of physics. What we mean by the unification of biology with the rest of science is only that there can be no freestanding principles of biology, any more than of geology. Any general principle of biology is what it is because of the fundamental principles of physics together with historical accidents, which by definition can never be explained.

The point of view described here is called (often disapprovingly) "reductionism." There is opposition to reductionism even within physics. Physicists who study fluids or solids often cite examples of "emergence," the appearance in the description of macroscopic phenomena of concepts like heat or phase transition that have no counterpart in elementary particle physics, and that do not depend on the details of elementary particles. For instance, thermodynamics, the science of heat, applies in a wide variety of systems: not just to those considered by Maxwell and Boltzmann, containing large numbers of molecules, but also to the surfaces of large black holes. But it does not apply to everything, and when we ask whether it applies to a given system and if so why, we must have reference to deeper, more truly fundamental, principles of physics. Reductionism in this sense is not a program for the reform of scientific practice; it is a view of why the world is the way it is.

We do not know how long science will continue on this reductive path. We may come to a point where further progress is impossible within the resources of our species. Right now, it seems that there is a scale of mass about a million trillion times larger than the mass of the hydrogen atom, at which gravity and other as yet undetected forces are unified with the forces of the Standard Model. (This is known as the "Planck mass"; it is the mass that particles would have to possess for their gravitational attraction to be as strong as the electrical repulsion between two electrons at the same separation.) Even if the economic resources of the human race were entirely at the disposal of physicists, we cannot now conceive of any way of creating particles with such huge masses in our laboratories.

We may instead run out of intellectual resources—humans may not be smart enough to understand the really fundamental laws of physics. Or we may encounter phenomena that in principle cannot be brought into a unified framework for all science. For instance, although we may well come to understand the processes in the brain responsible for consciousness, it is hard to see how we will ever describe conscious feelings themselves in physical terms.

Still, we have come a long way on this path, and are not yet at its end.[6] This is a grand story—how celestial and terrestrial physics were unified by Newton, how a unified theory of electricity and magnetism was developed that turned out to explain light, how the quantum theory of electromagnetism was expanded to include the weak and strong nuclear forces, and how chemistry and even biology were brought into a unified though incomplete view of nature based on physics. It is toward a more fundamental physical theory that the wide-ranging scientific principles we discover have been, and are being, reduced.

Acknowledgments

I was fortunate to have the help of several learned scholars: the classicist Jim Hankinson and the historians Bruce Hunt and George Smith. They read through most of the book, and I made many corrections based on their suggestions. I am deeply grateful for this help. I am indebted also to Louise Weinberg for invaluable critical comments, and for suggesting the lines of John Donne that now grace this book's front matter. Thanks, too, to Peter Dear, Owen Gingerich, Alberto Martinez, Sam Schweber, and Paul Woodruff for advice on specific topics. Finally, for encouragement and good advice, many thanks are due to my wise agent, Morton Janklow; and to my fine editors at HarperCollins, Tim Duggan and Emily Cunningham.

Technical Notes

The following notes describe the scientific and mathematical background for many of the historical developments discussed in this book. Readers who have learned some algebra and geometry in high school and have not entirely forgotten what they learned should have no trouble with the level of mathematics in these notes. But I have tried to organize this book so that readers who are not interested in technical details can skip these notes and still understand the main text.

A warning: The reasoning in these notes is not necessarily identical to that followed historically. From Thales to Newton, the style of the mathematics that was applied to physical problems was far more geometric and less algebraic than is common today. To analyze these problems in this geometric style would be difficult for me and tedious for the reader. In these notes I will show how the results obtained by the natural philosophers of the past do follow (or in some cases, do not follow) from the observations and assumptions on which they relied, but without attempting faithfully to reproduce the details of their reasoning.

Notes

1. Thales' Theorem

Thales' theorem uses simple geometric reasoning to derive a result about circles and triangles that is not immediately obvious. Whether or not Thales was the one who proved this result, it is useful to look at the theorem as an instance of the scope of Greek knowledge of geometry before the time of Euclid.

Consider any circle, and any diameter of it. Let *A* and *B* be the points where the diameter intersects the circle. Draw lines from A and B to any other point P on the circle. The diameter and the lines running from A to P and from B to P form a triangle, ABP. (We identify triangles by listing their three corner points.) Thales' Theorem tells us that this is a right triangle: the angle of triangle ABP at P is a right angle, or in other terms, 90°.

The trick in proving this theorem is to draw a line from the center C of the circle to point P. This divides triangle ABP into two triangles, ACP and BCP. (See Figure 1.) Both of these are isosceles triangles, that is, triangles with two sides equal. In triangle ACP, sides CA and CP are both radii of the circle, which have the same length according to the definition of a circle. (We label the sides of a triangle by the corner points they connect.) Likewise, in triangle BCP, sides CB and CP are equal. In an isosceles triangle the angles adjoining the two equal sides are equal, so angle α (alpha) at the intersection of sides AP and AC is equal to the angle at the intersection of sides AP and CP, while angle β (beta) at the intersection of sides BP and BC is equal to the angle at the intersection of sides BP and CP. The sum of the angles of any triangle is two right angles,* or in familiar terms 180°, so if we take α' as the third angle of triangle ACP, the angle at the intersection of sides AC and CP, and likewise take β' as the angle at the intersection of sides BC and CP, then

$$2\alpha + \alpha' = 180^\circ \qquad 2\beta + \beta' = 180^\circ$$

Adding these two equations and regrouping terms gives

$$2(\alpha + \beta) + (\alpha' + \beta') = 360^\circ$$

Now, $\alpha' + \beta'$ is the angle between AC and BC, which come together in a straight line, and is therefore half a full turn, or 180°, so

* This may not have been known in the time of Thales, in which case the proof must be of a later date.

$$2(\alpha + \beta) = 360° - 180° = 180°$$

and therefore $\alpha + \beta = 90°$. But a glance at Figure 1 shows that $\alpha + \beta$ is the angle between sides AP and BP of triangle ABP, with which we started, so we see that this is indeed a right triangle, as was to be proved.

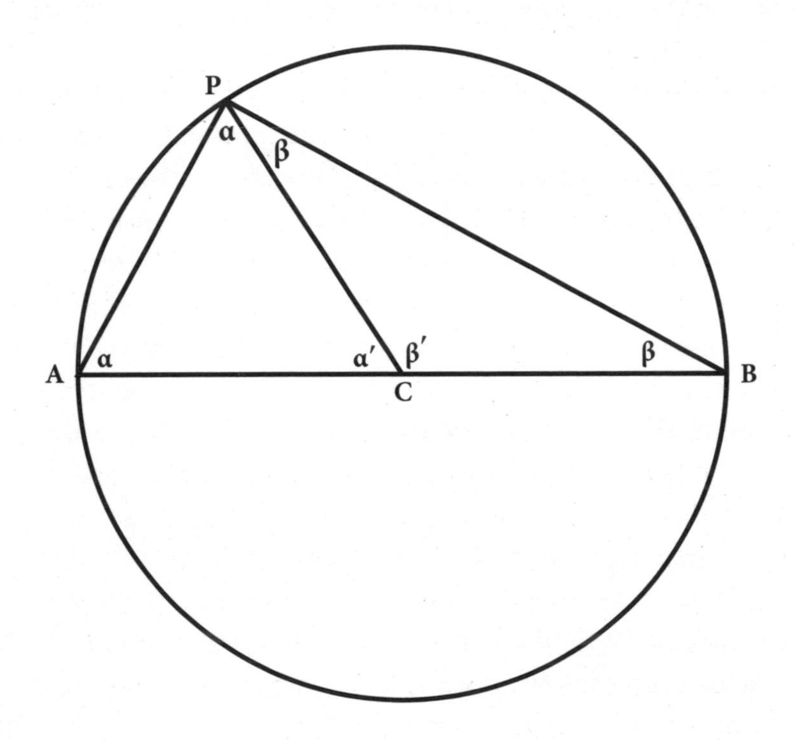

Figure 1. Proof of Thales' theorem. The theorem states that wherever point *P* is located on the circle, the angle between the lines from the ends of the diameter to *P* is a right angle.

2. *Platonic Solids*

In Plato's speculations about the nature of matter, a central role was played by a class of solid shapes known as regular polyhedrons, which have come also to be known as Platonic solids. The regular polyhedrons can be regarded as three-dimensional generalizations of the regular polygons of plane geometry, and are in a sense built up from regular polygons. A regular polygon is a plane figure bounded by some number *n* of straight lines, all of which are of the same length and meet at each of the *n* corners with the same angles. Examples are the equilateral triangle (a triangle with all sides equal) and the square. A regular polyhedron is a solid figure bounded by regular polygons, all of which are identical, with the same number *N* of polygons meeting with the same angles at every vertex.

The most familiar example of a regular polyhedron is the cube. A cube is bounded by six equal squares, with three squares meeting at each of its eight vertices. There is an even simpler regular polyhedron, the tetrahedron, a triangular pyramid bounded by four equal equilateral triangles, with three triangles meeting at each of the four vertices. (We will be concerned here only with polyhedrons that are convex, with every vertex pointed outward, as in the case of the cube and the tetrahedron.) As we read in the *Timaeus*, it had somehow became known to Plato that these regular polyhedrons come in only five possible shapes, which he took to be the shapes of the atoms of which all matter is composed. They are the tetrahedron, cube, octahedron, dodecahedron, and icosahedron, with 4, 6, 8, 12, and 20 faces, respectively.

The earliest attempt to prove that there are just five regular polyhedrons that has survived from antiquity is the climactic last paragraph of Euclid's *Elements*. In Propositions 13 through 17 of Book XIII Euclid had given geometric constructions of the tetrahedron, octahedron, cube, icosahedron, and dodecahedron. Then he states,* "I say next that no other figure, besides the said five figures, can be constructed which is contained by equilateral and equiangular figures equal to one another." In fact, what Euclid actually demonstrates after this statement is a weaker result, that there are only five combinations of the number *n* of sides of each polygonal face, and the number *N* of polygons meeting at each vertex, that are possible for a regular polyhedron. The proof

*This is from the standard translation by T. L. Heath (*Euclid's Elements*, Green Lion Press, Santa Fe, N.M., 2002, p. 480).

given below is essentially the same as Euclid's, expressed in modern terms.

The first step is to calculate the interior angle θ (theta) at each of the n vertices of an n-sided regular polygon. Draw lines from the center of the polygon to the vertices on the boundary. This divides the interior of the polygon into n triangles. Since the sum of the angles of any triangle is 180°, and each of these triangles has two vertices with angles $\theta/2$, the angle of the third vertex of each triangle (the one at the center of the polygon) must be $180° - \theta$. But these n angles must add up to 360°, so $n\,(180° - \theta) = 360°$. The solution is

$$\theta = 180° - \frac{360°}{n}$$

For instance, for an equilateral triangle we have $n = 3$, so $\theta = 180° - 120° = 60°$, while for a square $n = 4$, so $\theta = 180° - 90° = 90°$.

The next step is to imagine cutting all the edges and vertices of a regular polyhedron except at one vertex, and pushing the polyhedron down onto a plane at that vertex. The N polygons meeting at that vertex will then be lying in the plane, but there must be space left over or the N polygons would have formed a single face. So we must have $N\theta < 360°$. Using the above formula for θ and dividing both sides of the inequality by 360° then gives

$$N\left(\frac{1}{2} - \frac{1}{n}\right) < 1$$

or equivalently (dividing both sides by N),

$$\frac{1}{2} < \frac{1}{n} + \frac{1}{N}$$

Now, we must have $n \geq 3$ because otherwise there would be no area between the sides of the polygons, and we must have $N \geq 3$ because otherwise there would be no space between the faces coming together at a vertex. (For instance, for a cube $n = 4$ because the sides are squares, and $N = 3$.) Thus the above inequality does not allow either $1/n$ or $1/N$ to be as small as $1/2 - 1/3 = 1/6$, and consequently neither n nor N can be as large as 6. We can easily check every pair of values of whole numbers $5 \geq N \geq 3$ and $5 \geq n \geq 3$ to see if they satisfy the inequality, and find that there are only five pairs that do:

$$
\begin{array}{lll}
(a) & N = 3\,, & n = 3 \\
(b) & N = 4\,, & n = 3 \\
(c) & N = 5\,, & n = 3 \\
(d) & N = 3\,, & n = 4 \\
(e) & N = 3\,, & n = 5
\end{array}
$$

(In the cases $n = 3$, $n = 4$, and $n = 5$ the sides of the regular polyhedron are respectively equilateral triangles, squares, and regular pentagons.) These are the values of N and n that we find in the tetrahedron, octahedron, icosahedron, cube, and dodecahedron.

This much was proved by Euclid. But Euclid did not prove that there is only one regular polyhedron for each pair of n and N. In what follows, we will go beyond Euclid, and show that for each value of N and n we can find unique results for the other properties of the polyhedron: the number F of faces, the number E of edges, and the number V of vertices. There are three unknowns here, so for this purpose we need three equations. To derive the first, note that the total number of borders of all the polygons on the surface of the polyhedron is nF, but each of the E edges borders two polygons, so

$$2E = nF$$

Also, there are N edges coming together at each of the V vertices, and each of the E edges connects two vertices, so

$$2E = NV$$

Finally, there is a more subtle relation among F, E, and V. In deriving this relation, we must make an additional assumption, that the polyhedron is simply connected, in the sense that any path between two points on the surface can be continuously deformed into any other path between these points. This is the case for instance for a cube or a tetrahedron, but not for a polyhedron (regular or not) constructed by drawing edges and faces on the surface of a doughnut. A deep theorem states that any simply connected polyhedron can be constructed by adding edges, faces, and/or vertices to a tetrahedron, and then if necessary continuously squeezing the resulting polyhedron into some desired shape. Using this fact, we shall now show that any simply connected polyhedron (regular or not) satisfies the relation:

$$F - E + V = 2$$

It is easy to check that this is satisfied for a tetrahedron, in which case we have $F = 4$, $E = 6$, and $V = 4$, so the left-hand side is $4 - 6 + 4 = 2$. Now, if we add an edge to any polyhedron, running across a face from one edge to another, we add one new face and two new vertices, so F and V increase by one unit and two units, respectively. But this splits each old edge at the ends of the new edge into two pieces, so E increases by $1 + 2 = 3$, and the quantity $F - E + V$ is thereby unchanged. Likewise, if we add an edge that runs from a vertex to one of the old edges, then we increase F and V by one unit each, and E by two units, so the quantity $F - E + V$ is still unchanged. Finally, if we add an edge that runs from one vertex to another vertex, then we increase both F and E by one unit each and do not change V, so again $F - E + V$ is unchanged. Since any simply connected polyhedron can be built up in this way, all such polyhedrons have the same value for this quantity, which therefore must be the same value $F - E + V = 2$ as for a tetrahedron. (This is a simple example of a branch of mathematics known as topology; the quantity $F - E + V$ is known in topology as the "Euler characteristic" of the polyhedron.)

We can now solve these three equations for E, F, and V. It is simplest to use the first two equations to replace F and V in the third equation with $2E/n$ and $2E/N$, respectively, so that the third equation becomes $2E/n - E + 2E/N = 2$, which gives

$$E = \frac{2}{2/n - 1 + 2/N}$$

Then using the other two equations, we have

$$F = \frac{4}{2 - n + 2n/N} \qquad V = \frac{4}{2N/n - N + 2}$$

Thus for the five cases listed above, the numbers of faces, vertices, and edges are:

	F	V	E	
$N = 3, n = 3$	4	4	6	tetrahedron
$N = 4, n = 3$	8	6	12	octahedron
$N = 5, n = 3$	20	12	30	icosahedron
$N = 3, n = 4$	6	8	12	cube
$N = 3, n = 5$	12	20	30	dodecahedron

These are the Platonic solids.

3. Harmony

The Pythagoreans discovered that two strings of a musical instrument, with the same tension, thickness, and composition, will make a pleasant sound when plucked at the same time, if the ratio of the strings' lengths is a ratio of small whole numbers, such as 1/2, 2/3, 1/4, 3/4, etc. To see why this is so, we first need to work out the general relation between the frequency, wavelength, and velocity of any sort of wave.

Any wave is characterized by some sort of oscillating amplitude. The amplitude of a sound wave is the pressure in the air carrying the wave; the amplitude of an ocean wave is the height of the water; the amplitude of a light wave with a definite direction of polarization is the electric field in that direction; and the amplitude of a wave moving along the string of a musical instrument is the displacement of the string from its normal position, in a direction at right angles to the string.

There is a particularly simple kind of wave known as a sine wave. If we take a snapshot of a sine wave at any moment, we see that the amplitude vanishes at various points along the direction the wave is traveling. Concentrating for a moment on one such point, if we look farther along the direction of travel we will see that the amplitude rises and then falls again to zero, then as we look farther it falls to a negative value and rises again to zero, after which it repeats the whole cycle again and again as we look still farther along the wave's direction. The distance between points at the beginning and end of any one complete cycle is a length characteristic of the wave, known as its wavelength, and conventionally denoted by the symbol λ (lambda). It will be important in what follows that, since the amplitude of the wave vanishes not only at the beginning and end of a cycle but also in the middle, the distance between successive vanishing points is half a wavelength, $\lambda/2$. Any two points where the amplitude vanishes therefore must be separated by some whole number of half wavelengths.

There is a fundamental mathematical theorem (not made explicit until the early nineteenth century) that virtually any disturbance (that is, any disturbance having a sufficiently smooth dependence on distance along the wave) can be expressed as a sum of sine waves with various wavelengths. (This is known as "Fourier analysis.")

Each individual sine wave exhibits a characteristic oscillation in time, as well as in distance along the wave's direction of motion. If the wave is traveling with velocity v, then in time t it travels a distance vt. The number of wavelengths that pass a fixed point in time t will thus be vt/λ, so the number of cycles per second at a given point in which the amplitude and rate of change both keep going back to the same value is v/λ. This is known as the frequency, denoted by the symbol ν (nu), so $\nu = v/\lambda$. The velocity of a wave of vibration of a string is close to a constant, depending on the string tension and mass, but nearly independent of its wavelength or its amplitude, so for these waves (as for light) the frequency is simply inversely proportional to the wavelength.

Now consider a string of some musical instrument, with length L. The amplitude of the wave must vanish at the ends of the string, where the string is held fixed. This condition limits the wavelengths of the individual sine waves that can contribute to the total amplitude of the string's vibration. We have noted that the distance between points where the amplitude of any sine wave vanishes can be any whole number of half wavelengths. Thus the wave on a string that is fixed at both ends must contain a whole number N of half wavelengths, so that $L = N\lambda/2$. That is, the only possible wavelengths are $\lambda = 2L/N$, with $N = 1, 2, 3$, etc., and so the only possible frequencies are*

$$\nu = vN/2L$$

The lowest frequency, for the case $N = 1$, is $v/2L$; all the higher frequencies, for $N = 2$, $N = 3$, etc., are known as the "overtones." For instance, the lowest frequency of the middle C string of any instrument is 261.63 cycles per second, but it also vibrates at 523.26 cycles per second, 784.89 cycles per second, and so on. The intensities of the different overtones make the difference in the qualities of the sounds from different musical instruments.

Now, suppose that vibrations are set up in two strings that have different lengths L_1 and L_2, but are otherwise identical, and in particular have the same wave velocity v. In time t the modes of vibration of the lowest frequency of the first and second strings will go through $n_1 = \nu_1 t = vt/2L_1$ and $n_2 = \nu_2 t = vt/2L_2$ cycles or fractions of cycles, respectively. The ratio is

$$n_1/n_2 = L_2/L_1$$

* For a piano string there are small corrections due to the stiffness of the string; these corrections produce terms in ν proportional to $1/L^3$. I will ignore them here.

Thus, in order for the lowest vibrations of both strings each to go through whole numbers of cycles in the same time, the quantity L_2/L_1 must be a ratio of whole numbers—that is, a rational number. (In this case, in the same time each overtone of each string will also go through a whole number of cycles.) The sound produced by the two strings will thus repeat itself, just as if a single string had been plucked. This seems to contribute to the pleasantness of the sound.

For instance, if $L_2/L_1 = 1/2$, then the vibration of string 2 of lowest frequency will go through two complete cycles for every complete cycle of the corresponding vibration of string 1. In this case, we say that the notes produced by the two strings are an octave apart. All of the different C keys on the piano keyboard produce frequencies that are octaves apart. If $L_2/L_1 = 2/3$, the two strings make a chord called a fifth. For example, if one string produces middle C, at 261.63 cycles per second, then another string that is 2/3 as long will produce middle G, at a frequency $3/2 \times 261.63 = 392.45$ cycles per second.* If $L_2/L_1 = 3/4$, the chord is called a fourth.

The other reason for the pleasantness of these chords has to do with the overtones. In order for the N_1th overtone of string 1 to have the same frequency as the N_2th overtone of string 2, we must have $vN_1/2L_1 = vN_2/2L_2$, and so

$$L_2/L_1 = N_2/N_1$$

Again, the ratio of the lengths is a rational number, though for a different reason. But if this ratio is an irrational number, like π or the square root of 2, then the overtones of the two strings can never match, though the frequencies of high overtones will come arbitrarily close. This apparently sounds terrible.

* In some musical scales middle G is given a slightly different frequency in order to make possible other pleasant chords involving middle G. The adjustment of frequencies to make as many chords as possible pleasant is called "tempering" the scale.

4. The Pythagorean Theorem

The so-called Pythagorean theorem is the most famous result of plane geometry. Although it is believed to be due to a member of the school of Pythagoras, possibly Archytas, the details of its origin are unknown. What follows is the simplest proof, one that makes use of the notion of proportionality commonly used in Greek mathematics.

Consider a triangle with corner points A, B, and P, with the angle at P a right angle. The theorem states that the area of a square whose side is AB (the hypotenuse of the triangle) equals the sum of the areas of squares whose sides are the other two sides of the triangle, AP and BP. In modern algebraic terms, we can think of AB, AP, and BP as numerical quantities, equal to the lengths of these sides, and state the theorem as

$$AB^2 = AP^2 + BP^2$$

The trick in the proof is to draw a line from P to the hypotenuse AB, which intersects the hypotenuse at a right angle, say at point C. (See Figure 2.) This divides triangle ABP into two smaller right triangles, APC and BPC. It is easy to see that both of these smaller triangles are similar to triangle ABP—that is, all their corresponding angles are equal. If we take the angles at the corners A and B to be α (alpha) and β (beta), then triangle ABP has angles α, β, and 90°, so $\alpha + \beta + 90° = 180°$. Triangle APC has two angles equal to α and 90°, so to make the sum of the angles 180° its third angle must be β. Likewise, triangle BPC has two angles equal to β and 90°, so its third angle must be α.

Because these triangles are all similar, their corresponding sides are proportional. That is, AC must be in the same proportion to the hypotenuse AP of triangle ACP that AP has to the hypotenuse AB of the original triangle ABP, and BC must be in the same proportion to BP that BP has to AB. We can put this in more convenient algebraic terms as a statement about ratios of the lengths AC, AP, etc.:

$$\frac{AC}{AP} = \frac{AP}{AB} \qquad \frac{BC}{BP} = \frac{BP}{AB}$$

It follows immediately that $AP^2 = AC \times AB$, and $BP^2 = BC \times AB$. Adding these two equations gives

$$AP^2 + BP^2 = (AC + BC) \times AB$$

But $AC + BC = AB$, so this is the result that was to be proved.

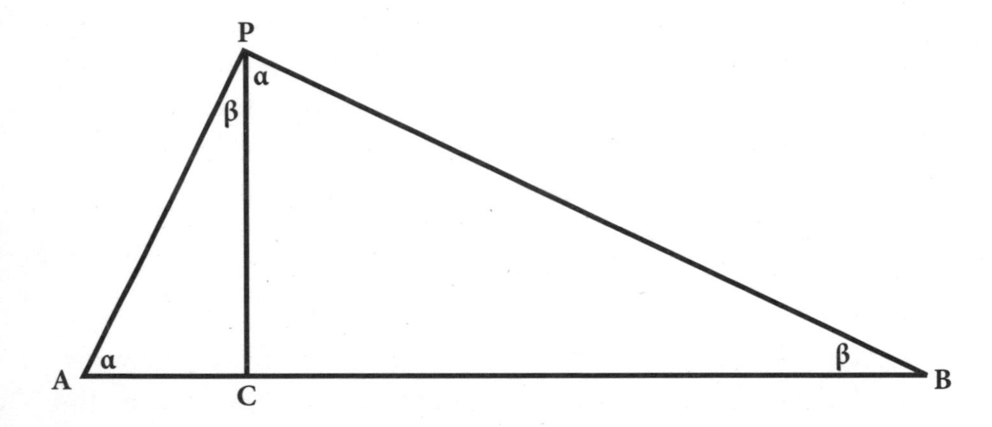

Figure 2. Proof of the Pythagorean theorem. This theorem states that the sum of the areas of two squares whose sides are *AP* and *BP* equals the area of a square whose sides are the hypotenuse *AB*. To prove the theorem, a line is drawn from *P* to a point *C*, which is chosen so that this line is perpendicular to the line from *A* to *B*.

5. *Irrational Numbers*

The only numbers that were familiar to early Greek mathematicians were rational. These are numbers that are either whole numbers, like 1, 2, 3, etc., or ratios of whole numbers, like 1/2, 2/3, etc. If the ratio of the lengths of two lines is a rational number, the lines were said to be "commensurable"—for instance, if the ratio is 3/5, then five times one line has the same length as three times the other. It was therefore shocking to learn that not all lines were commensurable. In particular, in a right isosceles triangle, the hypotenuse is incommensurable with either of the two equal sides. In modern terms, since according to the theorem of Pythagoras the square of the hypotenuse of such a triangle equals twice the square of either of the two equal sides, the length of the hypotenuse equals the length of either of the other sides times the square root of 2, so this amounts to the statement that the square root of 2 is not a rational number. The proof given by Euclid in Book X of the *Elements* consists of assuming the opposite, in modern terms that there is a rational number whose square is 2, and then deriving an absurdity.

Suppose that a rational number p/q (with p and q whole numbers) has a square equal to 2:

$$(p/q)^2 = 2$$

There will then be an infinity of such pairs of numbers, found by multiplying any given p and q by any equal whole numbers, but let us take p and q to be the smallest whole numbers for which $(p/q)^2 = 2$. It follows from this equation that

$$p^2 = 2q^2$$

This shows that p^2 is an even number, but the product of any two odd numbers is odd, so p must be even. That is, we can write $p = 2p'$, where p' is a whole number. But then

$$q^2 = 2p'^2$$

so by the same reasoning as before, q is even, and can therefore be written as $q = 2q'$, where q' is a whole number. But then $p/q = p'/q'$, so

$$(p'/q')^2 = 2$$

with p' and q' whole numbers that are respectively half p and q, contradicting the definition of p and q as the smallest whole numbers for which $(p/q)^2 = 2$. Thus the original assumption, that there are whole numbers p and q for which $(p/q)^2 = 2$, leads to a contradiction, and is therefore impossible.

This theorem has an obvious extension: any number like 3, 5, 6, etc. that is not itself the square of a whole number cannot be the square of a rational number. For instance, if $3 = (p/q)^2$, with p and q the smallest whole numbers for which this holds, then $p^2 = 3q^2$, but this is impossible unless $p = 3p'$ for some whole number p', but then $q^2 = 3p'^2$, so $q = 3q'$ for some whole number q, so $3 = (p'/q')^2$, contradicting the statement that p and q are the smallest whole numbers for which $p^2 = 3q^2$. Thus the square roots of 3, 5, 6 . . . are all irrational.

In modern mathematics we accept the existence of irrational numbers, such as the number denoted $\sqrt{2}$ whose square is 2. The decimal expansion of such numbers goes on forever, without ending or repeating; for example, $\sqrt{2} = 1.414215562\ldots$. The numbers of rational and irrational numbers are both infinite, but in a sense there are far more irrational than rational numbers, for the rational numbers can be listed in an infinite sequence that includes any given rational number:

$$1, 2, 1/2, 3, 1/3, 2/3, 3/2, 4, 1/4, 3/4, 4/3, \cdots$$

while no such list of all irrational numbers is possible.

6. *Terminal Velocity*

To understand how observations of falling bodies might have led Aristotle to his ideas about motion, we can make use of a physical principle unknown to Aristotle, Newton's second law of motion. This principle tells us that the acceleration a of a body (the rate at which its speed increases) equals the total force F acting on the body divided by the body's mass m:

$$a = F/m$$

There are two main forces that act on a body falling through the air. One is the force of gravity, which is proportional to the body's mass:

$$F_{grav} = mg$$

Here g is a constant independent of the nature of the falling body. It equals the acceleration of a falling body that is subject only to gravity, and has the value 32 feet/second per second on and near the Earth's surface. The other force is the resistance of the air. This is a quantity $f(v)$ proportional to the density of the air, which increases with velocity and also depends on the body's shape and size, but does not depend on its mass:

$$F_{air} = -f(v)$$

A minus sign is put in this formula for the force of air resistance because we are thinking of acceleration in a downward direction, and for a falling body the force of air resistance acts upward, so with this minus sign in the formula, $f(v)$ is positive. For instance, for a body falling through a sufficiently viscous fluid, the air resistance is proportional to velocity

$$f(v) = kv$$

with k a positive constant that depends on the body's size and shape. For a meteor or a missile entering the thin air of the upper atmosphere, we have instead

$$f(v) = Kv^2$$

with K another positive constant.

Using the formulas for these forces in the total force $F = F_{grav} + F_{air}$ and using the result in Newton's law, we have

$$a = g - f(v)/m$$

When a body is first released, its velocity vanishes, so there is no air resistance, and its acceleration downward is just g. As time passes its velocity increases, and air resistance begins to reduce its acceleration. Eventually the velocity approaches a value where the term $-f(v)/m$ just cancels the term g in the formula for acceleration, and the acceleration becomes negligible. This is the terminal velocity, defined as the solution of the equation:

$$f(v_{\text{terminal}}) = gm$$

Aristotle never spoke of terminal velocity, but the velocity given by this formula has some of the same properties that he attributed to the velocity of falling bodies. Since $f(v)$ is an increasing function of v, the terminal velocity increases with the mass m. In the special case where $f(v) = kv$, the terminal velocity is simply proportional to the mass and inversely proportional to the air resistance:

$$v_{\text{terminal}} = gm/k$$

But these are not general properties of the velocity of falling bodies; heavy bodies do not reach terminal velocity until they have fallen for a long time.

7. *Falling Drops*

Strato observed that falling drops get farther and farther apart as they fall, and concluded from this that these drops accelerate downward. If one drop has fallen farther than another, then it has been falling longer, and if the drops are separating, then the one that is falling longer must also be falling faster, showing that its fall is accelerating. Though Strato did not know it, the acceleration is constant, and as we shall see, this results in a separation between drops that is proportional to the time elapsed.

As mentioned in Technical Note 6, if air resistance is neglected, then the acceleration downward of any falling body is a constant g, which in the neighborhood of the Earth's surface has the value 32 feet/second per second. If a body falls from rest, then after a time interval τ (tau) its velocity downward will be $g\tau$. Hence if drops 1 and 2 fall from rest from the same downspout at times t_1 and t_2, then at a later time t the speed downward of these drops with be $v_1 = g(t - t_1)$ and $v_2 = g(t - t_2)$, respectively. The difference in their speeds will therefore be

$$v_1 - v_2 = g(t - t_1) - g(t - t_2) = g(t_2 - t_1)$$

Although both v_1 and v_2 are increasing with time, their difference is independent of the time t, so the separation s between the drops simply increases in proportion to the time:

$$s = (v_1 - v_2)t = gt(t_1 - t_2)$$

For instance, if the second drop leaves the downspout a tenth of a second after the first drop, then after half a second the drops will be 32 × 1/2 × 1/10 = 1.6 feet apart.

8. Reflection

The derivation of the law of reflection by Hero of Alexandria was one of the earliest examples of the mathematical deduction of a physical principle from a deeper, more general, principle. Suppose an observer at point A sees the reflection in a mirror of an object at point B. If the observer sees the image of the object at a point P on the mirror, the light ray must have traveled from B to P and then to A. (Hero probably would have said that the light traveled from the observer at A to the mirror and then to the object at *B*, as if the eye reached out to touch the object, but this makes no difference in the argument below.) The problem of reflection is: where on the mirror is *P*?

To answer this question, Hero assumed that light always takes the shortest possible path. In the case of reflection, this implies that *P* should be located so that the total length of the path from *B* to *P* and then to *A* is the shortest path that goes from *B* to anywhere on the mirror and then to *A*. From this, he concluded that angle θ_i (theta$_i$) between the mirror and the incident ray (the line from *B* to the mirror) equals angle θ_r between the mirror and the reflected ray (the line from the mirror to *A*).

Here is the proof of the equal-angles rule. Draw a line perpendicular to the mirror from *B* to a point *B′* that is as far behind the mirror as *B* is in front of the mirror. (See Figure 3.) Suppose that this line intersects the mirror at point *C*. Sides *B′C* and *CP* of right triangle *B′CP* have the same lengths as sides *BC* and *CP* of right triangle *BCP*, so the hypotenuses *B′P* and *BP* of these two triangles must also have the same length. The total distance traveled by the light ray from *B* to *P* and then to *A* is therefore the same as the distance that would be traveled by the light ray if it went from *B′* to *P* and then to *A*. The shortest distance between points *B′* and *A* is a straight line, so the path that minimizes the total distance between the object and the observer is the one for which *P* is on the straight line between *B′* and *A*. When two straight lines intersect, the angles on opposite sides of the intersection point are equal, so angle θ between line *B′P* and the mirror equals angle θ_r between the reflected ray and the mirror. But because the two right triangles *B′CP* and *BCP* have the same sides, angle θ must also equal angle θ_i between the incident ray *BP* and the mirror. So, since both θ_i and θ_r are equal to θ, they are equal to each other. This is the fundamental equal-angles rule that determines the location *P* on the mirror of the image of the object.

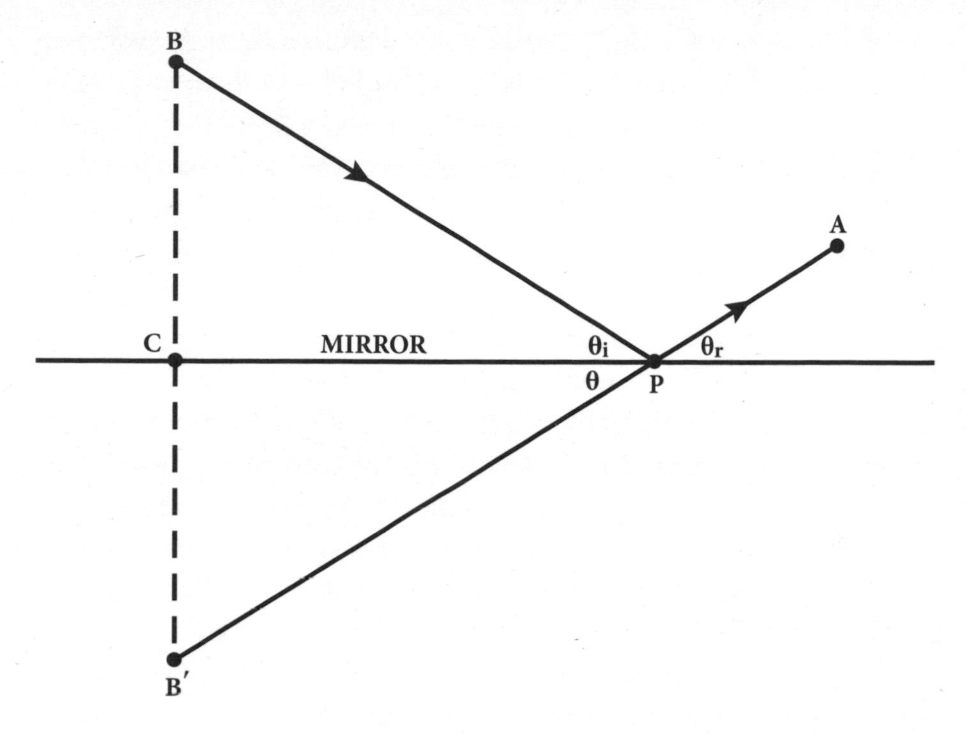

Figure 3. Proof of Hero's theorem. This theorem states that the shortest path from an object at B to the mirror and then to an eye at A is one for which angles θ_i and θ_r are equal. The solid lines marked with arrows represent the path of a light ray; the horizontal line is the mirror; and the dashed line is a line perpendicular to the mirror that runs from B to a point B' on the other side of the mirror at an equal distance from it.

9. Floating and Submerged Bodies

In his great work *On Floating Bodies*, Archimedes assumed that if bodies are floating or suspended in water in such a way that equal areas at equal depths in the water are pressed down by different weights, then the water and the bodies will move until all equal areas at any given depth are pressed down by the same weight. From this assumption, he derived general consequences about both floating and submerged bodies, some of which were even of practical importance.

First, consider a body like a ship whose weight is less than the weight of an equal volume of water. The body will float on the surface of the water, and displace some quantity of water. If we mark out a horizontal patch in the water at some depth directly below the floating body, with an area equal to the area of the body at its waterline, then the weight pressing down on this surface will be the weight of the floating body plus the weight of the water above that patch, but not including the water displaced by the body, as this water is no longer above the patch. We can compare this with the weight pressing down on an equal area at an equal depth, away from the location of the floating body. This of course does not include the weight of the floating body, but it does include all the water from this patch to the surface, with no water displaced. In order for both patches to be pressed down by the same weight, the weight of the water displaced by the floating body must equal the weight of the floating body. This is why the weight of a ship is referred to as its "displacement."

Next consider a body whose weight is greater than the weight of an equal volume of water. Such a body will not float, but it can be suspended in the water from a cable. If the cable is attached to one arm of a balance, then in this way we can measure the apparent weight $W_{apparent}$ of the body when submerged in water. The weight pressing down on a horizontal patch in the water at some depth directly below the suspended body will equal the true weight W_{true} of the suspended body, less the apparent weight $W_{apparent}$, which is canceled by the tension in the cable, plus the weight of the water above the patch, which of course does not include the water displaced by the body. We can compare this with the weight pressing down on an equal area at an equal depth, a weight that does not include W_{true} or $-W_{apparent}$, but does include the weight of all the water from this patch to the surface, with no water displaced. In order for both patches to be pressed down by the same weight, we must have

$$W_{true} - W_{apparent} = W_{displaced}$$

where $W_{\text{displaced}}$ is the weight of the water displaced by the suspended body. So by weighing the body when suspended in the water and weighing it when out of the water, we can find both W_{apparent} and W_{true}, and in this way find $W_{\text{displaced}}$. If the body has volume V, then

$$W_{\text{displaced}} = \rho_{\text{water}} V$$

where ρ_{water} ($\text{rho}_{\text{water}}$) is the density (weight per volume) of water, close to 1 gram per cubic centimeter. (Of course, for a body with a simple shape like a cube we could instead find V by just measuring the dimensions of the body, but this is difficult for an irregularly shaped body like a crown.) Also, the true weight of the body is

$$W_{\text{true}} = \rho_{\text{body}} V$$

where ρ_{body} is the density of the body. The volume cancels in the ratio of W_{true} and $W_{\text{displaced}}$, so from the measurements of both W_{apparent} and W_{true} we can find the ratio of the densities of the body and of water:

$$\frac{\rho_{\text{body}}}{\rho_{\text{water}}} = \frac{W_{\text{true}}}{W_{\text{displaced}}} = \frac{W_{\text{true}}}{W_{\text{true}} - W_{\text{apparent}}}$$

This ratio is called the "specific gravity" of the material of which the body is composed. For instance, if the body weighs 20 percent less in water than in air, then $W_{\text{true}} - W_{\text{apparent}} = 0.20 \times W_{\text{true}}$, so its density must be $1/0.2 = 5$ times the density of water. That is, its specific gravity is 5.

There is nothing special about water in this analysis; if the same measurements were made for a body suspended in some other liquid, then the ratio of the true weight of the body to the decrease in its weight when suspended in the liquid would give the ratio of the density of the body to the density of that liquid. This relation is sometimes used with a body of known weight and volume to measure the densities of various liquids in which the body may be suspended.

10. Areas of Circles

To calculate the area of a circle, Archimedes imagined that a polygon with a large number of sides was circumscribed outside the circle. For simplicity, let's consider a regular polygon, all of whose sides and angles are equal. The area of the polygon is the sum of the areas of all the right triangles formed by drawing lines from the center to the corners of the polygon, and lines from the center to the midpoints of the sides of the polygon. (See Figure 4, in which the polygon is taken to be a regular octagon.) The area of a right triangle is half the product of its two sides around the right angle, because two such triangles can be stacked on their hypotenuses to make a rectangle, whose area is the product of the sides. In our case, this means that the area of each triangle is half the product of the distance r to the midpoint of the side (which is just the radius of the circle) and the distance s from the midpoint of the side to the nearest corner of the polygon, which of course is half the length of that side of the polygon. When we add up all these areas, we find that the area of the whole polygon equals half of r times the total circumference of the polygon. If we let the number of sides of the polygon become infinite, its area approaches the area of the circle, and its circumference approaches the circumference of the circle. So the area of the circle is half its circumference times its radius.

In modern terms, we define a number $\pi = 3.14159\ldots$ such that the circumference of a circle of radius r is $2\pi r$. The area of the circle is thus

$$1/2 \times r \times 2\pi r = \pi r^2$$

The same argument works if we inscribe polygons within the circle, rather than circumscribing them outside the circle as in

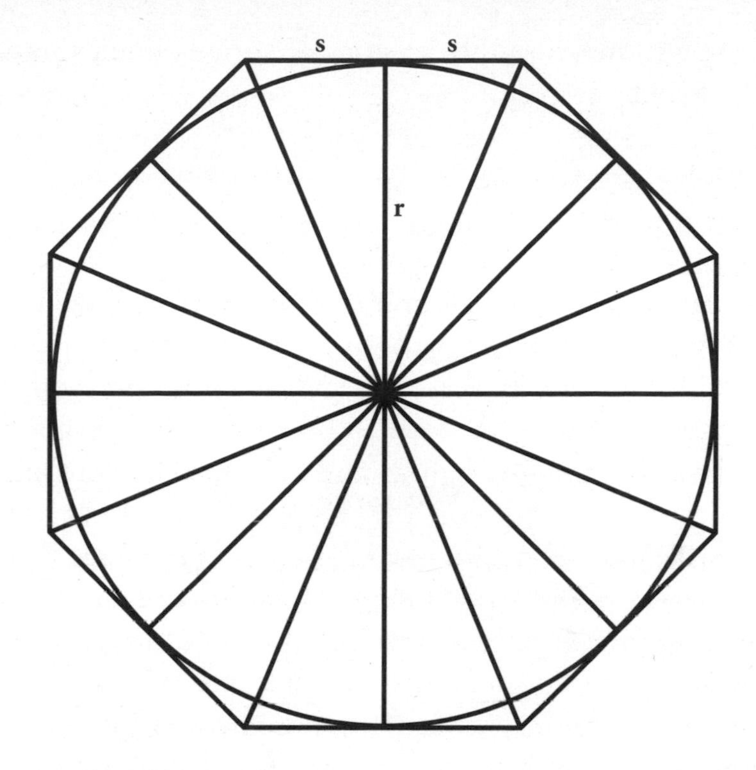

Figure 4. Calculation of the area of a circle. In this calculation a polygon with many sides is circumscribed about a circle. Here, the polygon has eight sides, and its area is already close to the area of the circle. As more sides are added to the polygon, its area becomes closer and closer to the area of the circle.

Figure 4. Since the circle is always between an outer polygon circumscribed around it and an inner polygon inscribed within it, using polygons of both sorts allowed Archimedes to give upper and lower limits for the ratio of the circumference of a circle to its radius—in other words, for 2π.

11. Sizes and Distances of the Sun and Moon

Aristarchus used four observations to determine the distances from the Earth to the Sun and Moon and the diameters of the Sun and Moon, all in terms of the diameter of the Earth. Let's look at each observation in turn, and see what can be learned from it. Below, d_s and d_m are the distances from the Earth to the Sun and Moon, respectively; and D_s, D_m, and D_e denote the diameters of the Sun, Moon, and Earth. We will assume that the diameters are negligible compared with the distances, so in talking of the distance from the Earth to the Moon or Sun, it is unnecessary to specify points on the Earth, Moon, or Sun from which the distances are measured.

Observation 1

When the Moon is half full, the angle between the lines of sight from the Earth to the Moon and to the Sun is 87°.

When the Moon is half full, the angle between the lines of sight from the Moon to the Earth and from the Moon to the Sun must be just 90° (see Figure 5a), so the triangle formed by the lines Moon–Sun, Moon–Earth, and Earth–Sun is a right triangle, with the Earth–Sun line as the hypotenuse. The ratio between the side adjacent to an angle θ (theta) of a right triangle and the hypotenuse is a trigonometric quantity known as the cosine of θ, abbreviated $\cos \theta$, which we can look up in tables or find on any scientific calculator. So we have

$$d_m/d_s = \cos 87° = 0.05234 = 1/19.11$$

and this observation indicates that the Sun is 19.11 times farther from the Earth than is the Moon. Not knowing trigonometry, Aristarchus could only conclude that this number is between 19 and 20. (The angle actually is not 87°, but 89.853°, and the Sun is really 389.77 times farther from the Earth than is the Moon.)

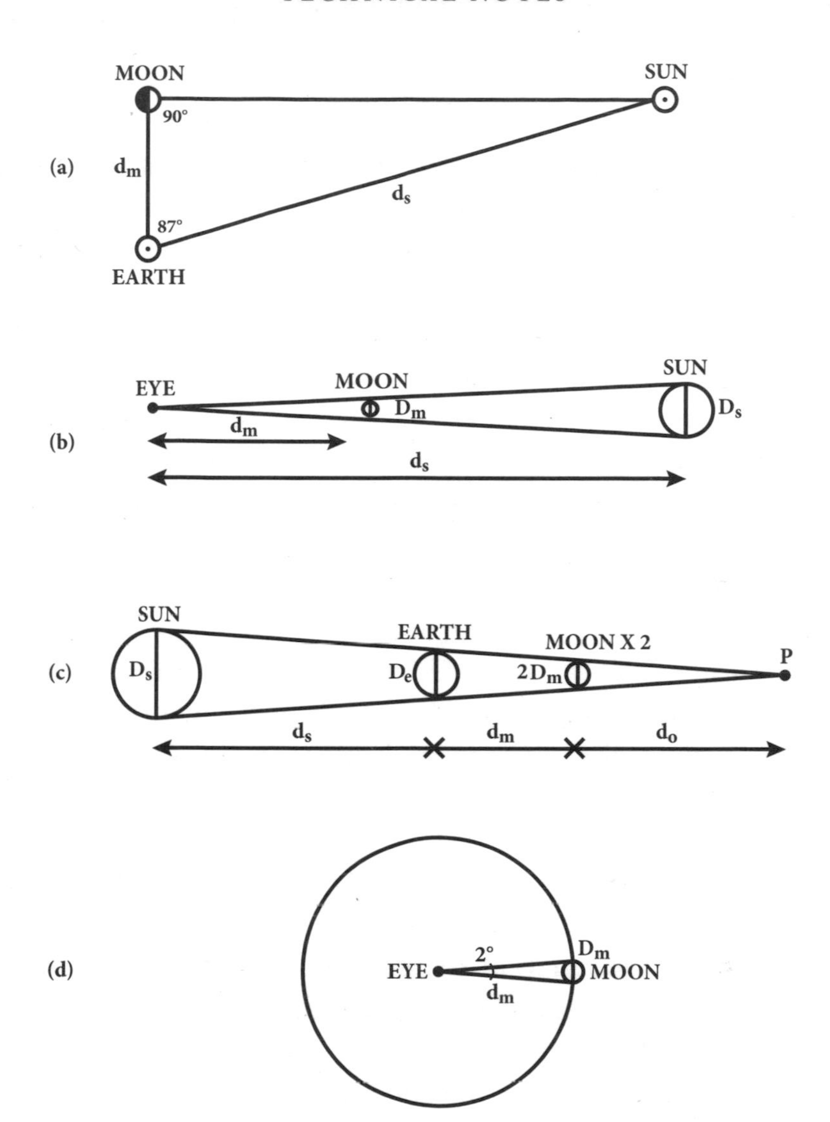

Figure 5. The four observations used by Aristarchus to calculate the sizes and distances of the Sun and Moon. (a) The triangle formed by the Earth, Sun, and Moon when the Moon is half full. (b) The Moon just blotting out the disk of the Sun during a total eclipse of the Sun. (c) The Moon passing into the shadow of the Earth during an eclipse of the Moon. The sphere that just fits into this shadow has a diameter twice that of the Moon, and *P* is the terminal point of the shadow cast by the Earth. (d) Lines of sight to the Moon spanning an angle of 2°; the actual angle is close to 0.5°.

with D_s, then all dependence of D_s cancels, and we have simply $D_e = 3D_m$, which is also not so far from the truth.

Of much greater historical importance is the fact that if we combine the results $D_s/D_m = 19.1$ and $D_e/D_m = 2.85$, we find $D_s/D_e = 19.1/2.85 = 6.70$. The actual value is $D_s/D_e = 109.1$, but the important thing is that the Sun is considerably bigger than the Earth. Aristarchus emphasized the point by comparing the volumes rather than the diameters; if the ratio of diameters is 6.7, then the ratio of volumes is $6.7^3 = 301$. It is this comparison that, if we believe Archimedes, led Aristarchus to conclude that the Earth goes around the Sun, not the Sun around the Earth.

The results of Aristarchus described so far yield values for all ratios of diameters of the Sun, Moon, and Earth, and the ratio of the distances to the Sun and Moon. But nothing so far gives us the ratio of any distance to any diameter. This was provided by the fourth observation:

Observation 4

The Moon subtends an angle of 2°.

(See Figure 5d.) Since there are 360° in a full circle, and a circle whose radius is d_m has a circumference $2\pi d_m$, the diameter of the Moon is

$$D_m = \left(\frac{2}{360}\right) \times 2\pi d_m = 0.035 d_m$$

Aristarchus calculated that the value of D_m/d_m is between 2/45 = 0.044 and 1/30 = 0.033. For unknown reasons Aristarchus in his surviving writings had grossly overestimated the true angular size of the Moon; it actually subtends an angle of 0.519°, giving $D_m/d_m = 0.0090$. As we noted in Chapter 8, Archimedes in *The Sand Reckoner* gave a value of 0.5° for the angle subtended by the Moon, which is quite close to the true value and would have given an accurate estimate of the ratio of the diameter and distance of the Moon.

With his results from observations 2 and 3 for the ratio D_e/D_m of the diameters of the Earth and Moon, and now with his result from observation 4 for the ratio D_m/d_m of the diameter and distance of the Moon, Aristarchus could find the ratio of the distance of the Moon to the diameter of the Earth. For instance, taking $D_e/D_m = 2.85$ and $D_m/d_m = 0.035$ would give

$$d_m/D_e = \frac{1}{D_e/D_m \times D_m/d_m} = \frac{1}{2.85 \times 0.035} = 10.0$$

(The actual value is about 30.) This could then be combined with the result of observation 1 for the ratio $d_s/d_m = 19.1$ of the distances to the Sun and Moon, giving a value of $d_s/D_e = 19.1 \times 10.0 = 191$ for the ratio of the distance to the Sun and the diameter of the Earth. (The actual value is about 11,600.) Measuring the diameter of the Earth was the next task.

12. *The Size of the Earth*

Eratosthenes used the observation that at noon on the summer solstice, the Sun at Alexandria is 1/50 of a full circle (that is, 360°/50 = 7.2°) away from the vertical, while at Syene, a city supposedly due south of Alexandria, the Sun at noon on the summer solstice was reported to be directly overhead. Because the Sun is so far away, the light rays striking the Earth at Alexandria and Syene are essentially parallel. The vertical direction at any city is just the continuation of a line from the center of the Earth to that city, so the angle between the lines from the Earth's center to Syene and to Alexandria must also be 7.2°, or 1/50 of a full circle. (See Figure 6.) Hence on the basis of the assumptions of Eratosthenes, the Earth's circumference must be 50 times the distance from Alexandria to Syene.

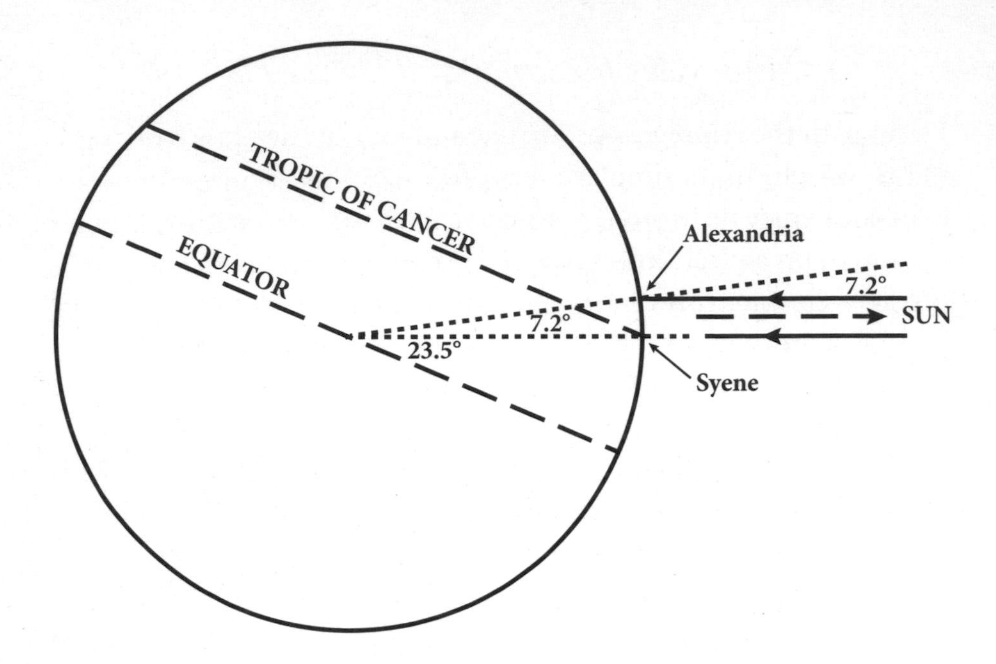

Figure 6. The observation used by Eratosthenes to calculate the size of the Earth. The horizontal lines marked with arrows indicate rays of sunlight at the summer solstice. The dotted lines run from the Earth's center to Alexandria and Syene, and mark the vertical direction at each place.

Syene is not on the Earth's equator, as might be suggested by the way the figure is drawn, but rather close to the Tropic of Cancer, the line at latitude 23½°. (That is, the angle between lines from the Earth's center to any point on the Tropic of Cancer and to a point due south on the equator is 23½°.) At the summer solstice the Sun is directly overhead at noon on the Tropic of Cancer rather than on the equator because the Earth's axis of rotation is not perpendicular to the plane of its orbit, but tilted from the perpendicular by an angle of 23½°.

13. *Epicycles for Inner and Outer Planets*

Ptolemy in the *Almagest* presented a theory of the planets according to which, in its simplest version, each planet goes on a circle called an epicycle around a point in space that itself goes around the Earth on a circle known as the planet's deferent. The question before us is why this theory worked so well in accounting for the apparent motions of the planets as seen from Earth. The answer for the inner planets, Mercury and Venus, is different from that for the outer planets, Mars, Jupiter, and Saturn.

First, consider the inner planets, Mercury and Venus. According to our modern understanding, the Earth and each planet go around the Sun at approximately constant distances from the Sun and at approximately constant speeds. If we do not concern ourselves with the laws of physics, we can just as well change our point of view to one centered on the Earth. From this point of view, the Sun goes around the Earth, and each planet goes around the Sun, all at constant speeds and distances. This is a simple version of the theory due to Tycho Brahe, which may also have been proposed by Heraclides. It gives the correct apparent motions of the planets, apart from small corrections due to the facts that planets actually move on nearly circular elliptical orbits rather than on circles, the Sun is not at the centers of these ellipses but at relatively small distances from the centers, and the speed of each planet varies somewhat as the planet goes around its orbit. It is also a special case of the theory of Ptolemy, though one never considered by Ptolemy, in which the deferent is nothing but the orbit of the Sun around the Earth, and the epicycle is the orbit of Mercury or Venus around the Sun.

Now, as far as the apparent position in the sky of the Sun and planets is concerned, we can multiply the changing distance of any planet from the Earth by a constant, without changing appearances. This can be done, for instance, by multiplying the radii of both the epicycle and the deferent by the same constant factor, chosen independently for Mercury and Venus. For instance, we could take the radius of the deferent of Venus to be half the distance of the Sun from the Earth, and the radius of its epicycle to be half the radius of the orbit of Venus around the Sun. This will not change the fact that the centers of the planets' epicycles always stay on the line between the Earth and the Sun. (See Figure 7a, which shows the epicycle and deferent for one of the inner planets, not drawn to scale.) The apparent motion of Venus and Mercury in the sky will be unchanged by this transformation, as long as we don't change the ratio of the radii of each planet's deferent and epicycle. This is a simple version of the theory proposed by Ptolemy for the inner planets. According to this theory, the planet goes around its epicycle in the same time that it actually takes to go around the Sun, 88 days for Mercury and 225 days for Venus, while the center of the epicycle follows the Sun around the Earth, taking one year for a complete circuit of the deferent.

Specifically, since we do not change the ratio of the radii of the deferent and epicycle, we must have

$$r_{\mathrm{EPI}}/r_{\mathrm{DEF}} = r_{\mathrm{P}}/r_{\mathrm{E}}$$

where r_{EPI} and r_{DEF} are the radii of the epicycle and deferent in Ptolemy's scheme, and r_P and r_E are the radii of the orbits of the planet and the Earth in the theory of Copernicus (or equivalently, the radii of the orbits of the planet around the Sun and the Sun around the Earth in the theory of Tycho). Of course, Ptolemy knew nothing of the theories of Tycho or Copernicus, and he did not obtain his theory in this way. The discussion above serves to show only why Ptolemy's theory worked so well, not how he derived it.

Now, let us consider the outer planets, Mars, Jupiter, and Saturn. In the simplest version of the theory of Copernicus (or Tycho) each planet keeps a fixed distance not only from the Sun, but also from a moving point C' in space, which keeps a fixed distance from the Earth. To find this point, draw a parallelogram (Figure 7b), whose first three vertices in order around it are the position S of the Sun, the position E of the Earth, and the position P' of one of the planets. The moving point C' is the empty

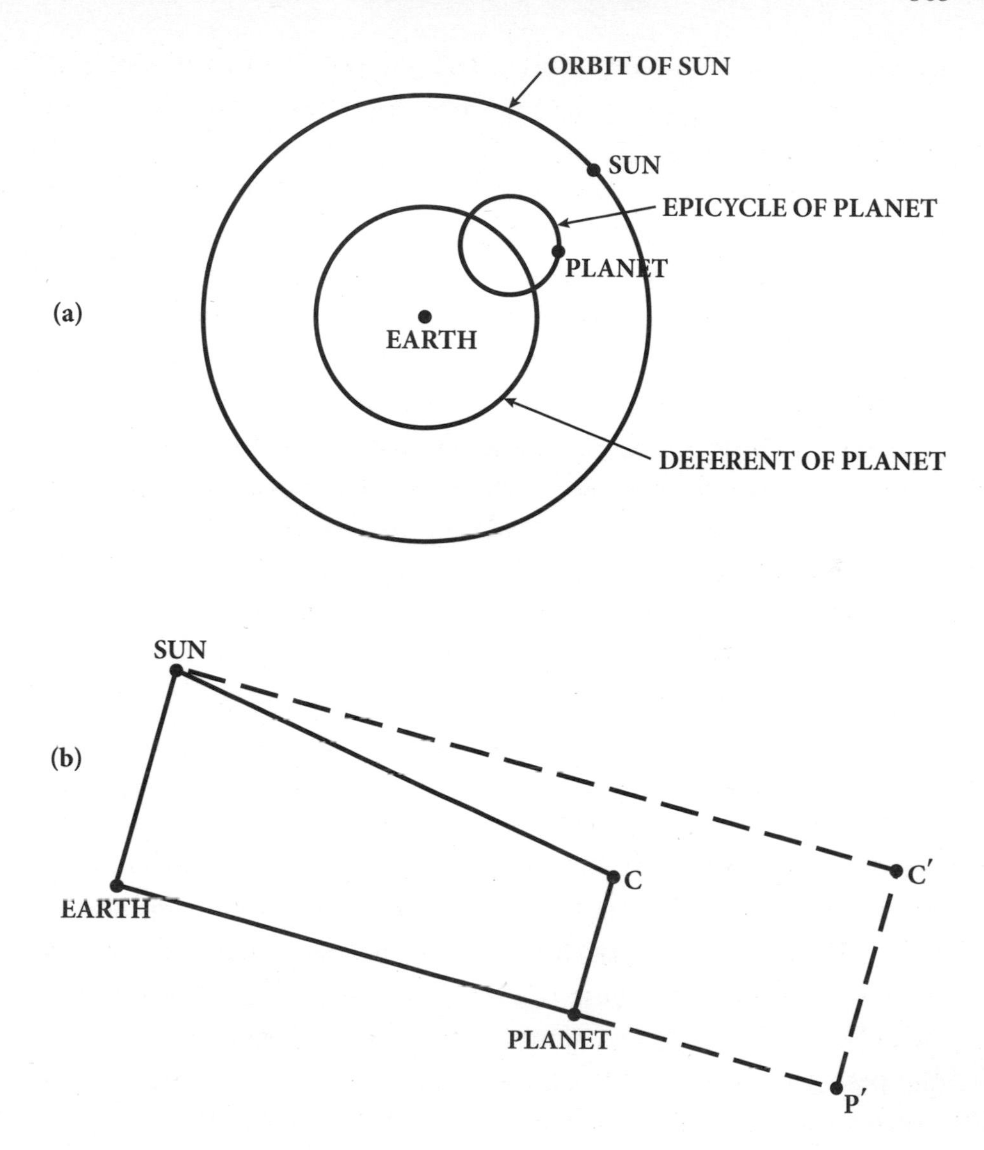

Figure 7. A simple version of the epicycle theory described by Ptolemy. (a) The supposed motion of one of the inner planets, Mercury or Venus. (b) The supposed motion of one of the outer planets Mars, Jupiter, or Saturn. Planet *P* goes on an epicycle around point *C* in one year, with the line from *C* to *P* always parallel to the line from the Earth to the Sun, while point *C* goes around the Earth on the deferent in a longer time. (The dashed lines indicate a special case of the Ptolemaic theory, for which it is equivalent to that of Copernicus.)

fourth corner of the parallelogram. Since the line between E and S has a fixed length, and the line between P' and C' is the opposite side of the parallelogram, it has an equal fixed length, so the planet stays at a fixed distance from C', equal to the distance of the Earth from the Sun. Likewise, since the line between S and P' has a fixed length, and the line between E and C' is the opposite side of the parallelogram, it has an equal fixed length, so point C' stays at a fixed distance from the Earth, equal to the distance of the planet from the Sun. This is a special case of the theory of Ptolemy, though a case never considered by him, in which the deferent is nothing but the orbit of point C' around the Earth, and the epicycle is the orbit of Mars, Jupiter, or Saturn around C'.

Once again, as far as the apparent position in the sky of the Sun and planets is concerned, we can multiply the changing distance of any planet from the Earth by a constant without changing appearances, by multiplying the radii of both the epicycle and the deferent by a constant factor, chosen independently for each outer planet. Although we no longer have a parallelogram, the line between the planet and C remains parallel to the line between the Earth and Sun. The apparent motion of each outer planet in the sky will be unchanged by this transformation, as long as we don't change the ratio of the radii of each planet's deferent and epicycle. This is a simple version of the theory proposed by Ptolemy for the outer planets. According to this theory, the planet goes around C on its epicycle in 1 year, while C goes around the deferent in the time that it actually takes the planet to go around the Sun: 1.9 years for Mars, 12 years for Jupiter, and 29 years for Saturn.

Specifically, since we do not change the ratio of the radii of the deferent and epicycle, we must now have

$$r_{\mathrm{EPI}}/r_{\mathrm{DEF}} = r_{\mathrm{E}}/r_{\mathrm{P}}$$

where r_{EPI} and r_{DEF} are again the radii of the epicycle and deferent in Ptolemy's scheme, and r_P and r_E are the radii of the orbits of the planet and the Earth in the theory of Copernicus (or equivalently, the radii of the orbits of the planet around the Sun and the Sun around the Earth in the theory of Tycho). Once again, the above discussion describes, not how Ptolemy obtained his theory, but only why this theory worked so well.

14. Lunar Parallax

Suppose that the direction to the Moon is observed from point O on the surface of the Earth to be at angle ζ' (zeta prime) to the zenith at O. The Moon moves in a smooth and regular way around the center of the Earth, so by using the results of repeated observations of the Moon it is possible to calculate the direction from the center C of the Earth to the Moon *M* at the same moment, and in particular to calculate angle ζ between the direction from C to the Moon and the direction of the zenith at O, which is the same as the direction of the line from the center of the Earth to O. Angles ζ and ζ' differ slightly because the radius r_e of the earth is not entirely negligible compared with the distance *d* of the Moon from the center of the Earth, and from this difference Ptolemy could calculate the ratio d/r_e.

Points C, O, and *M* form a triangle, in which the angle at C is ζ, the angle at O is $180° - \zeta'$, and (since the sum of the angles of any triangle is 180°) the angle at *M* is $180° - \zeta - (180° - \zeta') = \zeta' - \zeta$. (See Figure 8.) We can calculate the ratio d/r_e from these angles much more easily than Ptolemy did, by using a theorem of modern trigonometry: that in any triangle the lengths of sides are proportional to the sines of the opposite angles. (Sines are discussed in Technical Note 15.) The angle opposite the line of length r_e from C to O is $\zeta' - \zeta$, and the angle opposite the line of length *d* from C to *M* is $180° - \zeta'$, so

$$\frac{d}{r_e} = \frac{\sin(180° - \zeta')}{\sin(\zeta' - \zeta)} = \frac{\sin(\zeta')}{\sin(\zeta' - \zeta)}$$

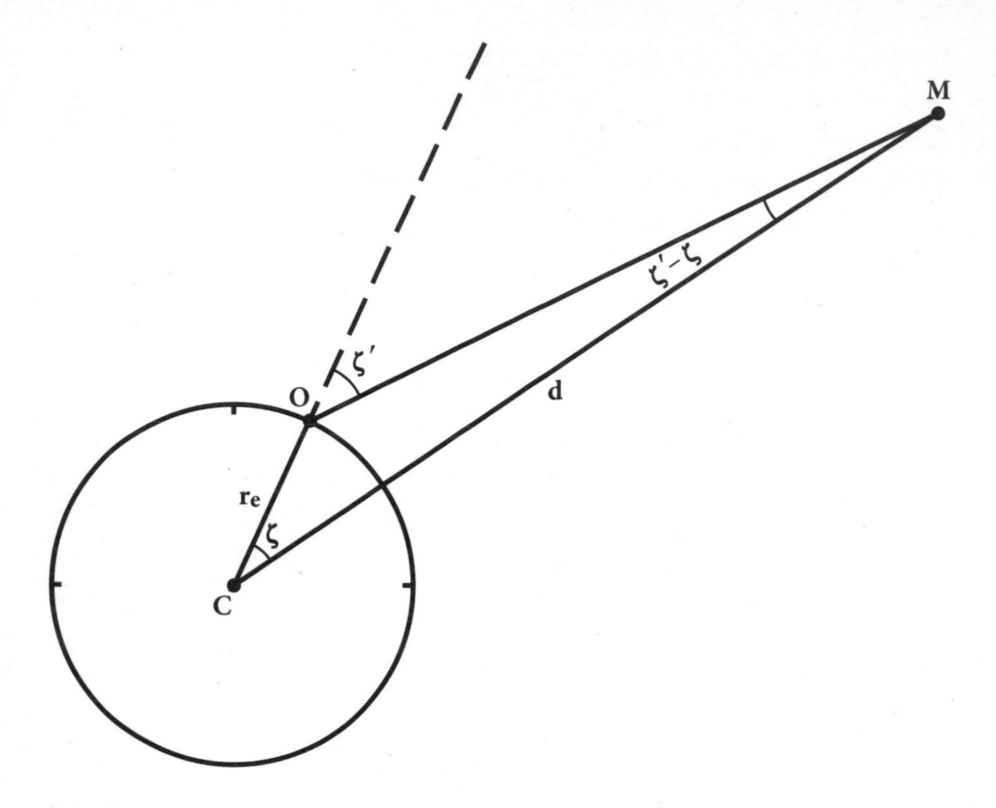

Figure 8. Use of parallax to measure the distance to the Moon. Here ζ' is the observed angle between the line of sight to the Moon and the vertical direction, and ζ is the value this angle would have if the Moon were observed from the center of the Earth.

On October 1, AD 135, Ptolemy observed that the zenith angle of the Moon as seen from Alexandria was $\zeta' = 50°55'$, and his calculations showed that at the same moment the corresponding angle that would be observed from the center of the Earth was $\zeta = 49°48'$. The relevant sines are

$$\sin\zeta' = 0.776 \qquad \sin(\zeta' - \zeta) = 0.0195$$

From this Ptolemy could conclude that the distance from the center of the Earth to the Moon in units of the radius of the Earth is

$$\frac{d}{r_e} = \frac{0.776}{0.0195} = 39.8$$

This was considerably less than the actual ratio, which on average is about 60. The trouble was that the difference $\zeta' - \zeta$ was not actually known accurately by Ptolemy, but at least it gave a good idea of the order of magnitude of the distance to the Moon.

Anyway, Ptolemy did better than Aristarchus, from whose values for the ratio of the diameters of the Earth and Moon and of the distance and diameter of the Moon he could have inferred that d/r_e is between 215/9 = 23.9 and 57/4 = 14.3. But if Aristarchus had used a correct value of about 1/2° for the angular diameter of the Moon's disk instead of his value of 2° he would have found d/r_e to be four times greater, between 57.2 and 95.6. That range includes the true value.

15. Sines and Chords

The mathematicians and astronomers of antiquity could have made great use of a branch of mathematics known as trigonometry, which is taught today in high schools. Given any angle of a right triangle (other than the right angle itself) trigonometry tells us how to calculate the ratios of the lengths of all the sides. In particular, the side opposite the angle divided by the hypotenuse is a quantity known as the "sine" of that angle, which can be found by looking it up in mathematical tables or typing the angle in a hand calculator and pressing "sin." (The side of the triangle adjacent to an angle divided by the hypotenuse is the "cosine" of the angle, and the side opposite divided by the side adjacent is the "tangent" of the angle, but it will be enough for us to deal here with sines.) Though no notion of a sine appears anywhere in Hellenistic mathematics, Ptolemy's *Almagest* makes use of a related quantity, known as the "chord" of an angle.

To define the chord of an angle θ (theta), draw a circle of radius 1 (in whatever units of length you find convenient), and draw two radial lines from the center to the circumference, separated by that angle. The chord of the angle is the length of the straight line, or chord, that connects the points where the two radial lines intersect the circumference. (See Figure 9.) The *Almagest* gives a table of chords* in a Babylonian sexigesimal notation, with angles expressed in degrees of arc, running from 1/2° to 180°. For instance, the chord of 45° is given as 45 15 19, or in modern notation

$$\frac{45}{60} + \frac{55}{60^2} + \frac{19}{60^3} = 0.7653658\ldots$$

while the true value is 0.7653669. . . .

The chord has a natural application to astronomy. If we imagine the stars as lying on a sphere of radius equal to 1, centered on the center of the Earth, then if the lines of sight to two stars are separated by angle θ, the apparent straight-line distance between the stars will be the chord of θ.

To see what these chords have to do with trigonometry, return to the figure used to define the chord of angle θ, and draw a line (the dashed line in Figure 9) from the center of the circle that just bisects the chord. We then have two right triangles, each with an angle at the center of the circle equal to $\theta/2$, and a side opposite this angle whose length is half the chord. The hypotenuse of each of these triangles is the radius of the circle, which we are taking to be 1, so the sine of $\theta/2$, in mathematical notation $\sin(\theta/2)$, is half the chord of θ, or:

$$\text{chord of } \theta = 2\sin(\theta/2)$$

Hence any calculation that can be done with sines can also be done with chords, though in most cases less conveniently.

* This table appears in the translation of the *Almagest* by G. J. Toomer (*Ptolemy's Almagest*, Duckworth, London, 1984, pp. 57–60).

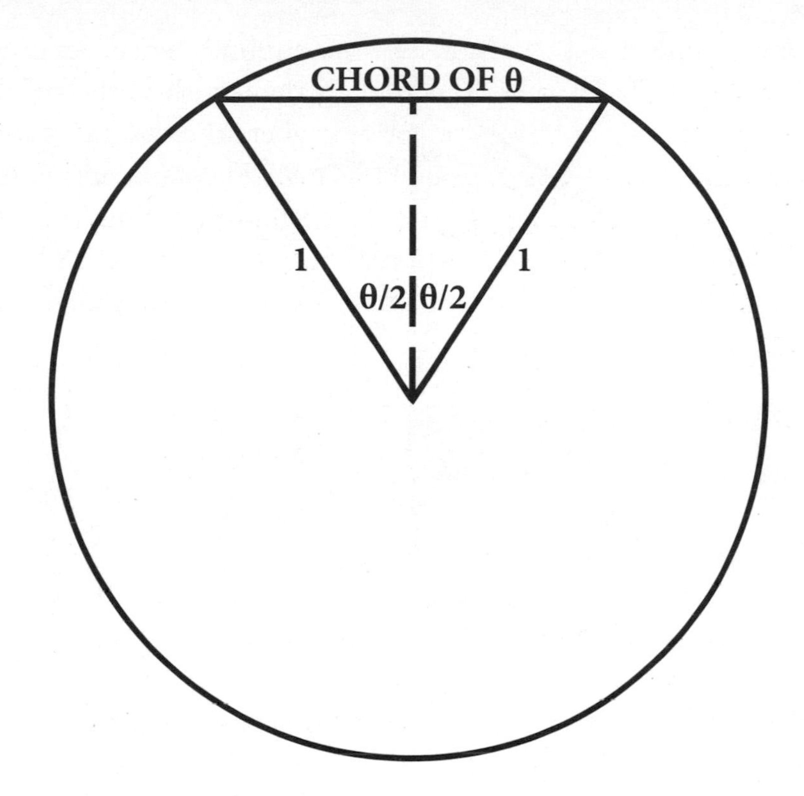

Figure 9. The chord of an angle *θ*. The circle here has a radius equal to 1. The solid radial lines make an angle θ at the center of the circle; the horizontal line runs between the intersections of these lines with the circle; and its length is the chord of this angle.

16. Horizons

Normally our vision outdoors is obstructed by nearby trees or houses or other obstacles. From the top of a hill, on a clear day, we can see much farther, but the range of our vision is still restricted by a horizon, beyond which lines of sight are blocked by the Earth itself. The Arab astronomer al-Biruni described a clever method for using this familiar phenomenon to measure the radius of the Earth, without needing to know any distances other than the height of the mountain.

An observer O on top of a hill can see out to a point H on the Earth's surface, where the line of sight is tangent to the surface.

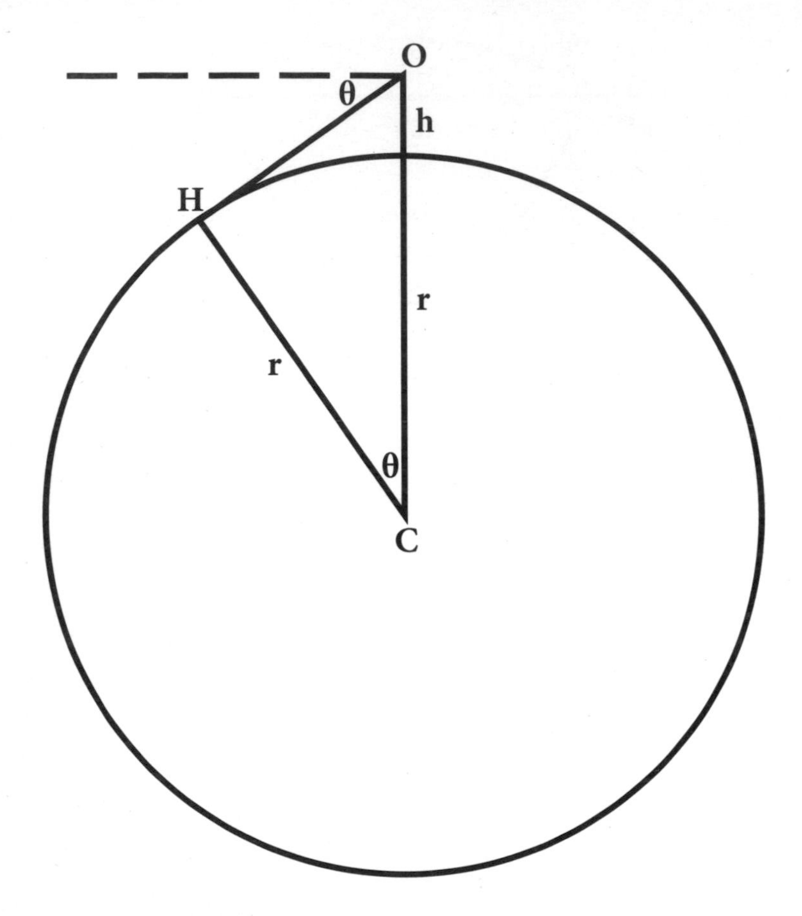

Figure 10. Al-Biruni's use of horizons to measure the size of the Earth. *O* is an observer on a hill of height *h*; *H* is the horizon as seen by this observer; the line from *H* to *O* is tangent to the Earth's surface at *H*, and therefore makes a right angle with the line from the center *C* of the Earth to *H*.

(See Figure 10.) This line of sight is at right angles to the line joining *H* to the Earth's center *C*, so triangle *OCH* is a right triangle. The line of sight is not in the horizontal direction, but below the horizontal direction by some angle θ, which is small because the Earth is large and the horizon far away. The angle between the line of sight and the vertical direction at the hill is then $90° - \theta$, so since the sum of the angles of any triangle must be 180°, the acute angle of the triangle at the center of the Earth is $180° - 90° - (90° - \theta) = \theta$. The side of the triangle adjacent to this angle is the line from C to H, whose length is the Earth's radius r, while the hypotenuse of the triangle is the distance from C to O, which is $r + h$, where h is the height of the mountain. According to the general definition of the cosine, the cosine of any angle is the ratio of the adjacent side to the hypotenuse, which here gives

$$\cos\theta = \frac{r}{r+h}$$

To solve this equation for r, note that its reciprocal gives $1 + h/r = 1/\cos\theta$, so by subtracting 1 from this equation and then taking the reciprocal again we have

$$r = \frac{h}{1/\cos\theta - 1}$$

For instance, on a mountain in India al-Biruni found $\theta = 34'$, for which $\cos\theta = 0.999951092$ and $1/\cos\theta - 1 = 0.0000489$. Hence

$$r = h/0.0000489 = 20{,}450\ h$$

Al-Biruni reported that the height of this mountain is 652.055 cubits (a precision much greater than he could possibly have achieved), which then actually gives r = 13.3 million cubits, while his reported result was 12.8 million cubits. I don't know the source of al-Biruni's error.

17. Geometric Proof of the Mean Speed Theorem

Suppose we make a graph of speed versus time during uniform acceleration, with speed on the vertical axis and time on the horizontal axis. The graph will be a straight line, rising from zero speed at zero time to the final speed at the final time. In each tiny interval of time, the distance traveled is the product of the speed at that time (this speed changes by a negligible amount during that interval if the interval is short enough) times the time inter-

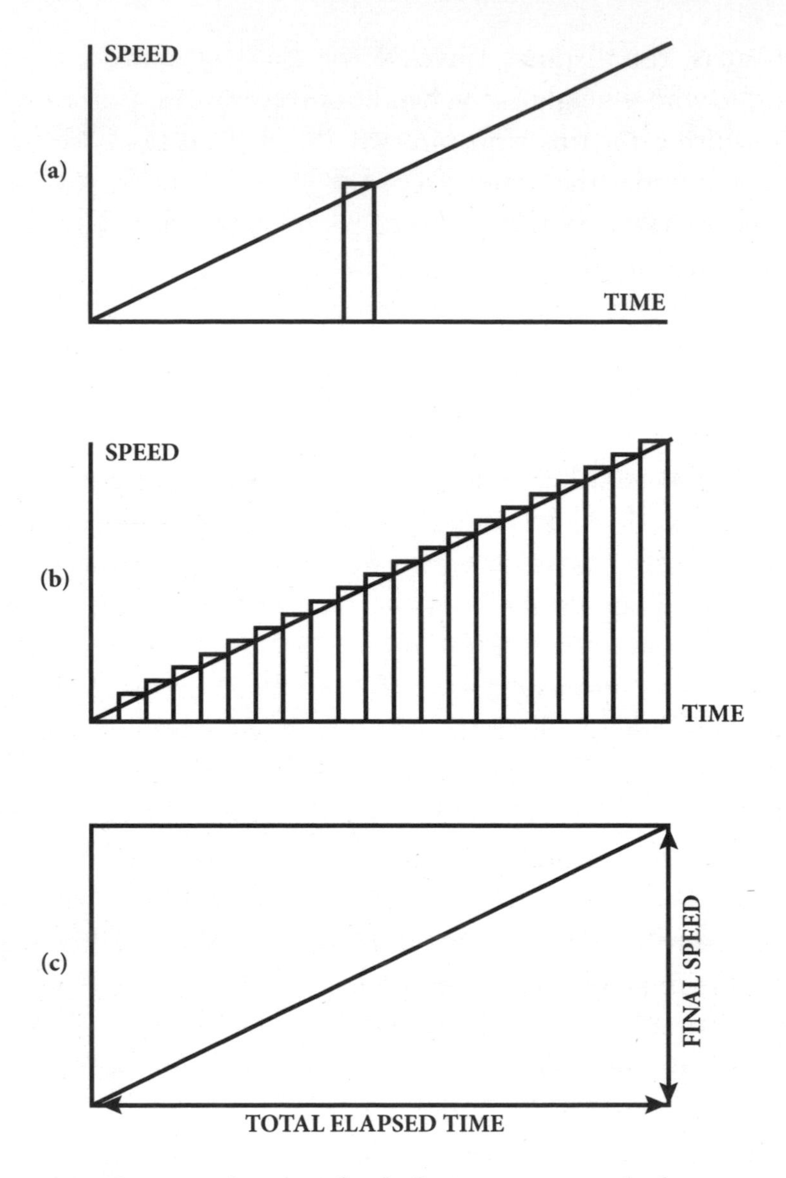

Figure 11. Geometric proof of the mean speed theorem. The slanted line is the graph of speed versus time for a body uniformly accelerated from rest. (a) The width of the small rectangle is a short time interval; its area is close to the distance traveled in that interval. (b) Time during a period of uniform acceleration, broken into short intervals; as the number of rectangles is increased the sum of the areas of the rectangles becomes arbitrarily close to the area under the slanted line. (c) The area under the slanted line is half the product of the elapsed time and the final speed.

val. That is, the distance traveled is equal to the area of a thin rectangle, whose height is the height of the graph at that time and whose width is the tiny time interval. (See Figure 11a.) We can fill up the area under the graph, from the initial to the final time, by such thin rectangles, and the total distance traveled will then be the total area of all these rectangles—that is, the area under the graph. (See Figure 11b.)

Of course, however thin we make the rectangles, it is only an approximation to say that the area under the graph equals the total area of the rectangles. But we can make the rectangles as thin as we like, and in this way make the approximation as good as we like. By imagining the limit of an infinite number of infinitely thin rectangles, we can conclude that the distance traveled equals the area under the graph of speed versus time.

So far, this argument would be unchanged if the acceleration was not uniform, in which case the graph would not be a straight line. In fact, we have just deduced a fundamental principle of integral calculus: that if we make a graph of the rate of change of any quantity versus time, then the change in this quantity in any time interval is the area under the curve. But for a uniformly increasing rate of change, as in uniform acceleration, this area is given by a simple geometric theorem.

The theorem says that the area of a right triangle is half the product of the two sides adjacent to the right angle—that is, the two sides other than the hypotenuse. This follows immediately from the fact that we can put two of these triangles together to form a rectangle, whose area is the product of its two sides. (See Figure 11c.) In our case, the two sides adjacent to the right angle are the final speed and the total time elapsed. The distance traveled is the area of a right triangle with those dimensions, or half the product of the final speed and the total time elapsed. But since the speed is increasing from zero at a constant rate, its mean value is half its final value, so the distance traveled is the mean speed multiplied by the time elapsed. This is the mean speed theorem.

18. Ellipses

An ellipse is a certain kind of closed curve on a flat surface. There are at least three different ways of giving a precise description of this curve.

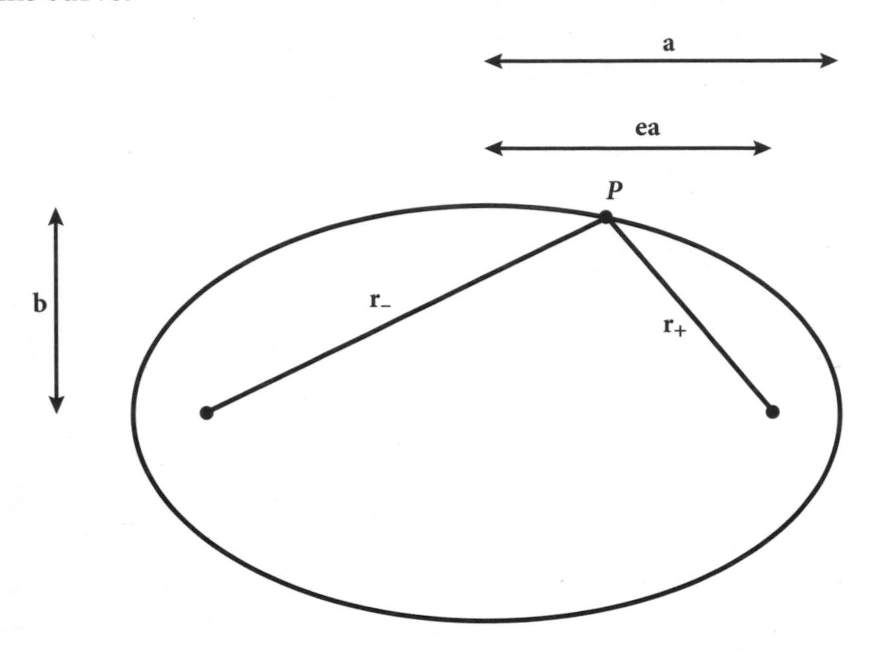

Figure 12. The elements of an ellipse. The marked points within the ellipse are its two foci; a and b are half the longer and shorter axes of the ellipse; and the distance from each focus to the center of the ellipse is ea. The sum of the lengths r_+ and r_- of the two lines from the foci to a point P equals $2a$ wherever P is on the ellipse. The ellipse shown here has ellipticity $e \simeq 0.8$.

First Definition

An ellipse is the set of points in a plane satisfying the equation

$$\frac{x^2}{a^2} + \frac{y^2}{b^2} = 1 \tag{1}$$

where x is the distance from the center of the ellipse of any point on the ellipse along one axis, y is the distance of the same point from the center along a perpendicular axis, and a and b are positive numbers that characterize the size and shape of the ellipse, conventionally defined so that $a \geq b$. For clarity of description it is convenient to think of the x-axis as horizontal and the y-axis as vertical, though of course they can lie along any two perpendicular directions. It follows from Eq. (1) that the distance $r = \sqrt{x^2 + y^2}$ of any point on the ellipse from the center at $x = 0$, $y = 0$ satisfies

$$\frac{r^2}{a^2} \leq \frac{x^2}{a^2} + \frac{y^2}{b^2} = 1 \text{ and } \frac{r^2}{b^2} \geq \frac{x^2}{a^2} + \frac{y^2}{b^2} = 1$$

so everywhere on the ellipse

$$b \leq r \leq a \tag{2}$$

Note that where the ellipse intersects the horizontal axis we have $y = 0$, so $x^2 = a^2$, and therefore $x = \pm a$; thus Eq. (1) describes an ellipse whose long diameter runs from $-a$ to $+a$ along the horizontal direction. Also, where the ellipse intersects the vertical axis we have $x = 0$, so $y^2 = b^2$, and therefore $y = \pm b$, and Eq. (1) therefore describes an ellipse whose short diameter runs along the vertical direction, from $-b$ to $+b$. (See Figure 12.) The parameter a is called the "semimajor axis" of the ellipse. It is conventional to define the eccentricity of an ellipse as

$$e \equiv \sqrt{1 - \frac{b^2}{a^2}} \tag{3}$$

The eccentricity is in general between 0 and 1. An ellipse with $e = 0$ is a circle, with radius $a = b$. An ellipse with $e = 1$ is so flattened that it just consists of a segment of the horizontal axis, with $y = 0$.

Second Definition

Another classic definition of an ellipse is that it is the set of points in a plane for which the sum of the distances to two fixed points (the foci of the ellipse) is a constant. For the ellipse defined by

Eq. (1), these two points are at $x = \pm ea$, $y = 0$, where e is the eccentricity as defined in Eq. (3). The distances from these two points to a point on the ellipse, with x and y satisfying Eq. (1), are

$$r_\pm = \sqrt{(x \mp ea)^2 + y^2} = \sqrt{(x \mp ea)^2 + (1 - e^2)(a^2 - x^2)}$$

$$= \sqrt{e^2x^2 \mp 2eax + a^2} = a \mp ex \qquad (4)$$

so their sum is indeed constant:

$$r_+ + r_- = 2a \qquad (5)$$

This can be regarded as a generalization of the classic definition of a circle, as the set of points that are all the same distance from a single point.

Since there is complete symmetry between the two foci of the ellipse, the average distances $\bar{r}_+$ and $\bar{r}_-$ of points on the ellipse (with every line segment of a given length on the ellipse given equal weight in the average) from the two foci must be equal: $\bar{r}_+ = \bar{r}_-$, and therefore Eq. (5) gives

$$\bar{r}_+ = \bar{r}_- = \frac{1}{2}(\bar{r}_+ + \bar{r}_-) = a \qquad (6)$$

This is also the average of the greatest and least distances of points on the ellipse from either focus:

$$\frac{1}{2}[(a + ea) + (a - ea)] = a \qquad (7)$$

Third Definition

The original definition of an ellipse by Apollonius of Perga is that it is a conic section, the intersection of a cone with a plane at a tilt to the axis of the cone. In modern terms, a cone with its axis in the vertical direction is the set of points in three dimensions satisfying the condition that the radii of the circular cross sections of the cone are proportional to distance in the vertical direction:

$$\sqrt{u^2 + y^2} = \alpha z \tag{8}$$

where u and y measure distance along the two perpendicular horizontal directions, z measures distance in the vertical direction, and α (alpha) is a positive number that determines the shape of the cone. (Our reason for using u instead of x for one of the horizontal coordinates will become clear soon.) The apex of this cone, where $u = y = 0$, is at $z = 0$. A plane that cuts the cone at an oblique angle can be defined as the set of points satisfying the condition that

$$z = \beta u + \gamma \tag{9}$$

where β (beta) and γ (gamma) are two more numbers, which respectively specify the tilt and height of the plane. (We are defining the coordinates so that the plane is parallel to the y-axis.) Combining Eq. (9) with the square of Eq. (8) gives

$$u^2 + y^2 = \alpha^2(\beta u + \gamma)^2$$

or equivalently

$$(1 - \alpha^2\beta^2)\left(u - \frac{\alpha^2\beta\gamma}{1 - \alpha^2\beta^2}\right)^2 + y^2 = \alpha^2\gamma^2\left(\frac{1}{1 - \alpha^2\beta^2}\right)$$

This is the same as the defining Eq. (1) if we identify

$$x = u - \frac{\alpha^2\beta\gamma}{1 - \alpha^2\beta^2} \qquad a = \frac{\alpha\gamma}{1 - \alpha^2\beta^2} \qquad b = \frac{\alpha\gamma}{\sqrt{1 - \alpha^2\beta^2}} \tag{10}$$

Note that this gives $e = \alpha\beta$, so the eccentricity depends on the shape of the cone and on the tilt of the plane cutting the cone, but not on the plane's height.

19. Elongations and Orbits of the Inner Planets

One of the great achievements of Copernicus was to work out definite values for the relative sizes of planetary orbits. A particularly simple example is the calculation of the radii of the orbits of the inner planets from the maximum apparent distance of these planets from the Sun.

Consider the orbit of one of the inner planets, Mercury or Venus, in the approximation that it and the Earth's orbit are both circles with the Sun at the center. At what is called "maximum

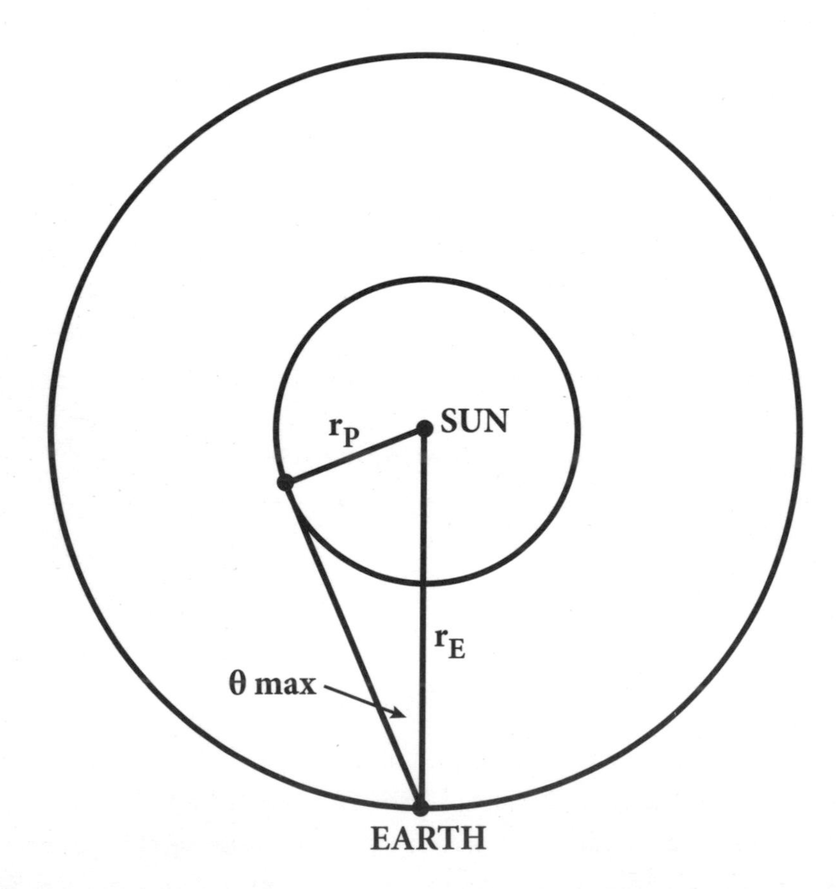

Figure 13. The positions of the Earth and an inner planet (Mercury or Venus) when the planet is at its maximum apparent distance from the Sun. The circles are the orbits of the Earth and planet.

elongation," the planet is seen at the greatest angular distance θ_{max} (theta$_{max}$) from the Sun. At this time, the line from the Earth to the planet is tangent to the planet's orbit, so the angle between this line and the line between the Sun and the planet is a right angle. These two lines and the line from the Sun to the Earth thus form a right triangle. (See Figure 13.) The hypotenuse of this triangle is the line between the Earth and the Sun, so the ratio of the distance r_P between the planet and the Sun to the distance r_E of the Earth from the Sun is the sine of θ_{max}. Here is a table of the angles of maximum elongation, their sines, and the actual orbital radii r_P of Mercury and Venus, in units of the radius r_E of the Earth's orbit:

	Maximum elongation θ_{max}	**Sine of θ_{max}**	**r_P/r_E**
Mercury	24°	0.41	0.39
Venus	45°	0.71	0.72

The small discrepancies between the sine of θ_{max} and the observed ratios r_P/r_E of the orbital radii of the inner planets and the Earth are due to the departure of these orbits from perfect circles with the Sun at the center, and to the fact that the orbits are not in precisely the same plane.

20. Diurnal Parallax

Consider a "new star" or another object that either is at rest with respect to the fixed stars or moves very little relative to the stars in the course of a day. Suppose that it is much closer to Earth than the stars. One can assume that the Earth rotates once a day on its axis from east to west, or that this object and the stars revolve around the Earth once a day from west to east; in either case, because we see the object in different directions at different times of night, its position will seem to shift during every evening relative to the stars. This is called the "diurnal parallax" of the object. A measurement of the diurnal parallax allows the determination of the distance of the object, or if it is found that the diurnal parallax is too small to be measured, this gives a lower limit to the distance.

To calculate the amount of this angular shift, consider the object's apparent position relative to the stars seen from a fixed observatory on Earth, when the object just rises above the horizon, and when it is highest in the sky. To facilitate the calculation, we will consider the case that is simplest geometrically: the observatory is on the equator, and the object is in the same plane as the equator. Of course, this does not accurately give the diurnal parallax of the new star observed by Tycho, but it will indicate the order of magnitude of that parallax.

The line to the object from this observatory when the object just rises above the horizon is tangent to the Earth's surface, so the angle between this line and the line from the observatory to the center of the Earth is a right angle. These two lines, together with the line from the object to the center of the Earth, thus form a right triangle. (See Figure 14.) Angle θ (theta) of this triangle at the object has a sine equal to the ratio of the opposite side, the radius r_E of the Earth, to the hypotenuse, the distance d of the object from the center of the Earth. As shown in the figure, this angle is also the apparent shift of the position of the object relative to the stars during the time between when it rises above the horizon and when it is highest in the sky. The total shift in its position from when it rises above the horizon to when it sets below the horizon is 2θ.

For instance, if we take the object to be at the distance of the Moon, then $d \simeq 250{,}000$ miles, while $r_E \simeq 4{,}000$ miles, so $\sin \theta \simeq 4/250$, and therefore $\theta \simeq 0.9°$, and the diurnal parallax is 1.8°. From a typical spot on Earth, such as Hven, to an object at a typical location in the sky like the new star of 1572, the diurnal parallax is smaller, but still of the same order of magnitude, in the neighborhood of 1°. This is more than large enough for it to have been detected by a naked-eye astronomer as expert as Tycho

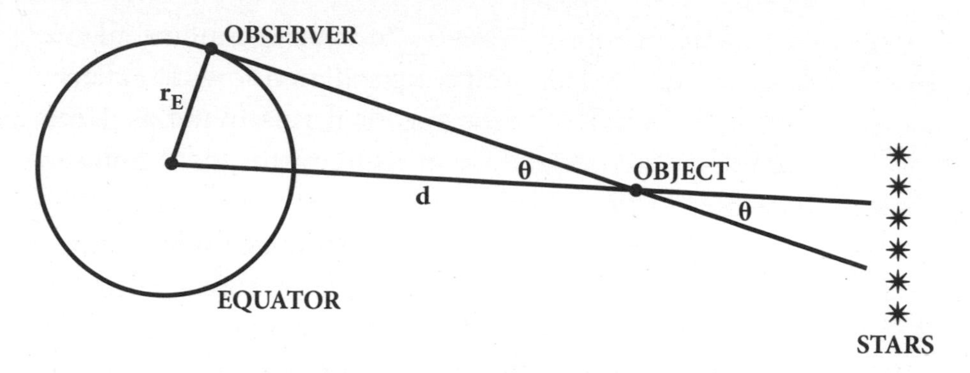

Figure 14. Use of diurnal parallax to measure the distance *d* from the Earth to some object. Here the view is from a point above the Earth's north pole. For simplicity, the observer is supposed to be on the equator, and the object is in the same plane as the equator. Thc two lines separated by an angle θ are the lines of sight to the object when it just arises above the horizon and six hours later, when it is directly above the observer.

Brahe, but Tycho could not detect any diurnal parallax, so he was able to conclude that the new star of 1572 is farther than the Moon. On the other hand, there was no difficulty in measuring the diurnal parallax of the Moon itself, and in this way finding the distance of the Moon from the Earth.

21. The Equal-Area Rule, and the Equant

According to Kepler's first law, the planets, including the Earth, each go around the Sun on an elliptical orbit, but the Sun is not at the center of the ellipse; it is at an off-center point on the major axis, one of the two foci of the ellipse. (See Technical Note 18.) The eccentricity e of the ellipse is defined so that the distance of each focus from the center of the ellipse is ea, where a is half the length of the major axis of the ellipse. Also, according to Kepler's second law, the speed of each planet in its orbit is not constant, but varies in such a way that the line from the Sun to the planet sweeps out equal areas in equal times.

There is a different approximate way of stating the second law, closely related to the old idea of the equant used in Ptolemaic astronomy. Instead of considering the line from the Sun to the planet, consider the line to the planet from the *other* focus of the ellipse, the empty focus. The eccentricity e of some planetary orbits is not negligible, but e^2 is very small for all planets. (The most eccentric orbit is that of Mercury, for which $e = 0.206$ and $e^2 = 0.042$; for the Earth, $e^2 = 0.00028$.) So it is a good approximation in calculating the motions of the planets to keep only terms that are independent of the eccentricity e or proportional to e, neglecting all terms proportional to e^2 or higher powers of e. In this approximation, Kepler's second law is equivalent to the statement that the line from the empty focus to the planet sweeps out equal *angles* in equal times. That is, the line between the empty focus of the ellipse and the planet rotates around that focus at a constant rate.

Specifically, we show below that if $\dot{A}$ is the rate at which area is swept out by the line from the Sun to the planet, and $\dot{\phi}$ (dotted phi) is the rate of change of the angle ϕ between the major axis of the ellipse and the line from the empty focus to the planet, then

$$\dot{\phi} = 2R\dot{A}/a^2 + O(e^2) \tag{1}$$

where $O(e^2)$ denotes terms proportional to e^2 or higher powers of e, and R is a number whose value depends on the units we use to measure angles. If we measure angles in degrees, then $R = 360° / 2\pi = 57.293\ldots°$, an angle known as a "radian." Or we can measure angles in radians, in which case we take $R = 1$. Kepler's second law tells us that in a given time interval the area swept out by the line from the Sun to the planet is always the same; this means that $\dot{A}$ is constant, so $\dot{\phi}$ is constant, up to terms proportional to e^2. So to a good approximation in a given time interval the angle swept out from the empty focus of the ellipse to the planet is also always the same.

Now, in the theory described by Ptolemy, the center of each planet's epicycle goes around the Earth on a circular orbit, the deferent, but the Earth is not at the center of the deferent. Instead, the orbit is eccentric—the Earth is at a point a small distance from the center. Furthermore, the velocity at which the center of the epicycle goes around the Earth is not constant, and the rate at which the line from the Earth to this center swivels around is not constant. In order to account correctly for the apparent motion of the planets, the device of the equant was introduced. This is a point on the other side of the center of the deferent from the Earth, and at an equal distance from the center. The line from the equant (rather than from the Earth) to the center of the epicycle was supposed to sweep out equal angles in equal times.

It will not escape the reader's notice that this is very similar to what happens according to Kepler's laws. Of course, the roles of the Sun and Earth are reversed in Ptolemaic and Copernican astronomy, but the empty focus of the ellipse in the theory of Kepler plays the same role as the equant in Ptolemaic astronomy, and Kepler's second law explains why the introduction of the equant worked well in explaining the apparent motion of the planets.

For some reason, although Ptolemy introduced an eccentric to describe the motion of the Sun around the Earth, he did not use an equant in this case. With this final equant included (and with some additional epicycles introduced to account for the large departure of Mercury's orbit from a circle), the Ptolemaic theory could account very well for the apparent motions of the planets.

Here is the proof of Eq. (1). Define θ as the angle between the major axis of the ellipse and the line from the Sun to the planet, and recall that ϕ is defined as the angle between the major axis and the line from the empty focus to the planet. As in Technical Note 18, define r_+ and r_- as the lengths of these lines—that is, the distances from the Sun to the planet and from the empty focus to the planet, respectively, given (according to that note) by

$$r_\pm = a \mp ex \qquad (2)$$

where x is the horizontal coordinate of the point on the ellipse—that is, it is the distance from this point to a line cutting through

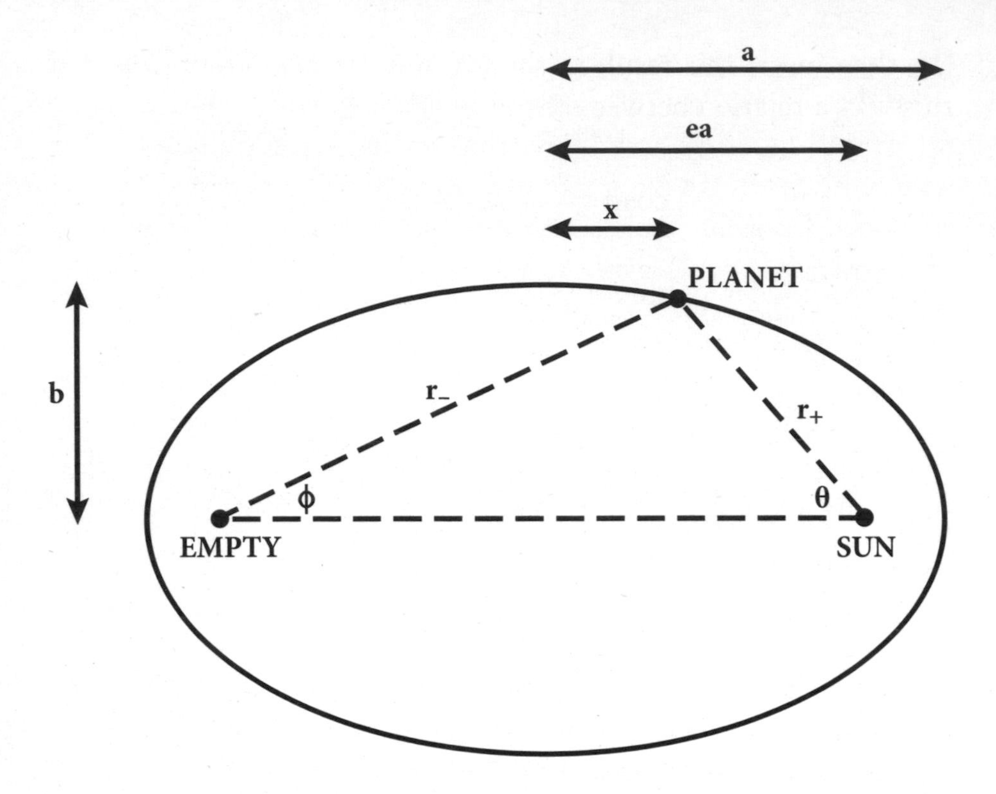

Figure 15. Elliptical motion of planets. The orbit's shape here is an ellipse, which (as in Figure 12) has ellipticity 0.8, much larger than the ellipticity of any planetary orbit in the solar system. The lines marked r_+ and r_- go respectively to the planet from the Sun and from the empty focus of the ellipse.

the ellipse along its minor axis. The cosine of an angle (symbolized cos) is defined in trigonometry by considering a right triangle with that angle as one of the vertices; the cosine of the angle is the ratio of the side adjacent to that angle to the hypotenuse of the triangle. Hence, referring to Figure 15,

$$\cos\theta = \frac{ea - x}{r_+} = \frac{ea - x}{a - ex} \qquad \cos\phi = \frac{ea + x}{r_-} = \frac{ea + x}{a + ex} \tag{3}$$

We can solve the equation at the left for x:

$$x = a\frac{e - \cos\theta}{1 - e\cos\theta} \tag{4}$$

We then insert this result in the formula for cos ϕ, obtaining in this way a relation between θ and ϕ:

$$\cos\phi = \frac{2e - (1 + e^2)\cos\theta}{1 + e^2 - 2e\cos\theta} \tag{5}$$

Since both sides of this equation are equal whatever the value of θ, the change in the left-hand side must equal the change in the right-hand side when we make any change in θ. Suppose we make an infinitesimal change $\delta\theta$ (delta theta) in θ. To calculate the change in ϕ, we make use of a principle in calculus, that when any angle α (such as θ or ϕ) changes by an amount $\delta\alpha$ (delta alpha), the change in cos α is $-(\delta\alpha/R)\sin\alpha$. Also, when any quantity f, such as the denominator in Eq. (5), changes by an infinitesimal amount δf, thc change in $1/f$ is $-\delta f/f^2$. Equating the changes in the two sides of Eq. (5) thus gives

$$\delta\phi\sin\phi = -\delta\theta\sin\theta\frac{(1 - e^2)^2}{(1 + e^2 - 2e\cos\theta)^2} \tag{6}$$

Now we need a formula for the ratio of sin ϕ and sin θ. For this purpose, note from Figure 15 that the vertical coordinate y of a point on the ellipse is given by $y = r_+ \sin\theta$ and also by $y = r_- \sin\phi$, so by climinating y,

$$\frac{\sin\theta}{\sin\phi} = \frac{r_-}{r_+} = \frac{a + ex}{a - ex} = \frac{1 - 2e\cos\theta + e^2}{1 - e^2} \tag{7}$$

Using this in Eq. (6), we have

$$\delta\phi = -\delta\theta\frac{1 - e^2}{1 + e^2 - 2e\cos\theta} \tag{8}$$

Now, what is the area swept out by the line from the Sun to the planet when the angle θ is changed by $\delta\theta$? If we measure angles in degrees, then it is the area of an isosceles triangle with two sides equal to r_+, and the third side equal to the fraction $2\pi r_+ \times \delta\theta/360°$ of the circumference $2\pi r_+$ of a circle of radius r_+. This area is

$$\delta A = -\frac{1}{2} \times r_+ \times 2\pi r_+ \times \delta\theta / 360° = -\frac{1}{2R} r_+^2 \delta\theta \tag{9}$$

$$= -\frac{a^2}{2R} \left(\frac{1 - e^2}{1 - e \cos\theta} \right)^2 \delta\theta$$

(A minus sign has been inserted here because we want δA to be positive when ϕ increases; but as we have defined them, ϕ increases when θ decreases, so $\delta\phi$ is positive when $\delta\theta$ is negative.) Thus Eq. (8) may be written

$$\delta\phi = \frac{2R}{a^2} \delta A \frac{(1 - e \cos\theta)^2}{(1 - e^2)(1 + e^2 - 2e \cos\theta)} \tag{10}$$

Taking δA and $\delta\phi$ to be the area and angle swept out in an infinitesimal time interval δt, and dividing Eq. (10) by δt, we find a corresponding relation between the rates of sweeping out areas and angles:

$$\dot{\phi} = \frac{2R}{a^2} \dot{A} \frac{(1 - e \cos\theta)^2}{(1 - e^2)(1 + e^2 - 2e \cos\theta)} \tag{11}$$

So far, this is all exact. Now let's consider how this looks when e is very small. The numerator of the second fraction in Eq. (11) is $(1 - e \cos\theta)^2 = 1 - 2e \cos\theta + e^2 \cos^2 \theta$, so the terms of zeroth and first order in the numerator and denominator of this fraction are the same, the difference between numerator and denominator appearing only in terms proportional to e^2. Equation (11) thus immediately yields the desired result, Eq. (1). To be a little more definite, we can keep the terms in Eq. (11) of order e^2:

$$\dot{\phi} = \frac{2R\dot{A}}{a^2} [1 + e^2 \cos^2\theta + O(e^3)] \tag{12}$$

where $O(e^3)$ denotes terms proportional to e^3 or higher powers of e.

22. *Focal Length*

Consider a vertical glass lens, with a convex curved surface in front and a plane surface in back, like the lens that Galileo and Kepler used as the front end of their telescopes. The curved surfaces that are easiest to grind are segments of spheres, and we will assume that the convex front of the lens is a segment of a sphere of radius r. We will also assume throughout that the lens is thin, with a maximum thickness much less than r.

Suppose that a ray of light traveling in the horizontal direction, parallel to the axis of the lens, strikes the lens at point P, and that the line from the center C of curvature (behind the lens) to P makes an angle θ (theta) with the centerline of the lens. The lens will bend the ray of light so that when it emerges from the back of the lens it makes a different angle ϕ with the centerline of the lens. The ray will then strike the centerline of the lens at some point F. (See Figure 16a.) We are going to calculate the distance f of this point from the lens, and show that it is independent of θ, so that all horizontal rays of light striking the lens reach the centerline at the same point F. Thus we can say that the light striking the lens is focused at point F; the distance f of this point from the lens is known as the "focal length" of the lens.

First, note that the arc on the front of the lens from the centerline to P is a fraction $\theta/360°$ of the whole circumference $2\pi r$ of a circle of radius r. On the other hand, the same arc is a fraction $\phi/360°$ of the whole circumference $2\pi f$ of a circle of radius f. Since these arcs are the same, we have

$$\frac{\theta}{360°} \times 2\pi r = \frac{\phi}{360°} \times 2\pi f$$

and therefore, canceling factors of 360° and 2π,

$$\frac{f}{r} = \frac{\theta}{\phi}$$

So to calculate the focal length, we need to calculate the ratio of ϕ to θ.

For this purpose, we need to look more closely at what happens to the light ray inside the lens. (See Figure 16b.) The line from the center of curvature *C* to the point *P* where a horizontal light ray strikes the lens is perpendicular to the convex spherical surface of the lens at *P*, so the angle between this perpendicular and the light ray (that is, the angle of incidence) is just θ. As was known to Claudius Ptolemy, if θ is small (as it will be for a thin

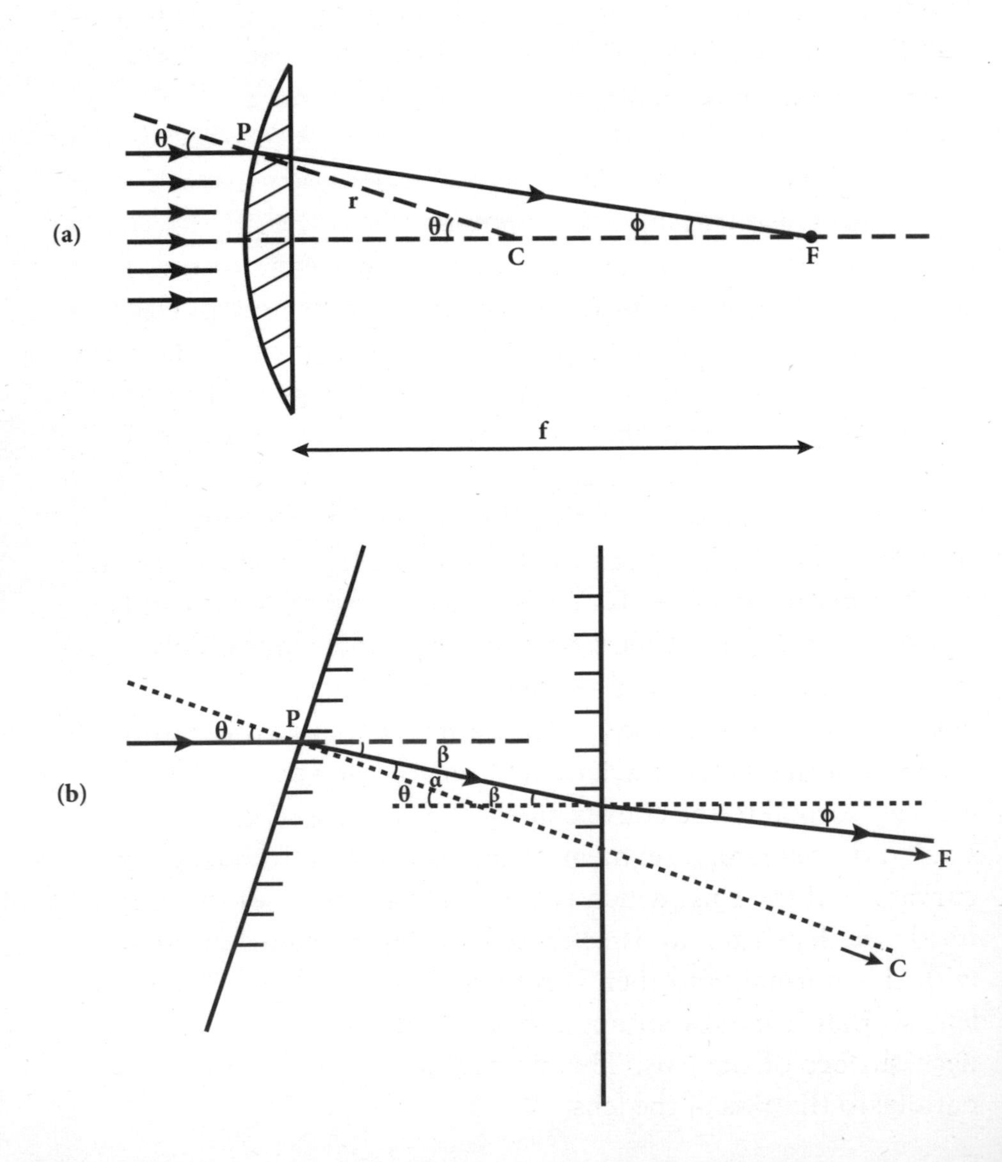

lens), then the angle α (alpha) between the ray of light inside the glass and the perpendicular (that is, the angle of refraction) will be proportional to the angle of incidence, so that

$$\alpha = \theta / n$$

where $n > 1$ is a constant, known as the "index of refraction," that depends on the properties of glass and the surrounding medium, typically air. (It was shown by Fermat that n is the speed of light in air divided by the speed of light in glass, but this information is not needed here.) The angle β (beta) between the light ray inside the glass and the centerline of the lens is then

Figure 16. Focal length. (a) Definition of focal length. The horizontal dashed line is the axis of the lens. Horizontal lines marked with arrows indicate rays of light that enter the lens parallel to this axis. One ray is shown entering the lens at point *P*, where the ray makes a small angle θ to a line from the center of curvature C that is perpendicular to the convex spherical surface at *P*; this ray is bent by the lens to make an angle ϕ to the lens axis and strikes this axis at focal point *F*, at a distance *f* from the lens. This is the focal length. With ϕ proportional to θ, all horizontal rays are focused to this point. (b) Calculation of focal length. Shown here is a small part of the lens, with the slanted hatched solid line on the left indicating a short segment of the convex surface of the lens. The solid line marked with an arrow shows the path of a ray of light that enters the lens at *P*, where it makes a small angle θ to the normal to the convex surface. This normal is shown as a slanted dotted line, a segment of the line from *P* to the center of curvature of the lens, which is beyond the borders of this figure. Inside the lens this ray is refracted so that it makes an angle α with this normal, and then is refracted again when it leaves the lens so that it makes an angle ϕ with the normal to the planar back surface of the lens. This normal is shown as a dotted line parallel to the axis of the lens.

$$\beta = \theta - \alpha = (1 - 1/n)\theta$$

This is the angle between the light ray and the normal to the flat back surface of the lens when the light ray reaches this surface. On the other hand, when the light ray emerges from the back of the lens it makes a different angle ϕ (phi) to the normal to the surface. The relation between ϕ and β is the same as if the light were going in the opposite direction, in which case ϕ would be the angle of incidence and β the angle of refraction, so that $\beta = \phi/n$, and therefore

$$\phi = n\beta = (n - 1)\theta$$

So we see that ϕ is simply proportional to θ, and therefore, using our previous formula for f/r, we have

$$f = \frac{r}{n - 1}$$

This is independent of θ, so as promised all horizontal light rays entering the lens are focused to the same point on the centerline of the lens.

If the radius of curvature r is very large, then the curvature of the front surface of the lens is very small, so that the lens is nearly the same as a flat plate of glass, with the bending of light on entering the lens being nearly canceled by its bending on leaving the lens. Similarly, whatever the shape of the lens, if the index of refraction n is close to 1, then the lens bends the light ray very little. In either case the focal length is very large, and we say that the lens is *weak*. A *strong* lens is one that has a moderate radius of curvature and an index of refraction appreciably different from 1, as for instance a lens made of glass, for which $n \simeq 1.5$.

A similar result holds if the back surface of the lens is not plane, but a segment of a sphere of radius r'. In this case the focal length is

$$f = \frac{rr'}{(r + r')(n - 1)}$$

This gives the same result as before if r' is much larger than r, in which case the back surface is nearly flat.

The concept of focal length can also be extended to concave lenses, like the lens that Galileo used as the eyepiece of his telescope. A concave lens can take rays of light that are converging and spread them out so that they are parallel, or even diverge. We can define the focal length of such a lens by considering converging rays of light that are made parallel by the lens; the focal length is the distance behind the lens of the point to which such rays *would* converge if not made parallel by the lens. Though its meaning is different, the focal length of a concave lens is given by a formula like the one we have derived for a convex lens.

23. Telescopes

As we saw in Technical Note 22, a thin convex lens will focus rays of light that strike it parallel to its central axis to a point F on this axis, at a distance behind the lens known as the focal length f of the lens. Parallel rays of light that strike the lens at a small angle γ (gamma) to the central axis will also be focused by the lens, but to a point that is a little off the central axis. To see how far off, we can imagine rotating the drawing of the ray path in Figure 16a around the lens by angle γ. The distance d of the focal point from the central axis of the lens will then be the same fraction of the circumference of a circle of radius f that γ is of 360°:

$$\frac{d}{2\pi f} = \frac{\gamma}{360^\circ}$$

and therefore

$$d = \frac{2\pi f \gamma}{360^\circ}$$

(This works only for thin lenses; otherwise d also depends on the angle θ introduced in Technical Note 22.) If the rays of light from some distant object strike the lens in a range $\Delta\gamma$ (delta gamma) of angles, they will be focused to a strip of height Δd, given by

$$\Delta d = \frac{2\pi f \Delta\gamma}{360^\circ}$$

(As usual, this formula is simpler if $\Delta\gamma$ is measured in radians, equal to $360^\circ / 2\pi$, rather than in degrees; in this case it reads simply $\Delta d = f\,\Delta\gamma$.) This strip of focused light is known as a "virtual image." (See Figure 17a).

We cannot see the virtual image just by peering at it, because after reaching this image the rays of light diverge again. To be focused to a point on the retina of a relaxed human eye, rays of light must enter the lens of the eye in more or less parallel directions. Kepler's telescope included a second convex lens, known as the eyepiece, to focus the diverging rays of light from the virtual image so that they left the telescope along parallel directions. By repeating the above analysis, but with the direction of the light rays reversed, we see that for the rays of light from a point on the light source to leave the telescope on parallel directions, the eyepiece must be placed at a distance f' from the virtual image, where f' is the focal length of the eyepiece. (See Figure 17b.) That is, the length L of the telescope must be the sum of the focal lengths

$$L = f + f'$$

The range $\Delta\gamma'$ of directions of the light rays from different points on the source entering the eye is related to the size of the virtual image by

$$\Delta d = \frac{2\pi f' \Delta\gamma'}{360^\circ}$$

The apparent size of any object is proportional to the angle subtended by rays of light from the object, so the magnification pro-

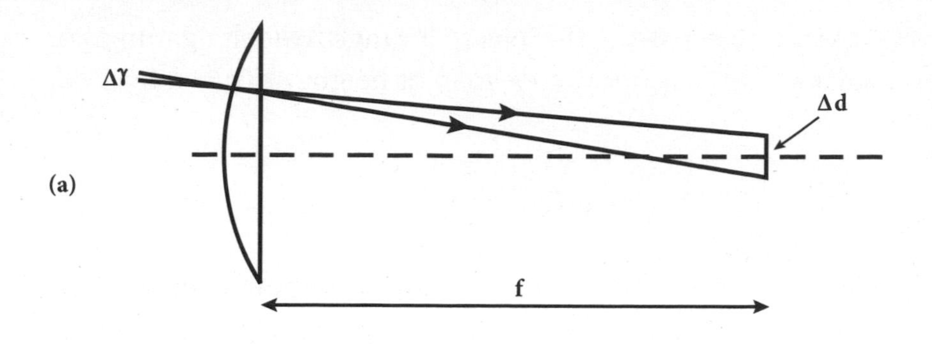

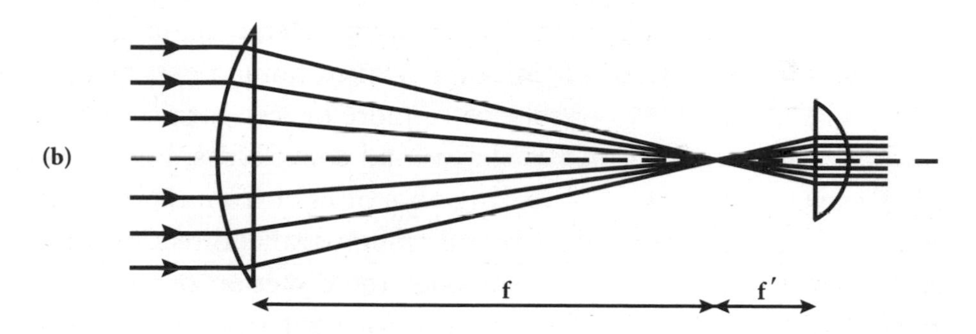

Figure 17. Telescopes. (a) Formation of a virtual image. The two solid lines marked with arrows are rays of light that enter the lens on lines separated by a small angle Δy. These lines (and others parallel to them) are focused to points at a distance f from the lens, with a vertical separation Δd proportional to Δy. (b) The lenses in Kepler's telescope. The lines marked with arrows indicate the paths of light rays that enter a weak convex lens from a distant object, on essentially parallel directions; are focused by the lens to a point at a distance f from the lens; diverge from this point; and are then bent by a strong convex lens so that they enter the eye on parallel directions.

duced by the telescope is the ratio of this angle when the rays enter the eye to the angle they would have spanned if there were no telescope:

$$\text{magnification} = \frac{\Delta\gamma'}{\Delta\gamma}$$

By taking the ratio of the two formulas we have derived for size Δd of the virtual image, we see that the magnification is

$$\frac{\Delta\gamma'}{\Delta\gamma} = \frac{f}{f'}$$

To get a significant degree of magnification, we need the lens at the front of the telescope to be much weaker than the eyepiece, with $f >> f'$.

This is not so easy. According to the formula for the focal length given in Technical Note 22, to have a strong glass eyepiece with short focal length f', it is necessary for it to have a small radius of curvature, which means either that it must be very small, or that it must not be thin (that is, with a thickness much less than the radius of curvature), in which case it does not focus the light well. We can instead make the lens at the front weak, with large focal length f, but in this case the length $L = f + f' \simeq f$ of the telescope must be large, which is awkward. It took some time for Galileo to refine his telescope to give it a magnification sufficient for astronomical purposes.

Galileo used a somewhat different design in his telescope, with a concave eyepiece. As mentioned in Technical Note 22, if a concave lens is properly placed, converging rays of light that enter it will leave on parallel directions; the focal length is the distance behind the lens at which the rays would converge if not for the lens. In Galileo's telescope there was a weak convex lens in front with focal length f, with a strong concave lens of focal length f' behind it, at a distance f' in *front* of the place where there would be a virtual image if not for the concave lens. The magnification of such a telescope is again the ratio f/f', but its length is only $f - f'$ instead of $f + f'$.

24. *Mountains on the Moon*

The bright and dark sides of the Moon are divided by a line known as the "terminator," where the Sun's rays are just tangent to the Moon's surface. When Galileo turned his telescope on the Moon he noticed bright spots on the dark side of the Moon near the terminator, and interpreted them as reflections from mountains high enough to catch the Sun's rays coming from the other side of the terminator. He could infer the height of these mountains by a geometrical construction similar to that used by al-Biruni to measure the size of the Earth. Draw a triangle whose vertices are the center C of the Moon, a mountaintop M on the dark side of the Moon that just catches a ray of sunlight, and the spot T on the terminator where this ray grazes the surface of the Moon. (See Figure 18.) This is a right triangle; line TM is tangent to the Moon's surface at T, so this line must be perpendicular to line CT. The length of CT is just the radius r of the moon, while the length of TM is the distance d of the mountain from the terminator. If the mountain has height h, then the length of CM (the hypotenuse of the triangle) is $r + h$. According to the Pythagorean theorem, we then have

$$(r + h)^2 = r^2 + d^2$$

and therefore

$$d^2 = (r + h)^2 - r^2 = 2rh + h^2$$

Since the height of any mountain on the Moon is much less than the size of the Moon, we can neglect h^2 compared with $2rh$. Dividing both sides of the equation by $2r^2$ then gives

$$\frac{h}{r} = \frac{1}{2}\left(\frac{d}{r}\right)^2$$

So by measuring the ratio of the apparent distance of a mountaintop from the terminator to the apparent radius of the Moon, Gal-

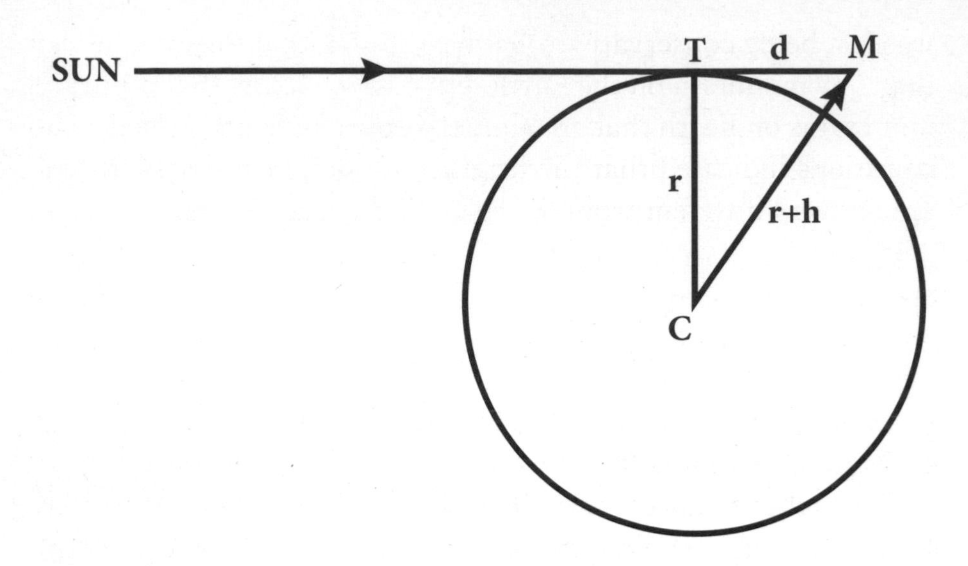

Figure 18. Galileo's measurement of the height of mountains on the Moon. The horizontal line marked with an arrow indicates a ray of light that grazes the Moon at the terminator *T* marking the boundary between the Moon's bright and dark sides, and then strikes the top *M* of a mountain of height *h* at a distance *d* from the terminator.

ileo could find the ratio of the mountain's height to the Moon's radius.

Galileo in *Siderius Nuncius* reported that he sometimes saw bright spots on the dark side of the Moon at an apparent distance from the terminator greater than 1/20 the apparent diameter of the Moon, so for these mountains $d/r > 1/10$, and therefore according to the above formula $h/r > (1/10)^2/2 = 1/200$. Galileo estimated the radius of the Moon to be 1,000 miles,* so these mountains would be at least 5 miles high. (For reasons that are not clear, Galileo gave a figure of 4 miles, but since he was trying only to set a lower bound on the mountain height, perhaps he

* Galileo used a definition of "mile" that is not very different from the modern English mile. In modern units, the radius of the Moon is actually 1,080 miles.

was just being conservative.) Galileo thought that this was higher than any mountain on the Earth, but now we know that there are mountains on Earth that are almost 6 miles high, so Galileo's observations indicated that the heights of mountains on the Moon are not very different from the heights of terrestrial mountains.

25. Gravitational Acceleration

Galileo showed that a falling body undergoes uniform acceleration—that is, its speed increases by the same amount in each equal interval of time. In modern terms, a body that falls from rest will after time t have a velocity v given by a quantity proportional to t:

$$v = gt$$

where g is a constant characterizing the gravitational field at the surface of the Earth. Although g varies somewhat from place to place on the Earth's surface, it is never very different from 32 feet/second per second, or 9.8 meters/second per second.

According to the mean speed theorem, the distance that such a body will fall from rest in time t is $v_{mean}t$, where v_{mean} is the average of gt and zero; in other words, $v_{mean} = gt/2$. Hence the distance fallen is

$$d = v_{mean}t = \frac{1}{2}gt^2$$

In particular, in the first second the body falls a distance $g(1 \text{ second})^2/2 = 16$ feet. The time required to fall distance d is in general

$$t = \sqrt{\frac{2d}{g}}$$

There is another, more modern way of looking at this result. The falling body has an energy equal to the sum of a *kinetic energy* term and a *potential energy* term. The kinetic energy is

$$E_{\text{kinetic}} = \frac{mv^2}{2} = \frac{mg^2t^2}{2}$$

where m is the body's mass. The potential energy is mg times the height (measured from any arbitrary altitude), so if the body is dropped from rest at an initial height h_0 and falls distance d, then

$$E_{\text{potential}} = mgh = mg(h_0 - d)$$

Hence with $d = gt^2/2$, the total energy is a constant:

$$E = E_{\text{kinetic}} + E_{\text{potential}} = mgh_0$$

We can turn this around, and derive the relation between velocity and distance fallen by *assuming* the conservation of energy. If we set E equal to the value mgh_0 that it has at $t = 0$, when $v = 0$ and $h = h_0$, then the conservation of energy gives at all times

$$\frac{mv^2}{2} + mg(h_0 - d) = mgh_0$$

from which it follows that $v^2/2 = gd$. Since v is the rate of increase of d, this is a differential equation that determines the relation between d and t. Of course, we know the solution of this equation: it is $d = gt^2/2$, for which $v = gt$. So by using the conservation of energy we can get these results without knowing in advance that the acceleration is uniform.

This is an elementary example of the conservation of energy, which makes the concept of energy useful in a wide variety of contexts. In particular, the conservation of energy shows the relevance of Galileo's experiments with balls rolling down inclined planes to the problem of free fall, though this is not an argument used by Galileo. For a ball of mass m rolling down a plane, the kinetic energy is $mv^2/2$, where v is now the velocity *along* the plane, and the potential energy is mgh, where h is again the height. In addition there is an energy of rotation of the ball, which takes the form

$$E_{\text{rotation}} = \frac{\zeta}{2} m r^2 (2\pi \nu)^2$$

where r is the radius of the ball, ν (nu) is the number of complete turns of the ball per second, and ζ (zeta) is a number that depends on the shape and mass-distribution of the ball. In the case that is probably relevant to Galileo's experiments, that of a solid uniform ball, ζ has the value $\zeta = 2/5$. (If the ball were hollow, we would have $\zeta = 2/3$.) Now, when the ball makes one complete turn it travels a distance equal to its circumference $2\pi r$, so in time t when it makes νt turns it travels distance $d = 2\pi r \nu t$, and therefore its velocity is $d/t = 2\pi \nu r$. Using this in the formula for the energy of rotation, we see that

$$E_{\text{rotation}} = \frac{\zeta}{2} m v^2 = \zeta E_{\text{kinetic}}$$

Dividing by m and $1 + \zeta$, the conservation of energy therefore requires that

$$\frac{v^2}{2} + \frac{gh}{1+\zeta} = \frac{gh_0}{1+\zeta}$$

This is the same relation between speed and distance fallen $d = h_0 - h$ that holds for a body falling freely, except that g has been replaced with $g/(1 + \zeta)$. Aside from this change, the dependence of the velocity of the ball rolling down the inclined plane on the vertical distance traveled is the same as that for a body in free fall. Hence the study of balls rolling down inclined planes could be used to verify that freely falling bodies experience uniform acceleration; but unless the factor $1/(1 + \zeta)$ is taken into account, it could not be used to measure the acceleration.

By a complicated argument, Huygens was able to show that the time it takes a pendulum of length L to swing through a small angle from one side to the other is

$$\tau = \pi \sqrt{\frac{L}{g}}$$

This equals π times the time required for a body to fall distance $d = L/2$, the result stated by Huygens.

26. Parabolic Trajectories

Suppose a projectile is shot horizontally with speed v. Neglecting air resistance, it will continue with this horizontal component of velocity, but it will accelerate downward. Hence after time t it will have moved a horizontal distance $x = vt$ and a downward distance z proportional to the square of the time, conventionally written as $z = gt^2/2$, with g = 32 feet/second per second, a constant measured after Galileo's death by Huygens. With $t = x/v$, it follows that

$$z = gx^2/2v^2$$

This equation, giving one coordinate as proportional to the square of the other, defines a parabola.

Note that if the projectile is fired from a gun at height h above the ground, then the horizontal distance x traveled when the projectile has fallen distance $z = h$ and reached the ground is $\sqrt{2v^2h/g}$. Even without knowing v or g, Galileo could have verified that the path of the projectile is a parabola, by measuring the distance traveled d for various heights fallen h, and checking that d is proportional to the square root of h. It is not clear whether Galileo ever did this, but there is evidence that in 1608 he did a closely related experiment, mentioned briefly in Chapter 12. A ball is allowed to roll down an inclined plane from various initial heights H, then rolls along the horizontal tabletop on which the inclined plane sits, and finally shoots off into the air from the table edge. As shown in Technical Note 25, the velocity of the ball at the bottom of the inclined plane is

$$v = \sqrt{\frac{2gH}{1 + \zeta}}$$

where as usual g = 32 feet/second per second, and ζ (zeta) is the ratio of rotational to kinetic energy of the ball, a number depending on the distribution of mass within the rolling ball. For a solid ball of uniform density, $\zeta = 2/5$. This is also the velocity of the ball when it shoots off horizontally into space from the edge of the tabletop, so the horizontal distance that the ball travels by the time it falls height h will be

$$d = \sqrt{2v^2 h/g} = 2\sqrt{\frac{Hh}{1+\zeta}}$$

Galileo did not mention the correction for rotational motion represented by ζ, but he may have suspected that some such correction could reduce the horizontal distance traveled, because instead of comparing this distance with the value $d = 2\sqrt{Hh}$ expected in the absence of ζ, he only checked that for fixed table height h, distance d was indeed proportional to $\sqrt{H}$, to an accuracy of a few percent. For one reason or another, Galileo never published the result of this experiment.

For many purposes of astronomy and mathematics, it is convenient to define a parabola as the limiting case of an ellipse, when one focus moves very far from the other. The equation for an ellipse with major axis $2a$ and minor axis $2b$ is given in Technical Note 18 as:

$$\frac{(z - z_0)^2}{a^2} + \frac{x^2}{b^2} = 1$$

in which for convenience later on we have replaced the coordinates x and y used in Note 18 with $z - z_0$ and x, with z_0 a constant that can be chosen as we like. The center of this ellipse is at $z = z_0$ and $x = 0$. As we saw in Note 18, there is a focus at $z - z_0 = -ae$, $x = 0$, where e is the eccentricity, with $e^2 \equiv 1 - b^2/a^2$, and the point of closest approach of the curve to this focus is at $z - z_0 = -a$ and $x = 0$. It will be convenient to give this point of closest approach the coordinates $z = 0$ and $x = 0$ by choosing $z_0 = a$ and in which case the nearby focus is at $z = z_0 - ea = (1 - e)a$. We want to let a and b become infinitely large, so that the other focus goes to infinity and the curve has no maximum x coordinate, but we want to keep the distance $(1 - e)a$ of closest approach to the nearer focus finite, so we set

$$1 - e = \ell/a$$

with ℓ held fixed as a goes to infinity. Since e approaches unity in this limit, the semiminor axis b is given by

$$b^2 = a^2(1 - e^2) = a^2(1 - e)(1 + e) \rightarrow 2a^2(1 - e) = 2\ell a$$

Using $z_0 = a$ and this formula for b^2, the equation for the ellipse becomes

$$\frac{z^2 - 2za + a^2}{a^2} + \frac{x^2}{2\ell a} = 1$$

The term a^2/a^2 on the left cancels the 1 on the right. Multiplying the remaining equation with a then gives

$$\frac{z^2}{a} - 2z + \frac{x^2}{2\ell} = 0$$

For a much larger than x, y, or ℓ, the first term may be dropped, so this equation becomes

$$z = \frac{x^2}{4\ell}$$

This is the same as the equation we derived for the motion of a projectile fired horizontally, provided we take

$$\frac{1}{4\ell} = \frac{g}{2v^2}$$

so the focus F of the parabola is at distance $\ell = v^2/2g$ below the initial position of the projectile. (See Figure 19.)

Parabolas can, like ellipses, be regarded as conic sections, but for parabolas the plane intersecting the cone is parallel to

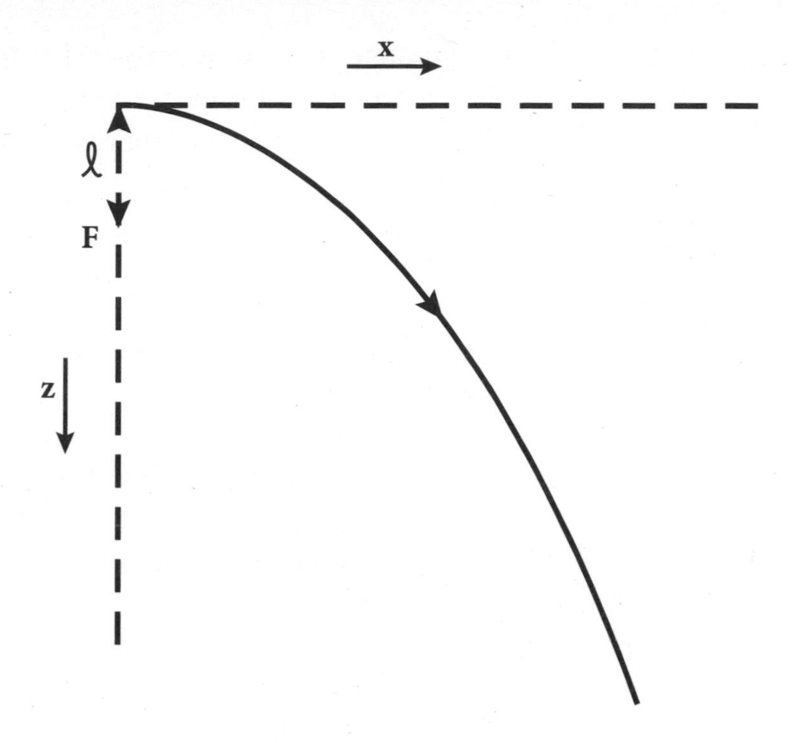

Figure 19. The parabolic path of a projectile that is fired from a hill in a horizontal direction. Point F is the focus of this parabola.

the cone's surface. Taking the equation of a cone centered on the z-axis as $\sqrt{x^2 + y^2} = \alpha(z + z_0)$, and the equation of a plane parallel to the cone as simply $y = \alpha(z - z_0)$, with z_0 arbitrary, the intersection of the cone and the plane satisfies

$$x^2 + \alpha^2(z^2 - 2zz_0 + z_0^2) = \alpha^2(z^2 + 2zz_0 + z_0^2)$$

After canceling the terms $\alpha^2 z^2$ and $\alpha^2 z^2{}_0$, this reads

$$z = \frac{x^2}{4\alpha^2 z_0}$$

which is the same as our previous result, provided we take $z_0 = \ell/\alpha^2$. Note that a parabola of a given shape can be obtained from any cone, with any value of the angular parameter α (alpha), because the shape of any parabola (as opposed to its location and orientation) is entirely determined by a parameter ℓ with the units of length; we do not need to know separately any unit-free parameter like α or the eccentricity of an ellipse.

27. Tennis Ball Derivation of the Law of Refraction

Descartes attempted a derivation of the law of refraction, based on the assumption that a ray of light is bent in passing from one medium to another in the same way that the trajectory of a tennis ball is bent in penetrating a thin fabric. Suppose a tennis ball with speed v_A strikes a thin fabric screen obliquely. It will lose some speed, so that after penetrating the screen its speed will be $v_B < v_A$, but we would not expect the ball's passage through the screen to make any change in the component of the ball's velocity *along* the screen. We can draw a right triangle whose sides are the components of the ball's initial velocity perpendicular to the screen and parallel to the screen, and whose hypotenuse is v_A. If the ball's original trajectory makes an angle i to the perpendicular to the screen, then the component of its velocity along the direction parallel to the screen is $v_A \sin i$. (See Figure 20.) Likewise, if after penetrating the screen the ball's trajectory makes an angle r with the perpendicular to the screen, then the component of its velocity along the direction parallel to the screen is $v_B \sin r$. Using Descartes' assumption that the passage of the ball through the screen could change only the component of velocity perpendicular to the interface, not the parallel component, we have

$$v_A \sin i = v_B \sin r$$

and therefore

$$\frac{\sin i}{\sin r} = n \tag{1}$$

where n is the quantity:

$$n = v_B / v_A \tag{2}$$

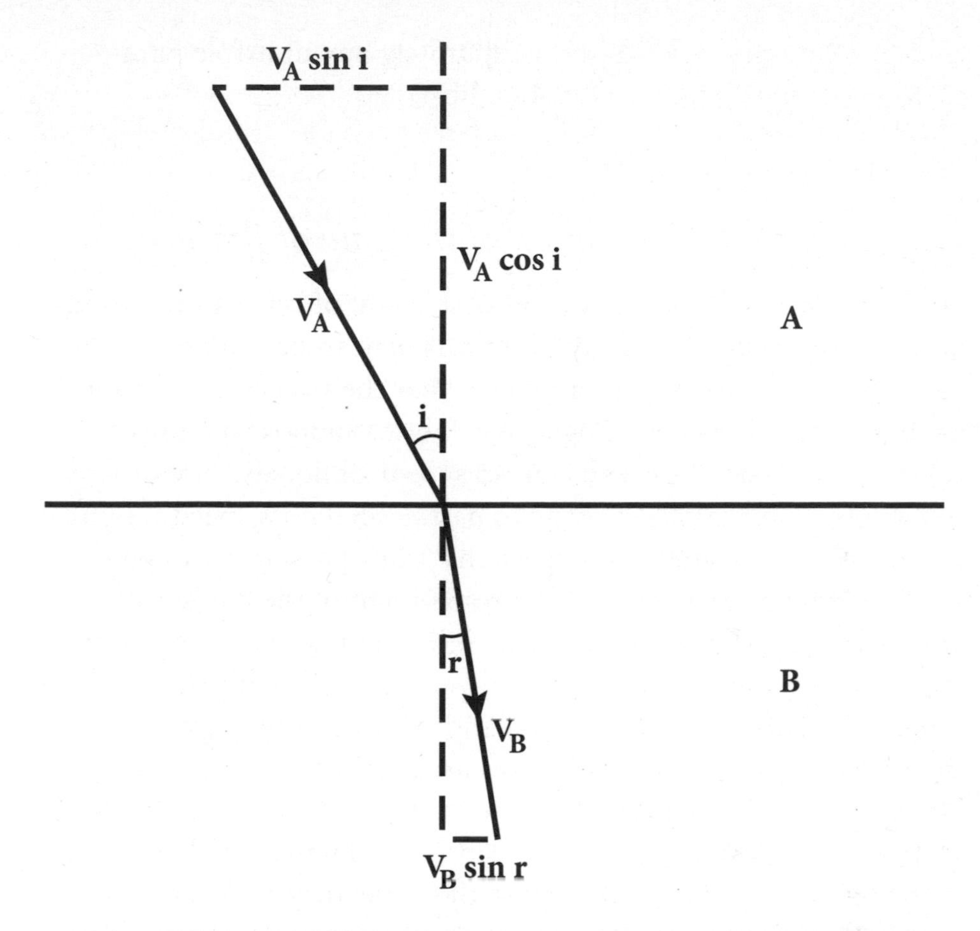

Figure 20. Tennis ball velocities. The horizontal line marks a screen penetrated by a tennis ball with initial speed vA and final speed vB. The solid lines marked with arrows indicate the magnitude and direction of the velocities of the ball before and after it penetrates the screen. This figure is drawn with the ball's path bent toward the perpendicular to the screen, as is the case for light rays entering a denser medium. It shows that in this case passage of the ball through the screen greatly reduces the component of its velocity along the screen, contrary to the assumption of Descartes.

Equation (1) is known as Snell's law, and it is the correct law of refraction for light. Unfortunately, the analogy between light and tennis balls breaks down when we come to Eq. (2) for n. Since for tennis balls, v_B is less than v_A, Eq. (2) gives $n < 1$, while when light passes from air to glass or water, we have $n > 1$. Not only that; there is no reason to suppose that for tennis balls v_B/v_A is actually independent of angles i and r, so that Eq. (1) is not useful as it stands.

As shown by Fermat, when light passes from a medium where its speed is v_A into one where its speed is v_B, the index of refraction n actually equals v_A/v_B, not v_B/v_A. Descartes did not know that light travels at a finite speed, and offered a hand-waving argument to explain why n is greater than unity when A is air and B is water. For seventeenth-century applications, like Descartes' theory of the rainbow, it didn't matter, because n was assumed to be angle-independent, which is correct for light if not for tennis balls, and its value was taken from observations of refraction, not from measurements of the speed of light in various media.

28. Least-Time Derivation of the Law of Refraction

Hero of Alexandria presented a derivation of the law of reflection, that the angle of reflection equals the angle of incidence, from the assumption that the path of the light ray from an object to the mirror and then to the eye is as short as possible. He might just as well have assumed that the time is as short as possible, since the time taken for light to travel any distance is the distance divided by the speed of light, and in reflection the speed of light does not change. On the other hand in refraction a light ray passes through the boundary between media (such as air and glass) in which the speed of light is different, and we have to distinguish between a principle of least distance and one of least time. Just from the fact that a ray of light is bent when passing from one medium to another, we know that light being refracted does not take the path of least distance, which would be a straight line. Rather, as shown by Fermat, the correct law of refraction can be derived by assuming that light takes the path of least time.

To carry out this derivation, suppose that a ray of light travels from point P_A in medium A in which the speed of light is v_A to point P_B in medium B in which the speed of light is v_B. To make this easy to describe, suppose that the surface separating the two media is horizontal. Let the angle between the light rays in media A and B and the vertical direction be i and r, respectively. If points P_A and P_B are at vertical distances d_A and d_B from the boundary surface, then the horizontal distance of these points from the point where the rays intersect this surface are $d_A \tan i$ and $d_B \tan_r$, respectively, where tan denotes the tangent of an angle, the ratio of the opposite to the adjacent side in a right triangle. (See Figure 21.) Although these distances are not fixed in advance, their sum is the fixed horizontal distance L between points P_A and P_B:

$$L = d_A \tan i + d_B \tan r$$

To calculate the time t elapsed when the light goes from P_A to P_B, we note that the distances traveled in media A and B are $d_A/\cos i$ and $d_B/\cos r$, respectively, where cos denotes the cosine of an angle, the ratio of the adjacent side to the hypotenuse in a right triangle. Time elapsed is distance divided by speed, so the total time elapsed here is

$$t = \frac{d_A}{v_A \cos i} + \frac{d_B}{v_B \cos r}$$

We need to find a general relation between angles i and r (independent of L, d_A, and d_B) that is satisfied by that angle i which makes time t a minimum, when r depends on i in such a way as to keep L fixed. For this purpose, consider δi, an infinitesimal variation δ (delta) of the angle of incidence i. The horizontal distance between P_A and P_B is fixed, so when i changes by amount δi, the angle of refraction r must also change, say by amount δr, given by the condition that L is unchanged. Also, at the minimum of t the graph of t versus i must be flat, for if t is increasing or decreasing at some i the minimum must be at some other value of i where t is smaller. This means that the change in t caused by a tiny change δi must vanish, at least to first order in δi. So to find the path of

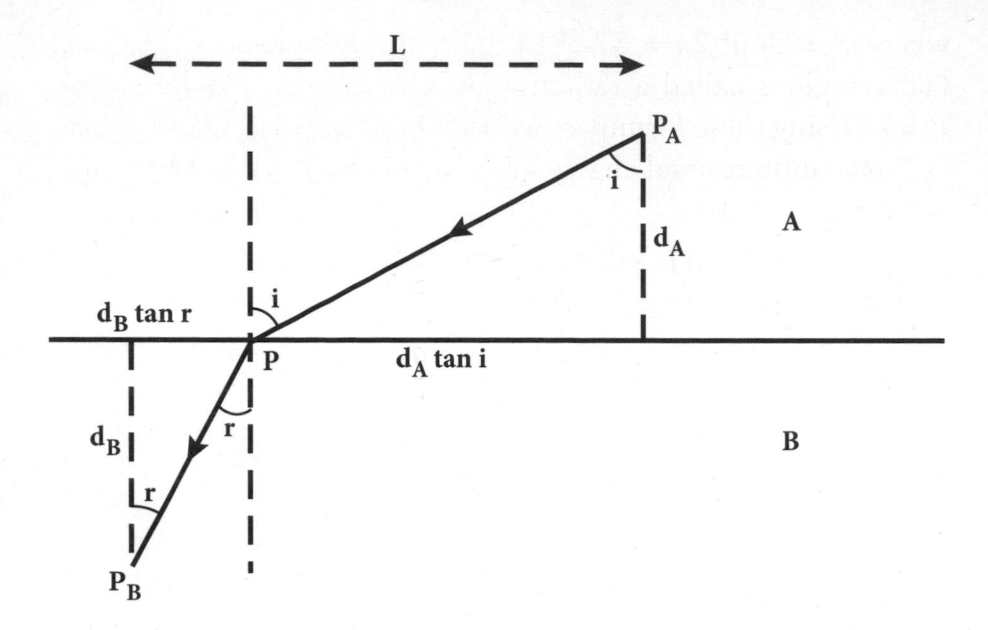

Figure 21. Path of a light ray during refraction. The horizontal line marks the interface between two transparent media *A* and *B*, in which light has different speeds v_A and v_B, and angles *i* and *r* are measured between the light ray and the dashed vertical line perpendicular to the interface. The solid line marked with arrows represents the path of a light ray that travels from point P_A in medium *A* to point P on the interface between the media and then to point P_B in medium *B*.

least time we can impose the condition that when we vary both *i* and *r* the changes δL and δt must both vanish at least to first order in δi and δr.

To implement this condition, we need standard formulas from differential calculus for the changes $\delta \tan \theta$ (theta) and $\delta(1/\cos\theta)$ when we make an infinitesimal change $\delta\theta$ in angle θ:

$$\delta \tan\theta = \frac{\delta\theta / R}{\cos^2\theta}$$

$$\delta(1/\cos\theta) = \frac{\sin\theta \ \delta\theta / R}{\cos^2\theta}$$

where $R = 360°/2\pi = 57.293\ldots°$ if θ is measured in degrees. (This angle is called a radian. If θ is measured in radians then $R = 1$.) Using these formulas, we find the changes in L and t when we make infinitesimal changes δi and δr in angles i and r:

$$\delta L = \frac{1}{R}\left(\frac{d_A}{\cos^2 i}\,\delta i + \frac{d_B}{\cos^2 r}\,\delta r\right)$$

$$\delta t = \frac{1}{R}\left(\frac{d_A \sin i}{v_A \cos^2 i}\,\delta i + \frac{d_B \sin r}{v_B \cos^2 r}\,\delta r\right)$$

The condition that $\delta L = 0$ tells us that

$$\delta r = -\frac{d_A/\cos^2 i}{d_B/\cos^2 r}\delta i$$

so that

$$\delta t = \left[\frac{d_A \sin i}{v_A \cos^2 i} - \frac{d_B \sin r}{v_B \cos^2 r}\frac{d_A/\cos^2 i}{d_B/\cos^2 r}\right]\frac{\delta i}{R} = \left[\frac{\sin i}{v_A} - \frac{\sin r}{v_B}\right]\frac{d_A}{\cos^2 i}\frac{\delta i}{R}$$

For this to vanish, wc must have

$$\frac{\sin i}{v_A} = \frac{\sin r}{v_B}$$

or in other words

$$\frac{\sin i}{\sin r} = n$$

with the index of refraction n given by the angle-independent ratio of velocities:

$$n = v_A/v_B$$

This is the correct law of refraction, with the correct formula for n.

29. *The Theory of the Rainbow*

Suppose that a ray of light arrives at a spherical raindrop at a point P, where it makes an angle i to the normal to the drop's surface. If there were no refraction, the light ray would continue straight through the drop. In this case, the line from the center C of the drop to the point Q of closest approach of the ray to the center would make a right angle with the light ray, so triangle PCQ would be a right triangle with hypotenuse equal to the radius R of the circle, and the angle at P equal to i. (See Figure 22a.) The impact parameter b is defined as the distance of closest approach of the unrefracted ray to the center, so it is the length of side CQ of the triangle, given by elementary trigonometry as

$$b = R \sin i$$

We can equally well characterize individual light rays by their value of b/R, as was done by Descartes, or by the value of the angle of incidence i.

Because of refraction, the ray will actually enter the drop at angle r to the normal, given by the law of refraction:

$$\sin r = \frac{\sin i}{n}$$

where $n \simeq 4/3$ is the ratio of the speed of light in air to its speed in water. The ray will cross the drop, and strike its back side at point P'. Since the distances from the center C of the drop to P and to P' are both equal to the radius R of the drop, the triangle with vertices C, P, and P' is isosceles, so the angles between the light ray and the normals to the surface at P and at P' must be equal, and hence both equal to r. Some of the light will be reflected from the back surface, and by the law of reflection the angle between the reflected ray and the normal to the surface at P' will again be r. The reflected ray will cross the drop and strike its front surface at a point P'', again making an angle r with the normal to the surface at P''. Some of the light will then emerge from the drop, and

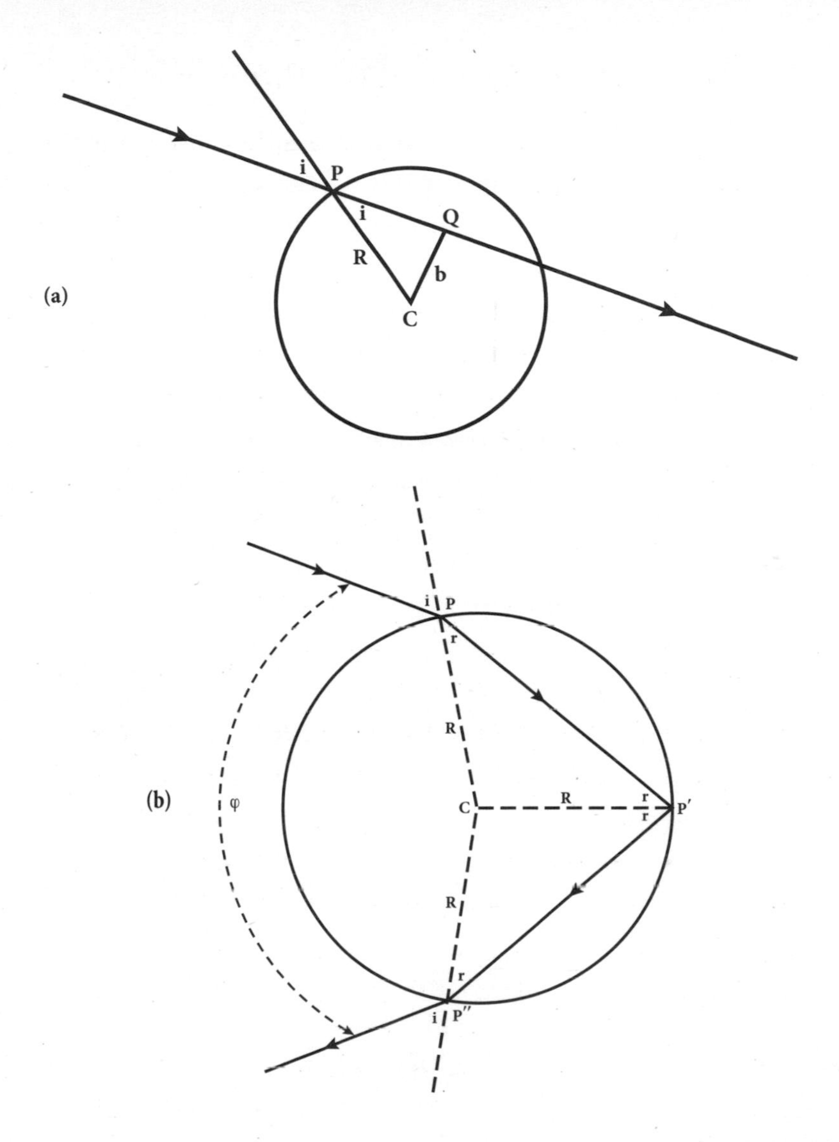

Figure 22. The path of a ray of sunlight in a spherical drop of water. The ray is indicated by solid lines marked with arrows, and enters the drop at point *P*, where it makes an angle *i* with the normal to the surface. (a) Path of the ray if there were no refraction, with *Q* the point of the ray's closest approach to the center *C* of the drop in this case. (b) The ray refracted on entering the drop at *P*, reflected from the back surface of the drop at *P*", and then refracted again on leaving the drop at *P*". Dashed lines run from the center *C* of the drop to points where the ray meets the surface of the drop.

by the law of refraction the angle between the emergent ray and the normal to the surface at P'' will equal the original incident angle i. (See Figure 22b. This figure shows the path of the light ray through a plane parallel to the ray's original direction that contains the center of the raindrop and the observer. Only rays that impinge on the drop's surface where it intersects this plane have a chance of reaching the observer.)

In the course of all this bouncing around, the light ray will have been bent toward the center of the drop by an angle $i - r$ twice on entering and leaving the drop, and by an angle $180° - 2r$ when reflected by the back surface of the drop, and hence by a total angle

$$2(i - r) + 180° - 2r = 180° - 4r + 2i$$

If the light ray bounced straight back from the drop (as is the case for $i = r = 0$) this angle would be 180°, and the initial and final light rays would be along the same line, so the actual angle φ (phi) between the initial and final light rays is

$$\varphi = 4r - 2i$$

We can express r in terms of i, as

$$r = \arcsin\left(\frac{\sin i}{n}\right)$$

where for any quantity x, the quantity $\arcsin x$ is the angle (usually taken between –90° and +90°) whose sine is x. The numerical calculation for $n = 4/3$ reported in Chapter 13 shows that φ rises from zero at $i = 0$ to a maximum value of about 42° and then drops to about 14° at $i = 90°$. The graph of φ versus i is flat at its maximum, so light tends to emerge from the drop at a deflection angle φ close to 42°.

If we look up at a misty sky with the sun behind us, we see light reflected back primarily from directions in the sky where the angle between our line of sight and the sun's rays is near 42°.

These directions form a bow, usually running from the Earth's surface up into the sky and then down again to the surface. Because n depends slightly on the color of light, so does the maximum value of the deflection angle φ, so this bow is spread out into different colors. This is the rainbow.

It is not difficult to derive an analytic formula that gives the maximum value of φ for any value of the index of refraction n. To find the maximum of φ, we use the fact that the maximum occurs at an incident angle i where the graph of φ versus i is flat, so that the variation $\delta\varphi$ (delta phi) in φ produced by a small variation δi in i vanishes to first order in δi. To use this condition, we use a standard formula of calculus, which tells us that when we make a change δx in x, the change in arcsin x is

$$\delta \arcsin x = \mathcal{R} \frac{\delta x}{\sqrt{1 - x^2}}$$

where if arcsin x is measured in degrees, then $\mathcal{R} = 360°/2\pi$. Thus when the angle of incidence varies by an amount δi, the angle of deflection changes by

$$\delta\varphi = 4\mathcal{R} \frac{\delta \sin i}{n\sqrt{1 - (\sin^2 i)/n^2}} - 2\delta i$$

or, since $\delta \sin i = \cos i \; \delta i/\mathcal{R}$,

$$\delta\varphi = \left[4 \frac{\cos i}{n\sqrt{1 - (\sin^2 i)/n^2}} - 2 \right] \delta i$$

Hence the condition for a maximum of φ is that

$$4 \frac{\cos i}{n\sqrt{1 - (\sin^2 i)/n^2}} = 2$$

Squaring both sides of the equation and using $\cos^2 i = 1 - \sin^2 i$ (which follows from the theorem of Pythagoras), we can then solve for sin i, and find

$$\sin i = \sqrt{\frac{1}{3}(4 - n^2)}$$

At this angle, φ takes its maximum value:

$$\varphi_{max} = 4 \arcsin\left(\frac{1}{n}\sqrt{\frac{1}{3}(4 - n^2)}\right) - 2 \arcsin\left(\sqrt{\frac{1}{3}(4 - n^2)}\right)$$

For $n = 4/3$, the maximum value of φ is reached for $b/\mathcal{R} = \sin i = 0.86$, for which $i = 59.4°$, where $r = 40.2°$, and $\varphi_{max} = 42.0°$.

30. Wave Theory Derivation of the Law of Refraction

The law of refraction, which as described in Technical Note 28 can be derived from an assumption that refracted light rays take the path of least time, can also be derived on the basis of the wave theory of light. According to Huygens, light is a disturbance in a medium, which may be some transparent material or space that is apparently empty. The front of the disturbance is a line, which moves forward in a direction at right angles to the front, at a speed characteristic of the medium.

Consider a segment of the front of such a disturbance, which is of length L in medium 1, traveling toward an interface with medium 2. Let us suppose that the direction of motion of the disturbance, which is at right angles to this front, makes angle i with the perpendicular to this interface. When the leading edge of the disturbance strikes the interface at point A, the trailing edge B is still at a distance (along the direction the disturbance is traveling) equal to $L \tan i$. (See Figure 23.) Hence the time required for the trailing edge to reach the interface at point D is $L \tan i/v_1$, where v_1 is the velocity of the disturbance in medium 1. During this time the leading edge of the front will have traveled in medium 2 at an angle r to the perpendicular, reaching a point C at a distance $v_2 L \tan i/v_1$ from A, where v_2 is the velocity in medium 2. At this time the wave front, which is at right angles to the direction of motion in medium 2, extends from C to D, so that the triangle with vertices A, C, and D is a right triangle, with a 90° angle at C. The distance $v_2 L \tan i/v_1$ from A to C is the side opposite angle r in this right triangle, while the hypote-

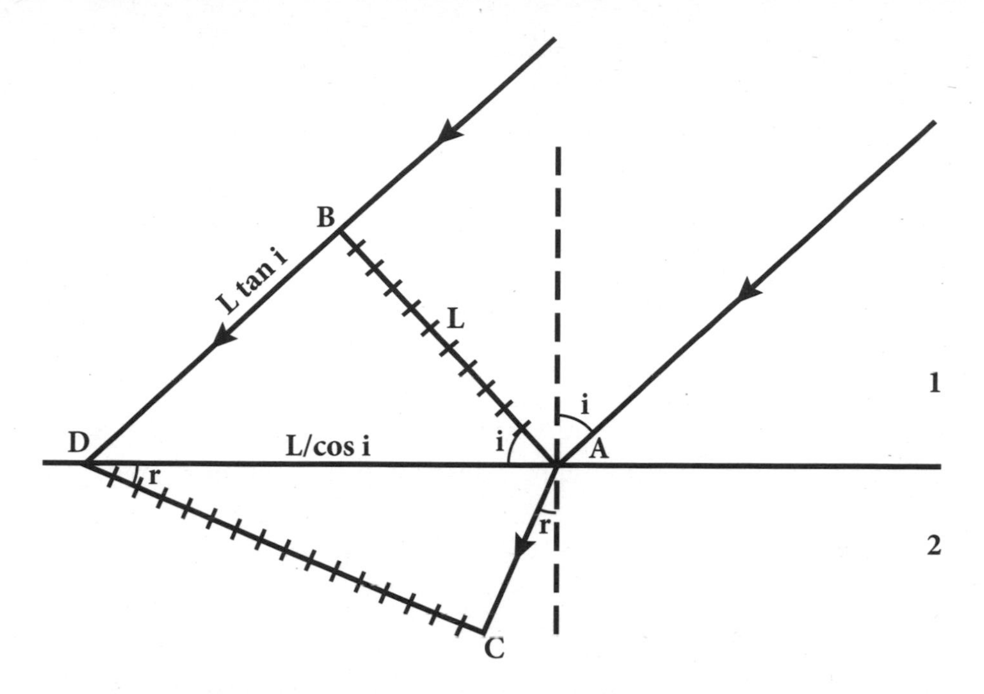

Figure 23. Refraction of a light wave. The horizontal line again marks the interface between two transparent media, in which light has different speeds. The crosshatched lines show a segment of a wave front at two different times—when the leading edge and when the trailing edge of the wave front just touch the interface. The solid lines marked with arrows show the paths taken by the leading and trailing edges of the wave front.

nuse is the line from A to D, which has length $L/\cos i$. (Again, see Figure 23.) Hence

$$\sin r = \frac{v_2 L \tan i / v_1}{L/\cos i}$$

Recalling that $\tan i = \sin i/\cos i$, we see that the factors of $\cos i$ and L cancel, so that

$$\sin r = v_2 \sin i / v_1$$

or in other words

$$\frac{\sin i}{\sin r} = \frac{v_1}{v_2}$$

which is the correct law of refraction.

It is not an accident that the wave theory, as worked out by Huygens, gives the same results for refraction as the least-time principle of Fermat. It can be shown that, even for waves passing through a heterogeneous medium in which the speed of light changes gradually in various directions, not just suddenly at a plane interface, the wave theory of Huygens will always give a light path that takes the shortest time to travel between any two points.

31. Measuring the Speed of Light

Suppose we observe some periodic process occurring at some distance from us. For definiteness we will consider a moon going around a distant planet, but the analysis below would apply to any process that repeats periodically. Suppose that the moon reaches the same stage in its orbit at two consecutive times t_1 and t_2; for instance these might be times that the moon consecutively emerges from behind the planet. If the intrinsic orbital period of the moon is T, then $t_2 - t_1 = T$. This is the period we observe, provided the distance between us and the planet is fixed. But if this distance is changing, then the period we observe will be shifted from T, by an amount that depends on the speed of light.

Suppose that the distances between us and the planet at two successive times when the moon is at the same stage in its orbit are d_1 and d_2. We then observe these stages in the orbit at times

$$t'_1 = t_1 + d_1/c \qquad t'_2 = t_2 + d_2/c$$

where c is the speed of light. (We are assuming here that the distance between the planet and its moon may be neglected.) If the distance between us and this planet is changing at rate v, either because it is moving or because we are or both, then $d_2 - d_1 = vT$, and so the observed period is

$$T' \equiv t'_2 - t'_1 = T + \frac{Tv}{c} = T\left[1 + \frac{v}{c}\right]$$

(This derivation depends on the assumption that v should change very little in a time T, which is typically true in the solar system, but v may be changing appreciably over longer time scales.) When the distant planet is moving toward or away from us, in which case v is respectively negative or positive, its moon's apparent period will be decreased or increased, respectively. We can measure T by observing the planet at a time when $v = 0$, and then measure the speed of light by observing the period again at a time when v has some known nonzero value.

This is the basis of the determination of the speed of light by Huygens, based on Rømer's observation of the changing apparent orbital period of Jupiter's moon Io. But with the speed of light known, the same calculation can tell us the relative velocity v of a distant object. In particular, the light waves from a specific line in the spectrum of a distant galaxy will oscillate with some characteristic period T, related to its frequency ν (nu) and wavelength λ (lambda) by $T = 1/\nu = \lambda/c$. This intrinsic period is known from observations of spectra in laboratories on Earth. Since the early twentieth century the spectral lines observed in very distant galaxies have been found to have longer wavelengths, and hence longer periods, from which we can infer that these galaxies are moving away from us.

32. Centripetal Acceleration

Acceleration is the rate of change of velocity, but the velocity of any body has both a magnitude, known as the speed, and a direction. The velocity of a body moving on a circle is continually changing its direction, turning toward the circle's center, so even at constant speed it undergoes a continual acceleration toward the center, known as its centripetal acceleration.

Let us calculate the centripetal acceleration of a body traveling on a circle of radius r, with constant speed v. During a short time interval from t_1 to t_2, the body will move along the circle by a small distance $v\Delta t$, where Δt (delta t) = $t_2 - t_1$, and the radial vector (the arrow from the center of the circle to the body) will swivel around by a small angle $\Delta\theta$ (delta theta). The velocity vector (an arrow with magnitude v in the direction of the body's motion) is always tangent to the circle, and hence at right angles to the radial vector, so while the radial vector's direction changes by an angle $\Delta\theta$, the velocity vector's direction will change by the same small angle. So we have two triangles: one whose sides are the radial vectors at times t_1 and t_2 and the chord connecting the positions of the body at these two times; and the other whose sides are the velocity vectors at times t_1 and t_2, and the change Δv in the velocity between these two times. (See Figure 24.) For small angles $\Delta\theta$, the difference in length between the chord and the arc connecting the positions of the bodies at times t_1 and t_2, is negligible, so we can take the length of the chord as $v\Delta t$.

Now, these triangles are similar (that is, they differ in size but not in shape) because they are both isosceles triangles (each has two equal sides) with the same small angle $\Delta\theta$ between the two equal sides. So the ratios of the short and long sides of each triangle should be the same. That is

$$\frac{v\Delta t}{r} = \frac{\Delta v}{v}$$

and therefore

$$\frac{\Delta v}{\Delta t} = \frac{v^2}{r}$$

This is Huygens' formula for the centripetal acceleration.

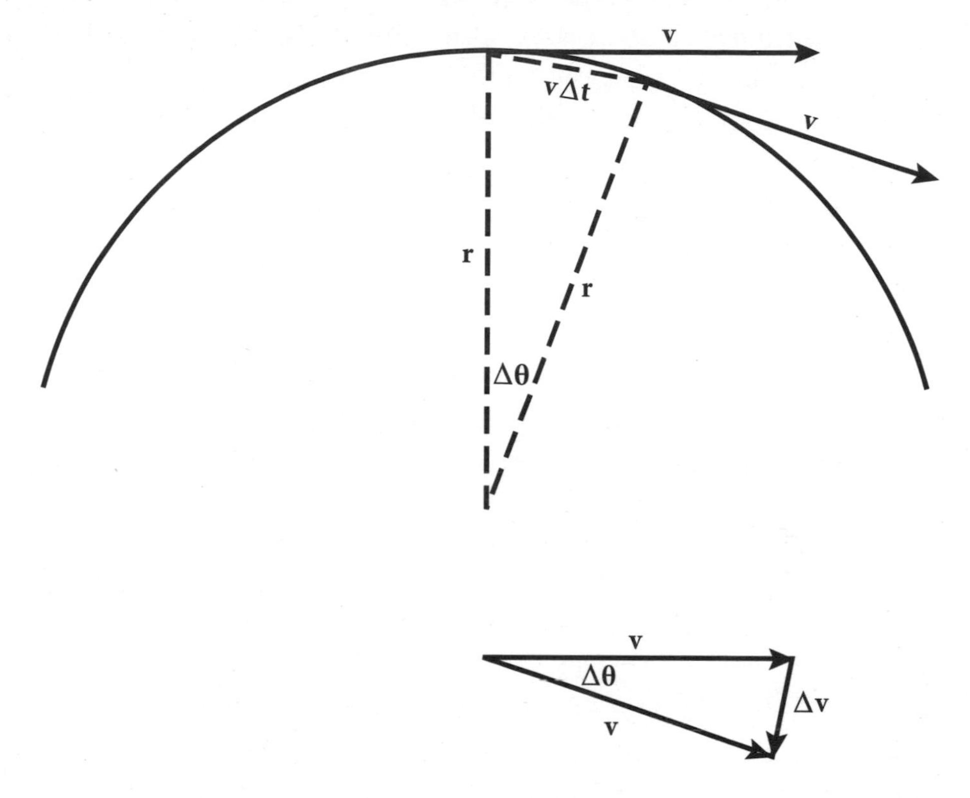

Figure 24. Calculation of centripetal acceleration. Top: Velocities of a particle moving on a circle at two times, separated by a short time interval Δt. Bottom: These two velocities, brought together into a triangle whose short side is the change of velocity in this time interval.

33. Comparing the Moon with a Falling Body

The ancient supposed distinction between phenomena in the heavens and on Earth was decisively challenged by Newton's comparison of the centripetal acceleration of the Moon in its orbit with the downward acceleration of a falling body near the surface of the Earth.

From measurements of the Moon's diurnal parallax, its average distance from the Earth was accurately known in Newton's time to be 60 times the radius of the Earth. (The actual ratio is

60.27.) To calculate the Earth's radius Newton took 1' (minute of arc) at the equator to be a mile of 5,000 feet, so with 360° for a circle and 60' to 1°, the earth's radius was

$$\frac{360 \times 60 \times 5{,}000 \text{ feet}}{2\pi} = 17{,}190{,}000 \text{ feet}$$

(The mean radius is actually 20,926,300 feet. This was the greatest source of error in Newton's calculation.) The orbital period of the Moon (the sidereal month) was accurately known to be 27.3 days, or 2,360,000 seconds. The velocity of the Moon in its orbit was then

$$\frac{60 \times 2\pi \times 17{,}190{,}000 \text{ feet}}{2{,}360{,}000 \text{ seconds}} = 2{,}745 \text{ feet per second}$$

This gives a centripetal acceleration of

$$\frac{(2{,}745 \text{ feet per second})^2}{60 \times 17{,}190{,}000 \text{ feet}} = 0.0073 \text{ foot/second per second}$$

According to the inverse square law, this should have equaled the acceleration of falling bodies on the surface of the Earth, 32 feet/second per second, divided by the square of the ratio of the radius of the Moon's orbit to the radius of the earth:

$$\frac{32 \text{ feet/second per second}}{60^2} = 0.0089 \text{ foot/second per second}$$

It is this comparison, of an "observed" lunar centripetal acceleration of 0.0073 foot/second per second with the value expected from the inverse square law, 0.0089 foot/second per second, to which Newton was referring when he said that they "answer pretty nearly." He did better later.

34. Conservation of Momentum

Suppose two moving objects with masses m_1 and m_2 collide head-on. If in a short time interval δt (delta t) object 1 exerts force F on object 2, then in this time interval object 2 will experience an acceleration a_2 that according to Newton's second law obeys the relation $m_2a_2 = F$. Its velocity v_2 will then change by an amount

$$\delta v_2 = a_2\,\delta t = F\,\delta t/m_2$$

According to Newton's third law, particle 2 will exert on particle 1 a force $-F$ that is equal in magnitude but (as indicated by the minus sign) opposite in direction, so in the same time interval the velocity v_1 of object 1 will undergo a change in the opposite direction to δv_2, given by

$$\delta v_1 = a_1\,\delta t = -F\,\delta t/m_1$$

The net change in the total momentum $m_1v_1 + m_2v_2$ is then

$$m_1\delta v_1 + m_2\delta v_2 = 0$$

Of course, the two objects may be in contact for an extended period, during which the force may not be constant, but since the momentum is conserved in every short interval of time, it is conserved during the whole period.

35. Planetary Masses

In Newton's time four bodies in the solar system were known to have satellites: Jupiter and Saturn as well as the Earth were known to have moons, and all the planets are satellites of the Sun. According to Newton's law of gravitation, a body of mass M exerts a force $F = GMm/r^2$ on a satellite of mass m at distance r (where G is a constant of nature), so according to Newton's second law of motion the centripetal acceleration of the satellite will be $a = F/m = GM/r^2$. The value of the constant G and the overall scale of the solar system were not known in Newton's time, but these unknown quantities do not appear in the *ratios* of masses calculated from ratios of distances and ratios of centripetal accelerations. If two satellites of bodies with masses M_1 and M_2 are found to be at distances from these bodies with a known ratio r_1/r_2 and to have centripetal accelerations with a known ratio a_1/a_2, then the ratio of the masses can be found from the formula

$$\frac{M_1}{M_2} = \left(\frac{r_1}{r_2}\right)^2 \frac{a_1}{a_2}$$

In particular, for a satellite moving at constant speed v in a circular orbit of radius r the orbital period is $T = 2\pi r/v$, so the centripetal acceleration v^2/r is $a = 4\pi^2 r/T^2$, the ratio of accelerations is $a_1/a_2 = (r_1/r_2)/(T_2/T_1)^2$, and the ratio of masses inferred from orbital periods and ratios of distances is

$$\frac{M_1}{M_2} = \left(\frac{r_1}{r_2}\right)^3 \left(\frac{T_2}{T_1}\right)^2$$

By 1687 all the *ratios* of the distances of the planets from the Sun were well known, and from the observation of the angular separation of Jupiter and Saturn from their moons Callisto and Titan (which Newton called the "Huygenian satellite") it was also possible to work out the ratio of the distance of Callisto from Jupiter to the distance of Jupiter from the Sun, and the ratio of the distance of Titan from Saturn to the distance of Saturn from the Sun. The distance of the Moon from the Earth was quite well known as a multiple of the size of the Earth, but *not* as a fraction of the distance of the Earth from the Sun, which was then not known. Newton used a crude estimate of the ratio of the distance of the Moon from the Earth and the distance of the Earth from the Sun, which turned out to be badly in error. Aside from this problem, the ratios of velocities and centripetal accelerations could be calculated from the known orbital periods of planets and moons. (Newton actually used the period of Venus rather than of Jupiter or Saturn, but this was just as useful because the ratios of the distances of Venus, Jupiter, and Saturn from the Sun were all well known.) As reported in Chapter 14, Newton's results for the ratios of the masses of Jupiter and Saturn to the mass of the Sun were reasonably accurate, while his result for the ratio of the mass of the Earth to the mass of the Sun was badly in error.

Endnotes

PART I: GREEK PHYSICS

1. Matter and Poetry

1. Aristotle, *Metaphysics*, Book I, Chapter 3, 983b 6, 20 (Oxford trans.). Here and below I follow the standard practice of citing passages from Aristotle by referring to their location in I. Bekker's 1831 Greek edition. By "Oxford trans.," I mean that the English language version is taken from *The Complete Works of Aristotle—The Revised Oxford Translation*, ed. J. Barnes (Princeton University Press, Princeton, N.J., 1984), which uses this convention in citing passages from Aristotle.
2. Diogenes Laertius, *Lives of the Eminent Philosophers*, Book I, trans. R. D. Hicks (Loeb Classical Library, Harvard University Press, Cambridge, Mass., 1972), p. 27.
3. From J. Barnes, *The Presocratic Philosophers*, rev. ed. (Routledge and Kegan Paul, London, 1982), p. 29. The quotations in this work, hereafter cited as *Presocratic Philosophers*, are translations into English of the fragmentary quotations in the standard source-book by Hermann Diels and Walter Kranz, *Die Fragmente der Vorsokratiker* (10th ed., Berlin, 1952).
4. *Presocratic Philosophers*, p. 53.
5. From J. Barnes, *Early Greek Philosophy* (Penguin, London, 1987), p. 97. Hereafter cited as *Early Greek Philosophy*. As in *Presocratic Philosophers*, these quotations are taken from Diels and Kranz, 10th ed.
6. From K. Freeman, *The Ancilla to the Pre-Socratic Philosophers* (Harvard University Press, Cambridge, Mass., 1966), p. 26. Here-

after cited as *Ancilla*. This is a translation into English of the quotations in Diels, *Fragmente der Vorsokratiker*, 5th ed.

7. *Ancilla*, p. 59.
8. *Early Greek Philosophy*, p. 166.
9. Ibid., p. 243.
10. *Ancilla*, p. 93.
11. Aristotle, *Physics*, Book VI, Chapter 9, 239b 5 (Oxford trans.).
12. Plato, *Phaedo*, 97C–98C. Here and below I follow the standard practice of citing passages from Plato's works by giving page numbers in the 1578 Stephanos Greek edition.
13. Plato, *Timaeus*, 54 A–B, from Desmond Lee, trans., *Timaeus and Critias* (Penguin Books, London, 1965).
14. For instance, in the Oxford translation of Aristotle's *Physics*, Book IV, Chapter 6, 213b 1–2.
15. *Ancilla*, p. 24.
16. *Early Greek Philosophy*, p. 253.
17. I have written about this point at greater length in the chapter "Beautiful Theories" in *Dreams of a Final Theory* (Pantheon, New York, 1992; reprinted with a new afterword, Vintage, New York, 1994).

2. Music and Mathematics

1. For the provenance of these stories, see Alberto A. Martínez, *The Cult of Pythagoras—Man and Myth* (University of Pittsburgh Press, Pittsburgh, Pa., 2012).
2. Aristotle, *Metaphysics*, Book I, Chapter 5, 985b 23–26 (Oxford trans.).
3. Ibid., 986a 2 (Oxford trans.).
4. Aristotle, *Prior Analytics*, Book I, Chapter 23, 41a 23–30.
5. Plato, *Theaetetus*, 147 D–E (Oxford trans.).
6. Aristotle, *Physics*, 215^{p} 1–5 (Oxford trans.).
7. Plato, *The Republic*, 529E, trans. Robin Wakefield (Oxford University Press, Oxford, 1993), p. 261.
8. E. P. Wigner, "The Unreasonable Effectiveness of Mathematics," *Communications in Pure and Applied Mathematics* 13 (1960): 1–14.

3. Motion and Philosophy

1. J. Barnes, in *The Complete Works of Aristotle—The Revised Oxford Translation* (Princeton University Press, Princeton, N.J., 1984).
2. R. J. Hankinson, in *The Cambridge Companion to Aristotle*, ed. J. Barnes (Cambridge University Press, Cambridge, 1995), p. 165.
3. Aristotle, *Physics*, Book II, Chapter 2, 194a 29–31 (Oxford trans., p. 331).
4. Ibid., Chapter 1, 192a 9 (Oxford trans., p. 329).
5. Aristotle, *Meteorology*, Book II, Chapter 9, 396b 7–11 (Oxford trans., p. 596).
6. Aristotle, *On the Heavens*, Book I, Chapter 6, 273b 30–31, 274a, 1 (Oxford trans., p. 455).
7. Aristotle, *Physics*, Book IV, Chapter 8, 214b 12–13 (Oxford trans., p. 365).
8. Ibid., 214b 32–34 (Oxford trans., p. 365).
9. Ibid., Book VII, Chapter 1, 242a 50–54 (Oxford trans., p. 408).
10. Aristotle, *On the Heavens*, Book III, Chapter 3, 301b 25–26 (Oxford trans., p. 494).
11. Thomas Kuhn, "Remarks on Receiving the Laurea," in *L'Anno Galileiano* (Edizioni LINT, Trieste, 1995).
12. David C. Lindberg, in *The Beginnings of Western Science* (University of Chicago Press, Chicago, Ill., 1992), pp. 53–54.
13. David C. Lindberg, in *The Beginnings of Western* Science, 2nd ed. (University of Chicago Press, Chicago, Ill., 2007), p. 65.
14. Michael R. Matthews, in Introduction to *The Scientific Background to Modern Philosophy* (Hackett, Indianapolis, Ind., 1989).

4. Hellenistic Physics and Technology

1. Here I borrow the title of the leading modern treatise on this age: Peter Green, *Alexander to Actium* (University of California Press, Berkeley, 1990).
2. I believe that this remark is originally due to George Sarton.
3. The description of Strato's work by Simplicius is presented in an English translation by M. R. Cohen and I. E. Drabkin, *A Source Book in Greek Science* (Harvard University Press, Cambridge, Mass., 1948), pp. 211–12.
4. H. Floris Cohen, *How Modern Science Came into the World* (Amsterdam University Press, Amsterdam, 2010), p. 17.
5. For the interaction of technology with physics research in modern

times, see Bruce J. Hunt, *Pursuing Power and Light: Technology and Physics from James Watt to Albert Einstein* (Johns Hopkins University Press, Baltimore, Md., 2010).

6. Philo's experiments are described in a letter quoted by G. I. Ibry-Massie and P. T. Keyser, *Greek Science of the Hellenistic Era* (Routledge, London, 2002), pp. 216–19.
7. The standard translation into English is Euclid, *The Thirteen Books of the Elements*, 2nd ed., trans. Thomas L. Heath (Cambridge University Press, Cambridge, 1925).
8. This is quoted in a Greek manuscript of the sixth century AD, and given in an English translation in Ibry-Massie and Keyser, *Greek Science of the Hellenistic Era*.
9. See Table V.1, p. 233, of the translation of Ptolemy's *Optics* by A. Mark Smith in "Ptolemy's Theory of Visual Perception," *Transactions of the American Philosophical Society* **86,** Part 2 (1996).
10. Quotes here are from T. L. Heath, trans., *The Works of Archimedes* (Cambridge University Press, Cambridge, 1897).

5. *Ancient Science and Religion*

1. Plato, *Timaeus*, 30A, trans. R. G. Bury, in *Plato*, Volume 9 (Loeb Classical Library, Harvard University Press, Cambridge, Mass., 1929), p. 55.
2. Erwin Schrödinger, Shearman Lectures at University College London, May 1948, published as *Nature and the Greeks* (Cambridge University Press, Cambridge, 1954).
3. Alexandre Koyré, *From the Closed World to the Infinite Universe* (Johns Hopkins University Press, Baltimore, Md., 1957), p. 159.
4. *Ancilla*, p. 22.
5. Thucydides, *History of the Peloponnesian War*, trans. Rex Warner (Penguin, New York, 1954, 1972), p. 511.
6. S. Greenblatt, "The Answer Man: An Ancient Poem Was Rediscovered and the World Swerved," *New Yorker*, August 8, 2011, pp. 28–33.
7. Edward Gibbon, *The Decline and Fall of the Roman Empire*, Chapter 23 (Everyman's Library, New York, 1991), p. 412. Hereafter cited as Gibbon, *Decline and Fall*.
8. Ibid., Chapter 2, p. 34.
9. Nicolaus Copernicus, *On the Revolutions of Heavenly Spheres*, trans. Charles Glenn Wallis (Prometheus, Amherst, N.Y., 1995), p. 7.
10. Lactantius, *Divine Institutes*, Book 3, Section 24, trans. A. Bowen and P. Garnsey (Liverpool University Press, Liverpool, 2003).

11. Paul, *Epistle to the Colossians* 2:8 (King James translation).
12. Augustine, *Confessions*, Book IV, trans. A. C. Outler (Dover, New York, 2002), p. 63.
13. Augustine, *Retractions*, Book I, Chapter 1, trans. M. I. Bogan (Catholic University of America Press, Washington, D.C., 1968), p. 10.
14. Gibbon, *Decline and Fall*, Chapter XL, p. 231.

PART II: GREEK ASTRONOMY

6. The Uses of Astronomy

1. This chapter is based in part on my article "The Missions of Astronomy," *New York Review of Books* **56**, 16 (October 22, 2009): 19–22; reprinted in *The Best American Science and Nature Writing*, ed. Freeman Dyson (Houghton Mifflin Harcourt, Boston, Mass., 2010), pp. 23–31, and in *The Best American Science Writing*, ed. Jerome Groopman (HarperCollins, New York, 2010), pp. 272–81.
2. Homer, *Iliad*, Book 22, 26–29. Quotation from Richmond Lattimore, trans., *The Iliad of Homer* (University of Chicago Press, Chicago, Ill., 1951), p. 458.
3. Homer, *Odyssey*, Book V, 280–87. Quotations from Robert Fitzgerald, trans., *The Odyssey* (Farrar, Straus and Giroux, New York, 1961), p. 89.
4. Diogenes Laertius, *Lives of the Eminent Philosophers*, Book I, 23.
5. This is the interpretation of some lines of Heraclitus argued by D. R. Dicks, *Early Greek Astronomy to Aristotle* (Cornell University Press, Ithaca, N.Y., 1970).
6. Plato's *Republic*, 527 D–E, trans. Robin Wakefield (Oxford University Press, Oxford, 1993).
7. Philo, *On the Eternity of the World*, I (1). Quotation from C. D. Yonge, trans., *The Works of Philo* (Hendrickson Peabody, Mass., 1993), 707.

7. Measuring the Sun, Moon, and Earth

1. The importance of Parmenides and Anaxagoras as founders of Greek scientific astronomy is emphasized by Daniel W. Graham, *Science Before Socrates—Parmenides, Anaxagoras, and the New Astronomy* (Oxford University Press, Oxford, 2013).
2. *Ancilla*, p. 18.
3. Aristotle, *On the Heavens*, Book II, Chapter 14, 297b 26–298a 5 (Oxford trans., pp. 488–89).

4. *Ancilla*, p. 23.
5. Aristotle, *On the Heavens*, Book II, Chapter 11.
6. Archimedes, *On Floating Bodies*, in T. L. Heath, trans., *The Works of Archimedes* (Cambridge University Press, Cambridge, 1897), p. 254. Hereafter cited as Archimedes, Heath trans.
7. A translation is given by Thomas Heath in *Aristarchus of Samos* (Clarendon, Oxford, 1923).
8. Archimedes, *The Sand Reckoner*, Heath trans., p. 222.
9. Aristotle, *On the Heavens*, Book II, 14, 296b 4–6 (Oxford trans.).
10. Aristotle, *On the Heavens*, Book II, 14, 296b 23–24 (Oxford trans.).
11. Cicero, *De Re Publica*, 1.xiv §21–22, in *Cicero, On the Republic and On the Laws*, trans. Clinton W. Keys (Loeb Classical Library, Harvard University Press, Cambridge, Mass., 1928), pp. 41, 43.
12. This work has been reconstructed by modern scholars; see Albert van Helden, *Measuring the Universe—Cosmic Dimensions from Aristarchus to Halley* (University of Chicago Press, Chicago, Ill., 1983), pp. 10–13.
13. *Ptolemy's Almagest*, trans. and annotated G. J. Toomer (Duckworth, London, 1984). The Ptolemy star catalog is on pages 341–99.
14. For a contrary view, see O. Neugebauer, *A History of Ancient Mathematical Astronomy* (Springer-Verlag, New York, 1975), pp. 288, 577.
15. Ptolemy, *Almagest*, Book VII, Chapter 2.
16. Cleomedes, *Lectures on Astronomy*, ed. and trans. A. C. Bowen and R. B. Todd (University of California Press, Berkeley and Los Angeles, 2004).

8. The Problem of the Planets

1. G. W. Burch, "The Counter-Earth," *Osiris* **11**, 267 (1954).
2. Aristotle, *Metaphysics*, Book I, Part 5, 986a 1 (Oxford trans.). But in Book II of *On the Heavens*, 293b 23–25, Aristotle says that the counter-Earth was supposed to explain why lunar eclipses are more common than solar eclipses.
3. The paragraph quoted here is as given by Pierre Duhem in *To Save the Phenomena—An Essay on the Idea of Physical Theory from Plato to Galileo*, trans. E. Dolan and C. Machler (University of Chicago Press, Chicago, Ill., 1969), p. 5, hereafter cited as Duhem, *To Save the Phenomena*. A more recent translation of this passage from Simplicius is given by I. Mueller: see Simplicius, *On*

Aristotle's "On the Heavens 2.10–14" (Cornell University Press, Ithaca, N.Y., 2005), 492.31–493.4, p. 33. We don't know if Plato ever actually proposed this problem. Simplicius was quoting Sosigenes the Peripatetic, a philosopher of the second century AD.

4. For very clear illustrations showing the model of Eudoxus, see James Evans, *The History and Practice of Ancient Astronomy* (Oxford University Press, Oxford, 1998), pp. 307–9.
5. Aristotle, *Metaphysics*, Book XII, Chapter 8, 1073b 1–1074a 1.
6. For a translation by I. Mueller, see Simplicius, *On Aristotle "On the Heavens 3.1–7"* (Cornell University Press, Ithaca, N.Y., 2005), 493.1–497.8, pp. 33–36.
7. This was the work, in 1956, of the physicists Tsung-Dao Lee and Chen-Ning Yang.
8. Aristotle, *Metaphysics*, Book XII, Section 8, 1073b 18–1074a 14 (Oxford trans.).
9. These references are given by D. R. Dicks, *Early Greek Astronomy to Aristotle* (Cornell University Press, Ithaca, N.Y., 1970), p. 202. Dicks takes a different view of what Aristotle was trying to accomplish.
10. Mueller, *Simplicius, On Aristotle's "On the Heavens 2.10–14,"* 519.9–11, p. 59.
11. Ibid., 504.19–30, p. 43.
12. See Book I of Otto Neugebauer, *A History of Ancient Mathematical Astronomy* (Springer-Verlag, New York, 1975).
13. G. Smith, private communication.
14. Ptolemy, *Almagest*, trans. G. J. Toomer (Duckworth, London, 1984), Book V, Chapter 13, pp. 247–51. Also see O. Neugebauer, *A History of Ancient Mathematical Astronomy, Part One* (Springer-Verlag, Berlin, 1975), pp. 100–3.
15. Barrie Fleet, trans., *Simplicius on Aristotle "Physics 2"* (Duckworth, London, 1997), 291.23–292.29, pp. 47–48.
16. Quoted by Duhem, *To Save the Phenomena*, pp. 20–21.
17. Ibid.
18. For comments on the meaning of explanation in science, and references to other articles on this subject, see S. Weinberg, "Can Science Explain Everything? Anything?" in *New York Review of Books* **48**, 9 (May 31, 2001): 47–50. Reprints: *Australian Review* (2001); in Portuguese, *Folha da S. Paolo* (2001); in French, *La Recherche* (2001); *The Best American Science Writing*, ed. M. Ridley and A. Lightman (HarperCollins, New York, 2002); *The Norton Reader* (W. W. Norton, New York, December 2003); *Explanations—Styles of Explanation in Science*, ed. John Cornwell (Oxford University

Press, London, 2004), 23–38; in Hungarian, *Akadeemia* **176**, No. 8: 1734–49 (2005); S. Weinberg, *Lake Views—This World and the Universe* (Harvard University Press, Cambridge, Mass., 2009).

19. This is not from the *Almagest* but from the *Greek Anthology*, verses compiled in the Byzantine Empire around AD 900. This translation is from Thomas L. Heath, *Greek Astronomy* (Dover, Mineola, N.Y., 1991), p. lvii.

PART III: THE MIDDLE AGES
9. The Arabs

1. This letter is quoted by Eutychius, then patriarch of Alexandria. The translation here is from E. M. Forster, *Pharos and Pharillon* (Knopf, New York, 1962), pp. 21–22. A less pithy translation is given by Gibbon, *Decline and Fall*, Chapter 51.
2. P. K. Hitti, *History of the Arabs* (Macmillan, London, 1937), p. 315.
3. D. Gutas, *Greek Thought, Arabic Culture—The Graeco-Arabic Translation Movement in Baghdad and Early 'Abbāsid Society* (Routledge, London, 1998), pp. 53–60.
4. Al-Biruni, *Book of the Determination at Coordinates of Localities*, Chapter 5, excerpted and trans. J. Lennart Berggren, in *The Mathematics of Egypt, Mesopotamia, China, India, and Islam*, ed. Victor Katz (Princeton University Press, Princeton, N.J., 2007).
5. Quoted in P. Duhem, *To Save the Phenomena*, p. 29.
6. Quoted by R. Arnaldez and A. Z. Iskandar in *The Dictionary of Scientific Biography* (Scribner, New York, 1975), Volume 12, p. 3.
7. G. J. Toomer, *Centaurus* **14**, 306 (1969).
8. Moses ben Maimon, *Guide to the Perplexed*, Part 2, Chapter 24, trans. M. Friedländer, 2nd ed. (Routledge, London, 1919), pp. 196, 198.
9. Ben Maimon is here quoting Psalms 115:16.
10. See E. Masood, *Science and Islam* (Icon, London, 2009).
11. N. M. Swerdlow, *Proceedings of the American Philosophical Society* **117**, 423 (1973).
12. The case that Copernicus learned of this device from Arab sources is made by F. J. Ragep, *History of Science* **14**, 65 (2007).
13. This is documented by Toby E. Huff, *Intellectual Curiosity and the Scientific Revolution* (Cambridge University Press, Cambridge, 2011), Chapter 5.
14. These are verses 13, 29, and 30 of the second version of Fitzgerald's translation.

15. Quoted in Jim al-Khalili, *The House of Wisdom* (Penguin, New York, 2011), p. 188.
16. *Al-Ghazali's Tahafut al-Falasifah*, trans. Sabih Ahmad Kamali (Pakistan Philosophical Congress, Lahore, 1958).
17. Al-Ghazali, *Fatihat al-'Ulum*, trans. I. Goldheizer, in *Studies on Islam*, ed. Merlin L. Swartz (Oxford University Press, 1981), quotation, p. 195.

10. Medieval Europe

1. See, e. g., Lynn White Jr., *Medieval Technology and Social Change* (Oxford University Press, Oxford, 1962), Chapter 2.
2. Peter Dear, *Revolutionizing the Sciences—European Knowledge and Its Ambitions, 1500–1700*, 2nd ed. (Princeton University Press, Princeton, N.J., and Oxford, 2009), p. 15.
3. The articles of the condemnation are given in a translation by Edward Grant in *A Source Book in Medieval Science*, ed. E. Grant (Harvard University Press, Cambridge, Mass., 1974), pp. 48–50.
4. Ibid., p. 47.
5. Quoted in David C. Lindberg, *The Beginnings of Western Science* (University of Chicago Press, Chicago, Ill., 1992), p. 241.
6. Ibid.
7. Nicole Oresme, *Le livre du ciel et du monde*, in French and trans. A. D. Menut and A. J. Denomy (University of Wisconsin Press, Madison, 1968), p. 369.
8. Quoted in "Buridan," in *Dictionary of Scientific Biography*, ed. Charles Coulston Gillespie (Scribner, New York, 1973), Volume 2, pp. 604–5.
9. See the article by Piaget in *The Voices of Time*, ed. J. T. Fraser (Braziller, New York, 1966).
10. Oresme, *Le livre*.
11. Ibid., pp. 537–39.
12. A. C. Crombie, *Robert Grosseteste and the Origins of Experimental Science—1100–1700* (Clarendon, Oxford, 1953).
13. For instance, see T. C. R. McLeish, *Nature* **507**, 161–63 (March 13, 2014).
14. Quoted in A. C. Crombie, *Medieval and Early Modern Science* (Doubleday Anchor, Garden City, N.Y., 1959), Volume 1, p. 53.
15. Translation by Ernest A. Moody, in *A Source Book in Medieval Science*, ed. E. Grant, p. 239. I have taken the liberty of changing the word "latitude" in Moody's translation to "increment of velocity," which I think more accurately indicates Heytesbury's meaning.

16. De Soto is quoted in an English translation by W. A. Wallace, *Isis* **59**, 384 (1968).
17. Quoted in Duhem, *To Save the Phenomena*, pp. 49–50.

PART IV: THE SCIENTIFIC REVOLUTION

1. Herbert Butterfield, *The Origins of Modern Science*, rev. ed. (Free Press, New York, 1957), p. 7.
2. For collections of essays on this theme, see *Reappraisals of the Scientific Revolution*, ed. D. C. Lindberg and R. S. Westfall (Cambridge University Press, Cambridge, 1990), and *Rethinking the Scientific Revolution*, ed. M. J. Osler (Cambridge University Press, Cambridge, 2000).
3. Steven Shapin, *The Scientific Revolution* (University of Chicago Press, Chicago, Ill., 1996), p. 1.
4. Pierre Duhem, *The System of the World: A History of Cosmological Doctrines from Plato to Copernicus* (Hermann, Paris, 1913).

11. The Solar System Solved

1. For an English translation, see Edward Rosen, *Three Copernican Treatises* (Farrar, Straus and Giroux, New York, 1939), or Noel M. Swerdlow, "The Derivation and First Draft of Copernicus's Planetary Theory: A Translation of the *Commentariolus* with Commentary," *Proceedings of the American Philosophical Society* **117**, 423 (1973).
2. For a review, see N. Jardine, *Journal of the History of Astronomy* **13**, 168 (1982).
3. O. Neugebauer, *Astronomy and History—Selected Essays* (Springer-Verlag, New York, 1983), essay 40.
4. The importance of this correlation for Copernicus is stressed by Bernard R. Goldstein, *Journal of the History of Astronomy* **33**, 219 (2002).
5. For an English translation, see *Nicolas Copernicus On the Revolutions*, trans. Edward Rosen (Polish Scientific Publishers, Warsaw, 1978; reprint, Johns Hopkins University Press, Baltimore, Md., 1978); or *Copernicus—On the Revolutions of the Heavenly Spheres*, trans. A. M. Duncan (Barnes and Noble, New York, 1976). Quotations here are from Rosen.
6. A. D. White, *A History of the Warfare of Science with Theology in Christendom* (Appleton, New York, 1895), Volume 1, pp. 126–28.

For a deflation of White, see D. C. Lindberg and R. L. Numbers, "Beyond War and Peace: A Reappraisal of the Encounter Between Christianity and Science," *Church History* **58**, 3 (September 1986): 338.

7. This paragraph has been quoted by Lindberg and Numbers, "Beyond War and Peace," and by T. Kuhn, *The Copernican Revolution* (Harvard University Press, Cambridge, Mass., 1957), p. 191. Kuhn's source is White, *A History of the Warfare of Science with Theology*. The German original is *Sämtliche Schriften*, ed. J. G. Walch (J. J. Gebauer, Halle, 1743), Volume 22, p. 2260.
8. Joshua 10:12.
9. This English translation of Osiander's preface is taken from Rosen, trans., *Nicolas Copernicus On the Revolutions*.
10. Quoted in R. Christianson, *Tycho's Island* (Cambridge University Press, Cambridge, 2000), p. 17.
11. On the history of the idea of hard celestial spheres, see Edward Rosen, "The Dissolution of the Solid Celestial Spheres," *Journal of the History of Ideas* **46**, 13 (1985). Rosen argues that Tycho exaggerated the extent to which this idea had been accepted before his time.
12. For claims to Tycho's system and for its variations, see C. Schofield, "The Tychonic and Semi-Tychonic World Systems," in *Planetary Astronomy from the Renaissance to the Rise of Astrophysics—Part A: Tycho Brahe to Newton*, ed. R. Taton and C. Wilson (Cambridge University Press, Cambridge, 1989).
13. For a photograph of this statue, taken by Owen Gingerich, see the frontispiece of my essay collection *Facing Up—Science and Its Cultural Adversaries* (Harvard University Press, Cambridge, Mass., 2001).
14. S. Weinberg, "Anthropic Bound on the Cosmological Constant," *Physical Review Letters* **59**, 2607 (1987); H. Martel, P. Shapiro, and S. Weinberg, "Likely Values of the Cosmological Constant," *Astrophysical Journal* **492**, 29 (1998).
15. J. R. Voelkel and O. Gingerich, "Giovanni Antonio Magini's 'Keplerian' Tables of 1614 and Their Implications for the Reception of Keplerian Astronomy in the Seventeenth Century," *Journal for the History of Astronomy* **32**, 237 (2001).
16. Quoted in Robert S. Westfall, *The Construction of Modern Science—Mechanism and Mechanics* (Cambridge University Press, Cambridge, 1977), p. 10.
17. This is the translation of William H. Donahue, in *Johannes*

Kepler—New Astronomy (Cambridge University Press, Cambridge, 1992), p. 65.
18. Johannes Kepler, *Epitome of Copernican Astronomy and Harmonies of the World*, trans. Charles Glenn Wallis (Prometheus, Amherst, N.Y., 1995), p. 180.
19. Quoted by Owen Gingerich in *Tribute to Galileo in Padua, International Symposium a cura dell'Universita di Padova, 2–6 dicembre 1992*, Volume 4 (Edizioni LINT, Trieste, 1995).
20. Quotations from Galileo Galilei, *Siderius Nuncius, or The Sidereal Messenger*, trans. Albert van Helden (University of Chicago Press, Chicago, Ill., 1989).
21. Galileo Galilei, *Discorse e Dimostrazione Matematiche*. For a facsimile of the 1663 translation by Thomas Salusbury, see Galileo Galilei, *Discourse on Bodies in Water*, with introduction and notes by Stillman Drake (University of Illinois Press, Urbana, 1960).
22. For a modern edition of a seventeenth-century translation, see Galileo, *Discourse on Bodies in Water*, trans. Thomas Salusbury, intro. and notes by Stillman Drake.
23. For details of this conflict, see J. L. Heilbron, *Galileo* (Oxford University Press, Oxford, 2010).
24. This letter is widely cited. The translation quoted here is from Duhem, *To Save the Phenomena*, p. 107. A fuller translation is given in Stillman Drake, *Discoveries and Opinions of Galileo* (Anchor, New York, 1957), pp. 162–64.
25. A translation of the entire letter is given in Drake, *Discoveries and Opinions of Galileo*, pp. 175–216.
26. Quoted in Stillman Drake, *Galileo* (Oxford University Press, Oxford, 1980), p. 64.
27. The letters of Maria Celeste to her father fortunately survive. Many are quoted in Dava Sobel, *Galileo's Daughter* (Walker, New York, 1999). Alas, Galileo's letters to his daughters are lost.
28. See Annibale Fantoli, *Galileo—For Copernicanism and for the Church*, 2nd ed., trans. G. V. Coyne (University of Notre Dame Press, South Bend, Ind., 1996); Maurice A. Finocchiaro, *Retrying Galileo, 1633–1992* (University of California Press, Berkeley and Los Angeles, 2005).
29. Quoted in Drake, *Galileo*, p. 90.
30. Quoted by Gingerich, *Tribute to Galileo*, p. 343.
31. I made a statement to this effect at the same meeting in Padua where Kuhn made the remarks about Aristotle cited in Chapter 4

and where Gingerich gave the talk about Galileo from which I have quoted here. See S. Weinberg, in *L'Anno Galileiano* (Edizioni LINT, Trieste, 1995), p. 129.

12. Experiments Begun

1. See G. E. R. Lloyd, *Proceedings of the Cambridge Philosophical Society*, N.S. **10**, 50 (1972), reprinted in *Methods and Problems in Greek Science* (Cambridge University Press, Cambridge, 1991).
2. Galileo Galilei, *Two New Sciences*, trans. Stillman Drake (University of Wisconsin Press, Madison, 1974), p. 68.
3. Stillman Drake, *Galileo* (Oxford University Press, Oxford, 1980), p. 33.
4. T. B. Settle, "An Experiment in the History of Science," *Science* **133**, 19 (1961).
5. This is Drake's conclusion in the endnote to p. 259 of Galileo Galilei, *Dialogue Concerning the Two Chief World Systems: Ptolemaic and Copernican*, trans. Stillman Drake (Modern Library, New York, 2001).
6. Our knowledge of this experiment is based on an unpublished document, folio 116v, in Biblioteca Nazionale Centrale, Florence. See Stillman Drake, *Galileo at Work—His Scientific Biography* (University of Chicago Press, Chicago, Ill., 1978), pp. 128–32; A. J. Hahn, "The Pendulum Swings Again: A Mathematical Reassessment of Galileo's Experiments with Inclined Planes," *Archive for the History of the Exact Sciences* **56**, 339 (2002), with a reproduction of the folio on p. 344.
7. Carlo M. Cipolla, *Clocks and Culture 1300–1700* (W. W. Norton, New York, 1978), pp. 59, 138.
8. Christiaan Huygens, *The Pendulum Clock or Geometrical Demonstrations Concerning the Motion of Pendula as Applied to Clocks*, trans. Richard J. Blackwell (Iowa State University Press, Ames, 1986), p. 171.
9. This measurement was described in detail by Alexandre Koyré in *Proceedings of the American Philosophical Society* **97**, 222 (1953) and **45**, 329 (1955). Also see Christopher M. Graney, "Anatomy of a Fall: Giovanni Battista Riccioli and the Story of *g*," *Physics Today*, September 2012, pp. 36–40.
10. On the controversy over these conservation laws, see G. E. Smith, "The Vis-Viva Dispute: A Controversy at the Dawn of Mathematics," *Physics Today*, October 2006, p. 31.

11. Christiaan Huygens, *Treatise on Light*, trans. Silvanus P. Thompson (University of Chicago Press, Chicago, Ill., 1945), p. vi.
12. Quoted by Steven Shapin in *The Scientific Revolution* (University of Chicago Press, Chicago, Ill., 1996), p. 105.
13. Ibid., p. 185.

13. Method Reconsidered

1. See articles on Leonardo in *Dictionary of Scientific Biography*, ed. Charles Coulston Gillespie (Scribner, New York, 1970), Volume 8, pp. 192–245.
2. Quotations are from René Descartes, *Principles of Philosophy*, trans. V. R. Miller and R. P. Miller (D. Reidel, Dordrecht, 1983), p. 15.
3. Voltaire, *Philosophical Letters*, trans. E. Dilworth (Bobbs-Merrill Educational Publishing, Indianapolis, Ind., 1961), p. 64.
4. It is odd that many modern English language editions of *Discourse on Method* leave out these supplements, as if they would not be of interest to philosophers. For an edition that does include them, see René Descartes, *Discourse on Method, Optics, Geometry, and Meteorology*, trans. Paul J. Olscamp (Bobbs-Merrill, Indianapolis, Ind., 1965). The Descartes quote and the numerical results below are from this edition.
5. It is argued that the tennis ball analogy fits well with Descartes' theory of light as arising from the dynamics of the tiny corpuscles that fill space; see John A. Schuster, "Descartes *Opticien*—The Construction of the Law of Refraction and the Manufacture of Its Physical Rationales, 1618–29," in *Descartes' Natural Philosophy*, ed. S. Graukroger, J. Schuster, and J. Sutton (Routledge, London and New York, 2000), pp. 258–312.
6. Aristotle, *Meteorology*, Book III, Chapter 4, 374a, 30–31 (Oxford trans., p. 603).
7. Descartes, *Principles of Philosophy*, trans. V. R. Miller and R. P. Miller, pp. 60, 114.
8. On this point, see Peter Dear, *Revolutionizing the Sciences—European Knowledge and Its Ambitions, 1500–1700*, 2nd ed. (Princeton University Press, Princeton, N.J., and Oxford, 2009), Chapter 8.
9. L. Laudan, "The Clock Metaphor and Probabilism: The Impact of Descartes on English Methodological Thought," *Annals of Science* 22, 73 (1966). Contrary conclusions were reached in G. A. J.

Rogers, "Descartes and the Method of English Science," *Annals of Science* **29**, 237 (1972).

10. Richard Watson, *Cogito Ergo Sum—The Life of René Descartes* (David R. Godine, Boston, Mass., 2002).

14. The Newtonian Synthesis

1. This is described in D. T. Whiteside, ed., General Introduction to Volume 20, *The Mathematical Papers of Isaac Newton* (Cambridge University Press, Cambridge, 1968), pp. xi–xii.
2. Ibid., Volume 2, footnote, pp. 206–7; and Volume 3, pp. 6–7.
3. See, for example, Richard S. Westfall, *Never at Rest—A Biography of Isaac Newton* (Cambridge University Press, Cambridge, 1980), Chapter 14.
4. Peter Galison, *How Experiments End* (University of Chicago Press, Chicago, Ill., 1987).
5. Quoted in Westfall, *Never at Rest*, p. 143.
6. Quoted in *Dictionary of Scientific Biography*, ed. Charles Coulston Gillespie (Scribner, New York, 1970), Volume 6, p. 485.
7. Quoted in James Gleick, *Isaac Newton* (Pantheon, New York, 2003), p. 120.
8. Quotations from I. Bernard Cohen and Anne Whitman, trans., *Isaac Newton—The Principia*, 3rd ed. (University of California Press, Berkeley and Los Angeles, 1999). Before this version, the standard translation was *The Principia—Mathematical Principles of Natural Philosophy* (University of California Press, Berkeley and Los Angeles, 1962), trans. Florian Cajori (1792), rev. trans. Andrew Motte.
9. G. E. Smith, "Newton's Study of Fluid Mechanics," *International Journal of Engineering Science* **36**, 1377 (1998).
10. Modern astronomical data in this chapter are from C. W. Allen, *Astrophysical Quantities*, 2nd ed. (Athlone, London, 1963).
11. The standard work on the history of the measurement of the size of the solar system is Albert van Helden, *Measuring the Universe—Cosmic Dimensions from Aristarchus to Halley* (University of Chicago Press, Chicago, Ill., 1985).
12. See Robert P. Crease, *World in the Balance—The Historic Quest for an Absolute System of Measurement* (W. W. Norton, New York, 2011).
13. See J. Z. Buchwald and M. Feingold, *Newton and the Origin of Civilization* (Princeton University Press, Princeton, N.J., 2014).

14. See S. Chandrasekhar, *Newton's* Principia *for the Common Reader* (Clarendon, Oxford, 1995), pp. 472–76; Westfall, *Never at Rest*, pp. 736–39.
15. R. S. Westfall, "Newton and the Fudge Factor," *Science* **179**, 751 (1973).
16. See G. E. Smith, "How Newton's *Principia* Changed Physics," in *Interpreting Newton: Critical Essays*, ed. A. Janiak and E. Schliesser (Cambridge University Press, Cambridge, 2012), pp. 360–95.
17. Voltaire, *Philosophical* Letters, trans. E. Dilworth (Bobbs-Merrill Educational Publishing, Indianapolis, Ind., 1961), p. 61.
18. The opposition to Newtonianism is described in articles by A. B. Hall, E. A. Fellmann, and P. Casini in "*Newton's Principia*: A Discussion Organized and Edited by D. G. King-Hele and A. R. Hall," *Monthly Notices of the Royal Astronomical Society* **42**, 1 (1988).
19. Christiaan Huygens, *Discours de la Cause de la Pesanteur* (1690), trans. Karen Bailey, with annotations by Karen Bailey and G. E. Smith, available from Smith at Tufts University (1997).
20. Shapin has argued that this conflict even had political implications: Steven Shapin, "Of Gods and Kings: Natural Philosophy and Politics in the Leibniz-Clarke Disputes," *Isis* **72**, 187 (1981).
21. S. Weinberg, *Gravitation and Cosmology* (Wiley, New York, 1972), Chapter 15.
22. G. E. Smith, to be published.
23. Quoted in *A Random Walk in Science*, ed. R. L. Weber and E. Mendoza (Taylor and Francis, London, 2000).
24. Robert K. Merton, "Motive Forces of the New Science," *Osiris* **4**, Part 2 (1938); reprinted in *Science, Technology, and Society in Seventeenth-Century England* (Howard Fertig, New York, 1970), and in *On Social Structure and Science*, ed. Piotry Sztompka (University of Chicago Press, Chicago, Ill., 1996), pp. 223–40.

15. Epilogue: The Grand Reduction

1. I have given a more detailed account of some of this progress in *The Discovery of Subatomic Particles*, rev. ed. (Cambridge University Press, Cambridge, 2003).
2. Isaac Newton, *Opticks, or A Treatise of the Reflections, Refractions, Inflections, and Colours of Light* (Dover, New York, 1952, based on 4th ed., London, 1730), p. 394.
3. Ibid., p. 376.

4. This is from Ostwald's *Outlines of General Chemistry*, and is quoted by G. Holton, in *Historical Studies in the Physical Sciences* **9**, 161 (1979), and I. B. Cohen, in *Critical Problems in the History of Science*, ed. M. Clagett (University of Wisconsin Press, Madison, 1959).
5. P. A. M. Dirac, "Quantum Mechanics of Many-Electron Systems," *Proceedings of the Royal Society* **A123**, 713 (1929).
6. To forestall accusations of plagiarism, I will acknowledge here that this last paragraph is a riff on the last paragraph of Darwin's *On the Origin of Species*.

Bibliography

This bibliography lists the modern secondary sources on the history of science on which I have relied, as well as original works of past scientists that I have consulted, from the fragments of the pre-Socratics to Newton's *Principia*, and more sketchily on to the present. The works listed are all in English or English translations; unfortunately, I have no Latin and less Greek, let alone Arabic. This is not intended to be a list of the most authoritative sources, or of the best editions of each source. These are simply the books that I have consulted in writing *To Explain the World*, in the best editions that happened to be available to me.

ORIGINAL SOURCES

Archimedes, *The Works of Archimedes*, trans. T. L. Heath (Cambridge University Press, Cambridge, 1897).

Aristarchus, *Aristarchus of Samos*, trans. T. L. Heath (Clarendon, Oxford, 1923).

Aristotle, *The Complete Works of Aristotle—The Revised Oxford Translation*, ed. J. Barnes (Princeton University Press, Princeton, N.J., 1984).

Augustine, *Confessions*, trans. Albert Cook Outler (Westminster, Philadelphia, Pa., 1955).

———, *Retractions*, trans. M. I. Bogan (Catholic University of America Press, Washington, D.C., 1968).

Cicero, *On the Republic and On the Laws*, trans. Clinton W. Keys

(Loeb Classical Library, Harvard University Press, Cambridge, Mass., 1928).

Cleomedes, *Lectures on Astronomy*, ed. and trans. A. C. Bowen and R. B. Todd (University of California Press, Berkeley and Los Angeles, 2004).

Copernicus, *Nicolas Copernicus On the Revolutions*, trans. Edward Rosen (Polish Scientific Publishers, Warsaw, 1978; reprint, Johns Hopkins University Press, Baltimore, Md., 1978).

———, *Copernicus—On the Revolutions of the Heavenly Spheres*, trans. A. M. Duncan (Barnes and Noble, New York, 1976).

———, *Three Copernican Treatises*, trans. E. Rosen (Farrar, Straus and Giroux, New York, 1939). Consists of *Commentariolus, Letter Against Werner*, and *Narratio prima of Rheticus*.

Charles Darwin, *On the Origin of Species by Means of Natural Selection*, 6th ed. (John Murray, London, 1885).

René Descartes, *Discourse on Method, Optics, Geometry, and Meteorology*, trans. Paul J. Olscamp (Bobbs-Merrill, Indianapolis, Ind., 1965).

———, *Principles of Philosophy*, trans. V. R. Miller and R. P. Miller (D. Reidel, Dordrecht, 1983).

Diogenes Laertius, *Lives of the Eminent Philosophers*, trans. R. D. Hicks (Loeb Classical Library, Harvard University Press, Cambridge, Mass., 1972).

Euclid, *The Thirteen Books of the Elements*, 2nd ed., trans. Thomas L. Heath (Cambridge University Press, Cambridge, 1925).

Galileo Galilei, *Dialogue Concerning the Two Chief World Systems: Ptolemaic and Copernican*, trans. Stillman Drake (Modern Library, New York, 2001).

———, *Discourse on Bodies in Water*, trans. Thomas Salusbury (University of Illinois Press, Urbana, 1960).

———, *Discoveries and Opinions of Galileo*, trans. Stillman Drake (Anchor, New York, 1957). Contains *The Starry Messenger, Letter to Christina*, and excerpts from *Letters on Sunspots* and *The Assayer*.

———, *The Essential Galileo*, trans. Maurice A. Finocchiaro (Hackett, Indianapolis, Ind., 2008). Includes *The Sidereal Messenger, Letter to Castelli, Letter to Christina, Reply to Cardinal Bellarmine*, etc.

———, *Siderius Nuncius, or The Sidereal Messenger*, trans. Albert van Helden (University of Chicago Press, Chicago, Ill., 1989).

———, *Two New Sciences, Including Centers of Gravity and Force of*

Percussion, trans. Stillman Drake (University of Wisconsin Press, Madison, 1974).

Galileo Galilei and Christoph Scheiner, *On Sunspots*, trans. and ed. Albert van Helden and Eileen Reeves (University of Chicago Press, Chicago, Ill., 1010).

Abu Hamid al-Ghazali, *The Beginnings of Sciences*, trans. I. Goldheizer, in *Studies on Islam*, ed. Merlin L. Swartz (Oxford University Press, Oxford, 1981).

———, *The Incoherence of the Philosophers*, trans. Sabih Ahmad Kamali (Pakistan Philosophical Congress, Lahore, 1958).

Herodotus, *The Histories*, trans. Aubery de Selincourt, rev. ed. (Penguin Classics, London, 2003).

Homer, *The Iliad*, trans. Richmond Lattimore (University of Chicago Press, Chicago, Ill., 1951).

———, *The Odyssey*, trans. Robert Fitzgerald (Farrar, Straus and Giroux, New York, 1961).

Horace, *Odes and Epodes*, trans. Niall Rudd (Loeb Classical Library, Harvard University Press, Cambridge, Mass., 2004).

Christiaan Huygens, *The Pendulum Clock or Geometrical Demonstrations Concerning the Motion of Pendula as Applied to Clocks*, trans. Richard J. Blackwell (Iowa State University Press, Ames, 1986).

———, *Treatise on Light*, trans. Silvanus P. Thompson (University of Chicago Press, Chicago, Ill., 1945).

Johannes Kepler, *Epitome of Copernican Astronomy and Harmonies of the World*, trans. C. G. Wallis (Prometheus, Amherst, N.Y., 1995).

———, *New Astronomy* (*Astronomia Nova*), trans. W. H. Donahue (Cambridge University Press, Cambridge, 1992).

Omar Khayyam, *The Rubáiyát, the Five Authorized Editions*, trans. Edward Fitzgerald (Walter J. Black, New York, 1942).

———, *The Rubáiyát, a Paraphrase from Several Literal Translations*, by Richard Le Gallienne (John Lan, London, 1928).

Lactantius, *Divine Institutes*, trans. A. Bowen and P. Garnsey (Liverpool University Press, Liverpool, 2003).

Gottfried Wilhelm Leibniz, *The Leibniz-Clarke Correspondence*, ed. H. G. Alexander (Manchester University Press, Manchester, 1956).

Martin Luther, *Table Talk*, trans. W. Hazlitt (H. G. Bohn, London, 1857).

Moses ben Maimon, *Guide to the Perplexed*, trans. M. Friedländer, 2nd ed. (Routledge, London, 1919).

Isaac Newton, *The Mathematical Papers of Isaac Newton*, ed. D. Thomas Whiteside (Cambridge University Press, Cambridge, 1968).

———, *Mathematical Principles of Natural Philosophy*, trans. Florian Cajori, rev. trans. Andrew Motte (University of California Press, Berkeley and Los Angeles, 1962).

———, *Opticks, or a Treatise of the Reflections, Refractions, Inflections, and Colours of Light* (Dover, New York, 1952, based on 4th ed., London, 1730).

———, *The Principia—Mathematical Principles of Natural Philosophy*, trans. I. Bernard Cohen and Anne Whitman, with "A Guide to Newton's *Principia*," by I. Bernard Cohen (University of California Press, Berkeley and Los Angeles, 1999).

Nicole Oresme, *The Book of the Heavens and the Earth*, trans. A. D. Menut and A. J. Denomy (University of Wisconsin Press, Madison, 1968).

Philo, *The Works of Philo*, trans. C. D. Yonge (Hendrickson, Peabody, Mass., 1993).

Plato, *Phaedo*, trans. Alexander Nehamas and Paul Woodruff (Hackett, Indianapolis, Ind., 1995).

———, *Plato*, Volume 9 (Loeb Classical Library, Harvard University Press, Cambridge, Mass., 1929). Includes *Phaedo*, etc.

———, *Republic*, trans. Robin Wakefield (Oxford University Press, Oxford, 1993).

———, *Timaeus and Critias*, trans. Desmond Lee (Penguin, New York, 1965).

———, *The Works of Plato*, trans. Benjamin Jowett (Modern Library, New York, 1928). Includes *Phaedo, Republic, Theaetetus*, etc.

Ptolemy, *Almagest*, trans. G. J. Toomer (Duckworth, London, 1984).

———, *Optics*, trans. A. Mark Smith, in "Ptolemy's Theory of Visual Perception—An English Translation of the *Optics* with Commentary," *Transactions of the American Philosophical Society* **86**, Part 2 (1996).

Simplicius, *On Aristotle "On the Heavens 2.10–14,"* trans. I. Mueller (Cornell University Press, Ithaca, N.Y., 2005).

———, *On Aristotle "On the Heavens 3.1–7,"* trans. I. Mueller (Cornell University Press, Ithaca, N.Y., 2005).

———, *On Aristotle "Physics 2,"* trans. Barrie Fleet (Duckworth, London, 1997).

Thucydides, *History of the Peloponnesian War*, trans. Rex Warner (Penguin, New York, 1954, 1972).

COLLECTIONS OF ORIGINAL SOURCES

J. Barnes, *Early Greek Philosophy* (Penguin, London, 1987).

———, *The Presocratic Philosophers*, rev. ed. (Routledge and Kegan Paul, London, 1982).

J. Lennart Berggren, "Mathematics in Medieval Islam," in *The Mathematics of Egypt, Mesopotamia, China, India, and Islam*, ed. Victor Katz (Princeton University Press, Princeton, N.J., 2007).

Marshall Clagett, *The Science of Mechanics in the Middle Ages* (University of Wisconsin Press, Madison, 1959).

M. R. Cohen and I. E. Drabkin, *A Source Book in Greek Science* (Harvard University Press, Cambridge, Mass., 1948).

Stillman Drake and I. E. Drabkin, *Mechanics in Sixteenth-Century Italy* (University of Wisconsin Press, Madison, 1969).

Stillman Drake and C. D. O'Malley, *The Controversy on the Comets of 1618* (University of Pennsylvania Press, Philadelphia, 1960). Translations of works of Galileo, Grassi, and Kepler.

K. Freeman, *The Ancilla to the Pre-Socratic Philosophers* (Harvard University Press, Cambridge, Mass., 1966).

D. W. Graham, *The Texts of Early Greek Philosophy—The Complete Fragments and Selected Testimonies of the Major Presocratics* (Cambridge University Press, New York, 2010).

E. Grant, ed., *A Source Book in Medieval Science* (Harvard University Press, Cambridge, Mass., 1974).

T. L. Heath, *Greek Astronomy* (J. M. Dent and Sons, London, 1932).

G. L. Ibry-Massie and P. T. Keyser, *Greek Science of the Hellenistic Era* (Routledge, London, 2002).

William Francis Magie, *A Source Book in Physics* (McGraw-Hill, New York, 1935).

Michael Matthews, *The Scientific Background to Modern Philosophy* (Hackett, Indianapolis, Ind., 1989).

Merlin L. Swartz, *Studies in Islam* (Oxford University Press, Oxford, 1981).

SECONDARY SOURCES

L'Anno Galileiano, International Symposium a cura dell'Universita di Padova, 2–6 dicembre 1992, Volume 1 (Edizioni LINT, Trieste, 1995). Speeches in English by T. Kuhn and S. Weinberg; see also *Tribute to Galileo*.

J. Barnes, ed., *The Cambridge Companion to Aristotle* (Cambridge

University Press, Cambridge, 1995). Articles by J. Barnes, R. J. Hankinson, and others.

Herbert Butterfield, *The Origins of Modern Science*, rev. ed. (Free Press, New York, 1957).

S. Chandrasekhar, *Newton's* Principia *for the Common Reader* (Clarendon, Oxford, 1995).

R. Christianson, *Tycho's Island* (Cambridge University Press, Cambridge, 2000).

Carlo M. Cipolla, *Clocks and Culture 1300–1700* (W. W. Norton, New York, 1978).

Marshall Clagett, ed., *Critical Studies in the History of Science* (University of Wisconsin Press, Madison, 1959). Articles by I. B. Cohen and others.

H. Floris Cohen, *How Modern Science Came into the World—Four Civilizations, One 17th-Century Breakthrough* (Amsterdam University Press, Amsterdam, 2010).

John Craig, *Newton at the Mint* (Cambridge University Press, Cambridge, 1946).

Robert P. Crease, *World in the Balance—The Historic Quest for an Absolute System of Measurement* (W. W. Norton, New York, 2011).

A. C. Crombie, *Medieval and Early Modern Science* (Doubleday Anchor, Garden City, N.Y., 1959).

———, *Robert Grosseteste and the Origins of Experimental Science—1100–1700* (Clarendon, Oxford, 1953).

Olivier Darrigol, *A History of Optics from Greek Antiquity to the Nineteenth Century* (Oxford University Press, Oxford, 2012).

Peter Dear, *Revolutionizing the Sciences—European Knowledge and Its Ambitions, 1500–1700*, 2nd ed. (Princeton University Press, Princeton, N.J., and Oxford, 2009).

D. R. Dicks, *Early Greek Astronomy to Aristotle* (Cornell University Press, Ithaca, N.Y., 1970).

The Dictionary of Scientific Biography, ed. Charles Coulston Gillespie (Scribner, New York, 1970).

Stillman Drake, *Galileo at Work—His Scientific Biography* (University of Chicago Press, Chicago, Ill., 1978).

Pierre Duhem, *The Aim and Structure of Physical Theory*, trans. Philip K. Weiner (Athenaeum, New York, 1982).

———, *Medieval Cosmology—Theories of Infinity, Place, Time, Void, and the Plurality of Worlds*, trans. Roger Ariew (University of Chicago Press, Chicago, Ill., 1985).

———, *To Save the Phenomena—An Essay on the Idea of Physical*

Theory from Plato to Galileo, trans. E. Dolan and C. Machler (University of Chicago Press, Chicago, Ill., 1969).

James Evans, *The History and Practice of Ancient Astronomy* (Oxford University Press, Oxford, 1998).

Annibale Fantoli, *Galileo—For Copernicanism and for the Church*, 2nd ed., trans. G. V. Coyne (University of Notre Dame Press, South Bend, Ind., 1996).

Maurice A. Finocchiaro, *Retrying Galileo, 1633–1992* (University of California Press, Berkeley and Los Angeles, 2005).

E. M. Forster, *Pharos and Pharillon* (Knopf, New York, 1962).

Kathleen Freeman, *The Pre-Socratic Philosophers*, 3rd ed. (Basil Blackwell, Oxford, 1953).

Peter Galison, *How Experiments End* (University of Chicago Press, Chicago, Ill., 1987).

Edward Gibbon, *The Decline and Fall of the Roman Empire* (Everyman's Library, New York, 1991).

James Gleick, *Isaac Newton* (Pantheon, New York, 2003).

Daniel W. Graham, *Science Before Socrates—Parmenides, Anaxagoras, and the New Astronomy* (Oxford University Press, Oxford, 2013).

Edward Grant, *The Foundations of Modern Science in the Middle Ages* (Cambridge University Press, Cambridge, 1996).

———, *Planets, Stars, and Orbs—The Medieval Cosmos, 1200–1687* (Cambridge University Press, Cambridge, 1994).

Stephen Graukroger, ed., *Descartes—Philosophy, Mathematics, and Physics* (Harvester, Brighton, 1980).

Stephen Graukroger, John Schuster, and John Sutton, eds., *Descartes' Natural Philosophy* (Routledge, London and New York, 2000).

Peter Green, *Alexander to Actium* (University of California Press, Berkeley, 1990).

Dimitri Gutas, *Greek Thought, Arabic Culture—The Graeco-Arabic Translation Movement in Baghdad and Early ʻAbbāsid Society* (Routledge, London, 1998).

Rupert Hall, *Philosophers at War: The Quarrel Between Newton and Leibniz* (Cambridge University Press, Cambridge, 1980).

Charles Homer Haskins, *The Rise of Universities* (Cornell University Press, Ithaca, N.Y., 1957).

J. L. Heilbron, *Galileo* (Oxford University Press, Oxford, 2010).

Albert van Helden, *Measuring the Universe—Cosmic Dimensions from Aristarchus to Halley* (University of Chicago Press, Chicago, Ill., 1983).

P. K. Hitti, *History of the Arabs* (Macmillan, London, 1937).

J. P. Hogendijk and A. I. Sabra, eds., *The Enterprise of Science in Islam = New Perspectives* (MIT Press, Cambridge, Mass., 2003).

Toby E. Huff, *Intellectual Curiosity and the Scientific Revolution* (Cambridge University Press, Cambridge, 2011).

Jim al-Khalifi, *The House of Wisdom* (Penguin, New York, 2011).

Henry C. King, *The History of the Telescope* (Charles Griffin, Toronto, 1955; reprint, Dover, New York, 1979).

D. G. King-Hele and A. R. Hale, eds., "Newton's *Principia* and His Legacy," *Notes and Records of the Royal Society of London* **42**, 1–122 (1988).

Alexandre Koyré, *From the Closed World to the Infinite Universe* (Johns Hopkins University Press, Baltimore, Md., 1957).

Thomas S. Kuhn, *The Copernican Revolution* (Harvard University Press, Cambridge, Mass., 1957).

———, *The Structure of Scientific Revolutions* (University of Chicago Press, Chicago, Ill., 1962; 2nd ed. 1970).

David C. Lindberg, *The Beginnings of Western Science* (University of Chicago Press, Chicago, Ill., 1992; 2nd ed. 2007).

D. C. Lindberg and R. S. Westfall, eds., *Reappraisals of the Scientific Revolution* (Cambridge University Press, Cambridge, 2000).

G. E. R. Lloyd, *Methods and Problems in Greek Science* (Cambridge University Press, Cambridge, 1991).

Peter Machamer, ed., *The Cambridge Companion to Galileo* (Cambridge University Press, Cambridge, 1998).

Alberto A. Martínez, *The Cult of Pythagoras—Man and Myth* (University of Pittsburgh Press, Pittsburgh, Pa., 2012).

E. Masood, *Science and Islam* (Icon, London, 2009).

Robert K. Merton, "Motive Forces of the New Science," *Osiris* **4**, Part 2 (1938); reprinted in *Science, Technology, and Society in Seventeenth-Century England* (Howard Fertig, New York, 1970), and *On Social Structure and Science*, ed. Piotry Sztompka (University of Chicago Press, Chicago, Ill., 1996), pp. 223–40.

Otto Neugebauer, *Astronomy and History—Selected Essays* (Springer-Verlag, New York, 1983).

——— *A History of Ancient Mathematical Astronomy* (Springer-Verlag, New York, 1975).

M. J. Osler, ed., *Rethinking the Scientific Revolution* (Cambridge University Press, Cambridge, 2000). Articles by M. J. Osler, B. J. T. Dobbs, R. S. Westfall, and others.

Ingrid D. Rowland, *Giordano Bruno—Philosopher and Heretic* (Farrar, Straus and Giroux, New York, 2008).

George Sarton, *Introduction to the History of Science*, Volume 1,

From Homer to Omar Khayyam (Carnegie Institution of Washington, Washington, D.C., 1927).

Erwin Schrödinger, *Nature and the Greeks* (Cambridge University Press, Cambridge, 1954).

Steven Shapin, *The Scientific Revolution* (University of Chicago Press, Chicago, Ill., 1996).

Dava Sobel, *Galileo's Daughter* (Walker, New York, 1999).

Merlin L. Swartz, *Studies in Islam* (Oxford University Press, Oxford 1981).

N. M. Swerdlow and O. Neugebauer, *Mathematical Astronomy in Copernicus's De Revolutionibus* (Springer-Verlag, New York, 1984).

R. Taton and C. Wilson, eds., *Planetary Astronomy from the Renaissance to the Rise of Astrophysics—Part A: Tycho Brahe to Newton* (Cambridge University Press, Cambridge, 1989).

Tribute to Galileo in Padua, International Symposium a cura dell'Universita di Padova, 2–6 dicembre 1992, Volume 4 (Edizioni LINT, Trieste, 1995). Articles in English by J. MacLachlan, I. B. Cohen, O. Gingerich, G. A. Tammann, L. M. Lederman, C. Rubbia, and Steven Weinberg; see also *L'Anno Galileiano.*

Gregory Vlastos, *Plato's Universe* (University of Washington Press, Seattle, 1975).

Voltaire, *Philosophical Letters*, trans. E. Dilworth (Bobbs-Merrill Educational Publishing, Indianapolis, Ind., 1961).

Richard Watson, *Cogito Ergo Sum—The Life of René Descartes* (David R. Godine, Boston, Mass., 2002).

Steven Weinberg, *Discovery of Subatomic Particles*, rev. ed. (Cambridge University Press, Cambridge, 2003).

———, *Dreams of a Final Theory* (Pantheon, New York, 1992; reprinted with a new afterword, Vintage, New York, 1994).

———, *Facing Up—Science and Its Cultural Adversaries* (Harvard University Press, Cambridge, Mass., 2001).

———, *Lake Views—This World and the Universe* (Harvard University Press, Cambridge, Mass., 2009).

Richard S. Westfall, *The Construction of Modern Science—Mechanism and Mechanics* (Cambridge University Press, Cambridge, 1977).

———, *Never at Rest—A Biography of Isaac Newton* (Cambridge University Press, Cambridge, 1980).

Andrew Dickson White, *A History of the Warfare of Science with Theology in Christendom* (Appleton, New York, 1895).

Lynn White, *Medieval Technology and Social Change* (Oxford University Press, Oxford, 1962).

Index